*Radio and Line
Transmission (A)*

Radio and Line Transmission (A)

*A textbook covering the second-year requirements
of the Telecommunication Technicians' Course*

D. C. Green
ASSOCIATE MEMBER IEE
C & G FULL TECHN CERT (TELECOMM)

*Lecturer in Telecommunication Engineering
at Willesden College of Technology*

London

Sir Isaac Pitman & Sons Ltd

First published 1968

SIR ISAAC PITMAN & SONS LTD
PITMAN HOUSE, PARKER STREET, KINGSWAY, LONDON, WC2
THE PITMAN PRESS, BATH
PITMAN HOUSE, BOUVERIE STREET, CARLTON, VICTORIA 3053
P.O. BOX 7721, JOHANNESBURG, TRANSVAAL
P.O. BOX 6038, PORTAL STREET, NAIROBI, KENYA

ASSOCIATED COMPANIES

PITMAN MEDICAL PUBLISHING COMPANY LTD
46 CHARLOTTE STREET, LONDON, W1

PITMAN PUBLISHING CORPORATION
20 EAST 46TH STREET, NEW YORK, N.Y. 10017

SIR ISAAC PITMAN & SONS (CANADA) LTD
PITMAN HOUSE, 381–383 CHURCH STREET, TORONTO

SBN: 273 40444 X

Made in Great Britain at the Pitman Press, Bath

Preface

The Telecommunication Technicians' Course of the City and Guilds
of London Institute requires a student to study Telecommunication
Principles, Mathematics, and one of several optional subjects.
Among the subjects available are Radio and Line Transmission.
In the second and third years of the course a combined Radio and
Line Transmission syllabus is studied, but in the fourth year the two
subjects are treated separately. This book has been written to cover
the second-year Radio and Line Transmission "A" syllabus and
provides a textbook for students taking this subject. It is thought
that the book will also help O.N.C. students who wish to gain some
insight into the field of radio.

Numerous worked examples are given in the text, and each chapter
concludes with a number of past City and Guilds examination questions. The source and date of each City and Guilds question is
indicated, the following source key being used—

A	Radio and Line Transmission "A"
B	Radio and Line Transmission "B"
E	Elementary Telecommunication Practice
P1	Telecommunication Principles "A"
P2	Telecommunication Principles "B"
1	Radio 1

Concisely worked answers to the numerical questions will be
found at the end of the book. These answers are the responsibility
of the author and are not necessarily endorsed by the City and Guilds
Institute.

Acknowledgement is made to the City and Guilds Institute for
permission to use past examination questions, and also to the
Institution of Post Office Electrical Engineers and S.G.S. Fairchild
Ltd. for permission to use diagrams from their publications. Individual acknowledgement is given beneath the diagrams concerned.

My thanks are due to the Ministry of Aviation for permission to
publish.

D.C.G.

Contents

List of Symbols and Abbreviations

Symbols for quantities are in *italic* type, and abbreviations for the names of units (unit symbols) are in ordinary type.

A_p	power gain
A_v	voltage gain
A	ampere
B	magnetic flux density
c	speed of light
C	capacitance
C_{ac}	anode–cathode capacitance
C_{ga}	grid–anode capacitance
C_{gc}	grid–cathode capacitance
dB	decibel
dBm dBr dBW	} see p. 76
E_{cb}	collector–base bias voltage
E_{cc}	collector supply voltage
E_{eb}	emitter–base bias voltage
f	frequency
f_0	resonant frequency
f_1	lowest frequency in modulating signal
f_2	highest frequency in modulating signal
f_c	carrier frequency
f_m	frequency of modulating signal
f.d.m.	frequency division multiplex
F	farad
g_m	mutual conductance
GHz	gigahertz (10^9 Hz)
H	henry
h	parameters (see p. 171)

h_{fb}	common–base short-circuit current gain (alternative, α)
h_{FB}	common–base d.c. current gain
h_{fe}	common–emitter short-circuit current gain (alternatives, β and α')
Hz	hertz
I_a	anode current
I_b	base current
I_c	collector current
I_e	emitter current
I_k	cathode current
I_{CBO}	common–base collector leakage current
I_{CEO}	common–emitter collector leakage current
kHz	kilohertz (10^3 Hz)
kW	kilowatt (10^3 W)
L	inductance
m	modulation factor
m	metre
mA	milliampere (10^{-3} A)
m/s	metre per second
mW	milliwatt (10^{-3} W)
MHz	megahertz (10^6 Hz)
MΩ	megohm ($10^6\ \Omega$)
N	newton
N/m^2	newton per square metre
N_p	number of turns on primary winding
N_s	number of turns on secondary winding
pF	picofarad (10^{-12} F)
P	power
Q	charge
r_a	anode a.c. resistance
r_{ac}	a.c. resistance of a diode
r_b	base resistance
r_c	a.c. resistance of reverse-biased collector–base junction
r_e	a.c. resistance of forward-biased emitter–base junction
R	resistance
r.m.s.	root mean square

s	second
T	period
t.d.m.	time-division multiplex
v	instantaneous voltage
v_p	phase velocity of propagation
V	peak value or amplitude of a voltage wave
V_a	anode voltage
V_{be}	base–emitter voltage
V_c	amplitude of carrier wave
V_{cb}	collector–base voltage
V_{ce}	collector–emitter voltage
V_{eb}	emitter–base voltage
V_g	grid voltage
V_m	amplitude of modulating signal
V_p	voltage across primary winding
V_s	screen voltage
V_s	voltage across secondary winding
V	volt
W	watt
X	reactance
Z	impedance
α	common–base short-circuit current gain (alternative, h_{fb})
α'	common-emitter short-circuit current gain (alternatives, h_{fe} and β)
β	base transmission factor common-emitter short-circuit current gain (alternatives, h_{fe} and α')
γ	emitter injection ratio
ϵ_r	relative permittivity
θ	phase angle of wave
λ	wavelength
μ	amplification factor
μF	microfarad (10^{-6} F)
μV	microvolt (10^{-6} V)
ω	angular velocity
Ω	ohm

1 Sound

Sound may be said to consist of mechanical vibrations of the air, or some other substance, at frequencies lying within the frequency range over which the ear can respond. Thus the source of a sound must always be the movement of a body although this movement may often be so small that the eye cannot detect it.

Sounds are produced by anything that will cause the air to vibrate sufficiently rapidly, such as the firing of a gun, the starting of a car, or the squeaking of a rusty gate. Some vibrations, however, are too slow to produce sounds, for example, the movement of a person's legs when walking or the fluttering of a flag in the wind.

Fig. 1.1

Vibration of a tuning fork

Consider, as an example, that a tuning fork has been caused to vibrate as shown in Fig. 1.1.

Before commencing to vibrate the tuning fork is in its stationary position, A, and it does not exert any pressure upon the layers of air adjacent to it. The adjacent layers of air remain in the positions marked x. Once the tuning fork begins to vibrate it will oscillate about its stationary position as shown; consider what happens when a prong of the fork is, instantaneously, in the position B. The prong will exert pressure upon the neighbouring layer of air causing it to move to the new position x'; this layer, in turn, will exert pressure on the next layer of air, y, moving it to a new position y'. Again, this displaced layer of air will disturb the air adjacent to it, and so on, and in this way the initial disturbance of the air, caused by the movement of the prong of the tuning fork, travels through the air. Each layer is said to have been compressed.

Conversely, when a prong moves from position B to position C, via position A, the pressure exerted on the adjacent layer x is reduced and the air is said to be rarefied. As a result the layer will move to the position x''. This movement of air causes the air layer y to move to position y'', and so on for succeeding layers of air. Thus another disturbance of the air, but this time a rarefaction, is transmitted through the air. The vibrations of the tuning fork thus cause a series of disturbances to be propagated through the air with a velocity of approximately 340 m/s (at an air temperature of 15°C).

Suppose the tuning fork to be vibrating sinusoidally; then at any given distance from the fork the displacement of the air layer at that point also varies sinusoidally, see Fig. 1.2.

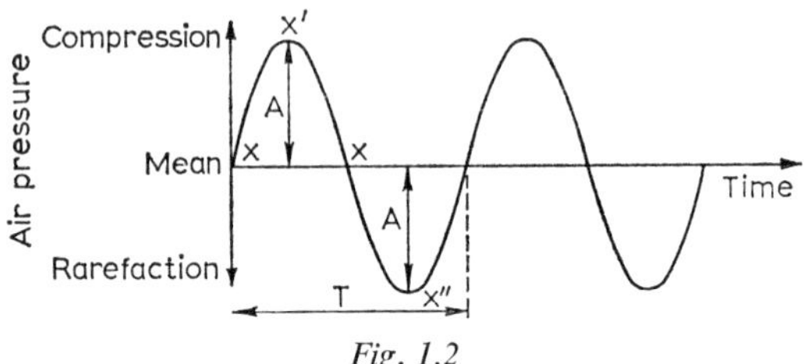

Fig. 1.2

Variation with time of the air pressure at a point

The frequency f of the sound wave is the number of cycles which occur in one second and is determined solely by the rate of vibration of the body producing the sound. The amplitude A of the wave is the maximum distance *either* side of its mean value that is reached during a cycle, while the intensity of the wave is proportional to the square of the amplitude. Finally, the period T of the wave is the time taken for the wave to go through one cycle and since the frequency f is the number of cycles per second, the relationship between the frequency and the period, or the periodic time, is

$$f = \frac{1}{T} \quad \text{Hz*} \tag{1.1}$$

Sound waves travel away from their source with a finite velocity; hence if a plot is made of the magnitude of the air layer displacement at different distances from the source, at a particular instant in time, a graph such as that shown in Fig. 1.3 is obtained. Three instants have been selected, (*a*) the instant when the air layer at the source is at its stationary position but about to be compressed, (*b*) an instant when the air layer at the source has maximum rarefaction and (*c*) an instant when the air layer at the source is in its stationary position but is about to be rarefied. It can be seen that the peak values, labelled p' and p'', of the waveform travel outwards from the source of the sound. The distance between two consecutive points of the

---

* The Hertz (Hz) was formerly known as the cycle per second (c/s).

same phase, e.g. two compression peaks, is known as the wavelength λ of the wave.

The velocity with which a sound wave travels depends upon the nature of the medium in which it is being propagated, and on the temperature. Table 1.1 gives some typical values at a temperature of $0°C$.

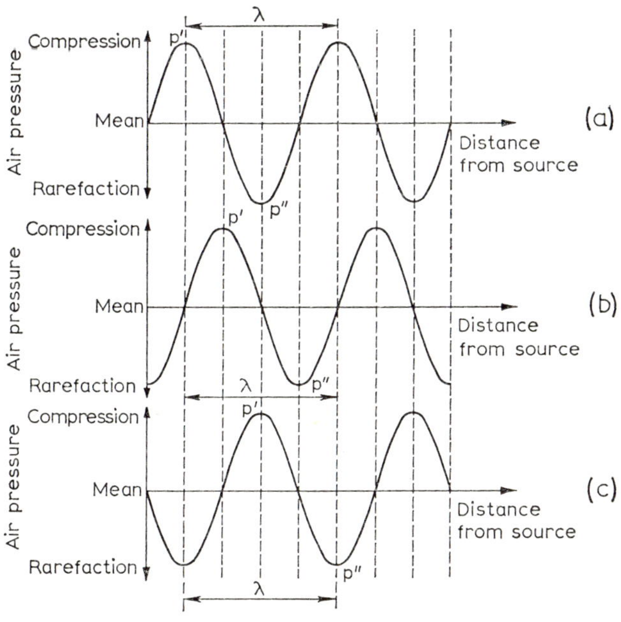

Fig. 1.3

Variation with distance from source of the air pressure at a given instant in time

Table 1.1

MATERIAL	VELOCITY OF SOUND
Air	332 m/s
Water	1,432 m/s
Brick	3,658 m/s
Iron	4,877 m/s

THE RELATIONSHIP BETWEEN FREQUENCY, WAVELENGTH AND VELOCITY OF PROPAGATION

The distance occupied by one complete cycle of a sound wave, i.e. its wavelength λ, is equal to the velocity of propagation of the wave, v_p, divided by the number of cycles f occurring in one second; hence

$$\lambda = \frac{v_p}{f} \tag{1.2}$$

where λ and v_p are in metres and metres/second respectively.

This relationship is very important since it also holds good for electromagnetic waves, e.g. radio waves.

Example 1.1
A sound wave is propagated with a velocity of 400 m/s. If its frequency is found to be 400 Hz, what is its wavelength?

Solution. From equation (1.2)

$$\lambda = \frac{v_p}{f}$$

Therefore, substituting the values given in the question,

$$\lambda = \frac{400}{400} = 1 \text{ metre} \qquad Ans.$$

NON-SINUSOIDAL WAVEFORMS

The discussion so far has been in terms of the sinusoidal sound wave produced by a vibrating tuning fork. In practice, however, the vast majority of the sounds heard by the ear have a waveform which is far from sinusoidal. This fact is illustrated by Fig. 1.4 which shows

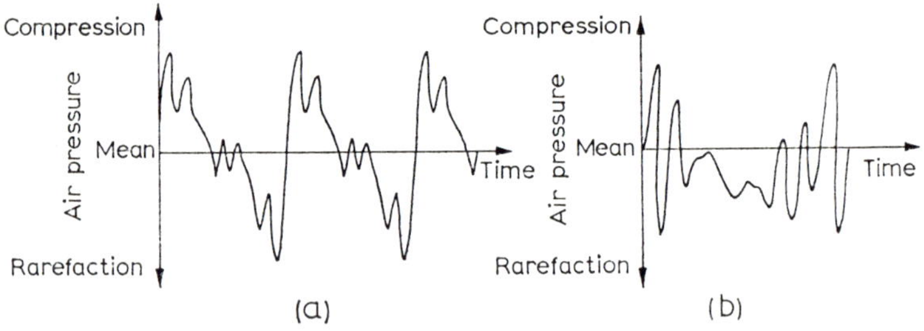

Fig. 1.4

Two typical sound waveforms
(*a*) "Ah" sung by human voice
(*b*) Note played on a clarinet

the waveforms of the sound "Ah" sung by the human voice and played on a clarinet. It should be noted that the waveforms are repetitive, and indeed, all sounds that are pleasant to the ear have a repetitive waveform.

It can be shown, however, mathematically or graphically, that any repetitive waveshape consists of a fundamental frequency component plus a number of other components whose frequencies are integral multiples of the fundamental frequency, and are known as harmonics of the fundamental frequency. It follows from this that the previous discussion of the propagation of a sinusoidal sound wave may be extended to include other, more practical, waveforms.

It should, perhaps, be noted at this point that sound produced by a non-repetitive waveform is discordant to the ear and is known as noise.

The Subjective Properties of Sound

The characteristics of sound mentioned so far are all objective; this means that they are capable of measurement with a measuring instrument of some kind. The characteristics which matter to the person hearing a sound, however, are subjective and cannot be measured with an instrument but must be assessed by that person. The subjective properties of sound are (*a*) the pitch, (*b*) the loudness and (*c*) the quality or timbre.

(A) PITCH

The pitch of a sound is its frequency or note as judged by the listener. Pitch is mainly determined by frequency but depends to some extent upon intensity also. If the sound observed is complex, i.e. comprises two or more frequencies, the pitch is that of the fundamental frequency contained in the sound waveform. Differences in pitch depend upon the *ratio* of the fundamental frequencies of the sounds heard and not on the *difference* between the frequencies. For example, if a person hears three sounds, one after the other, and their fundamental frequencies are 800 Hz, 1,600 Hz and 3,200 Hz the difference in pitch is the same in each case since the frequency ratio is constant at 2:1. A frequency ratio of 2:1 is known as an octave.

(B) LOUDNESS

Loudness, like pitch, is a sensation experienced by an observer and is dependent upon the intensity of the sound listened to and the sensitivity of the ear of the listener. Experiments have shown that sounds at different frequencies appearing equally loud to an observer have different intensities. Fig. 1.5 shows this fact graphically; it can be seen that the intensity required to produce a given loudness sensation varies with the frequency of the sound and that the minimum intensity is required at a frequency of approximately 1,000 Hz.

Because the minimum sound intensity for a given loudness is required at 1,000 Hz this frequency has been adopted as the frequency at which loudness measurements are (usually) carried out. A zero noise point has been agreed upon and it is the loudness, as

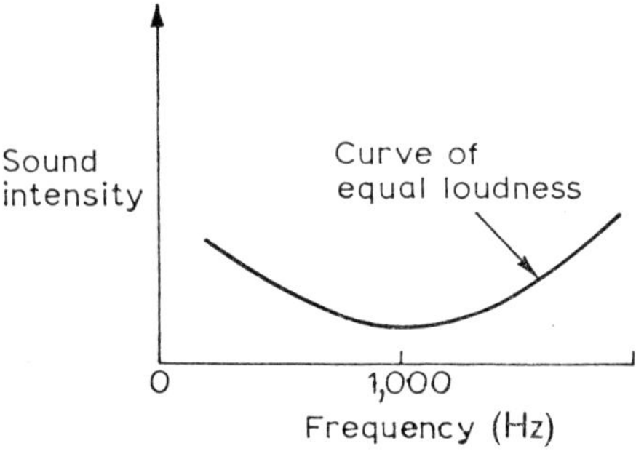

Fig. 1.5

Intensity/frequency curve for equal loudness

heard by the average observer, of a 1,000 Hz tone having a pressure of 0·00002 N/m².

The unit of loudness is the phon and this is defined thus:

A 1,000 Hz tone has a loudness of x phons if

$$x = \log_{10} \frac{\text{sound pressure in N/m}^2}{0 \cdot 00002} \tag{1.3}$$

Should the loudness of a sound at a frequency other than 1,000 Hz be required then an observer is employed to compare the loudnesses of the sound in question and a 1,000 Hz tone. The loudness of the latter is varied until the observer considers both sounds to be equally loud, when the loudness of the sound is x phons. It then remains to obtain the value of x by measurement of the sound pressure produced by the tone and the use of equation (1.3).

(c) QUALITY OR TIMBRE

If the same note is sounded on a violin and then on a clarinet the ear can easily tell that two different instruments have been employed. The ear is able to distinguish between different musical instruments sounding the same note because the quality of the note is different in each case. This is because the waveform of a note produced by a musical instrument is never sinusoidal but always contains a number of harmonics and these provide the quality of the note. The quality of any sound is determined by the presence of the various harmonics in its waveform and their amplitudes relative to one another.

The Transmission of Sound

In telecommunication engineering, media are provided for the transmission of intelligence from one point to another; for example, telephone conversations pass over telephone cables and radio programmes are broadcast through the atmosphere. It is necessary to ensure that sufficient information is available at the receiving end of a system to allow the intelligence to be understood and appreciated by the person receiving it. The requirements demanded of the transmission media depend upon the type of intelligence to be transmitted, but for the transmission of speech and music the main requirement is that sufficient frequencies are retained in the transmitted sound waveform to permit the received sound to be understandable and, in the case of music, enjoyable also. It is therefore necessary to have an appreciation of the range of frequencies produced by the human voice, and by musical instruments, and the frequency range over which the human ear is capable of responding.

THE VOICE AND SPEECH

A current of air expelled by the lungs passes through a narrow slit between the vocal cords in the larynx and causes them to vibrate. This vibration is then communicated to the air via various cavities in the mouth, throat and nose. The shape and size of the nose cavities are more or less fixed, but the mouth and throat cavities can have their shapes and dimensions considerably changed by the action of the tongue, lips, teeth and the throat muscles. The frequency at which a particular cavity allows the air to vibrate most freely depends upon the shape and dimensions of the cavity and these can readily be adjusted by the movement of the lips, tongue and teeth. The pitch of the spoken sound depends upon both the length and tension of the vocal cords and the width of the slit between them. The length of the vocal cords varies from person to person; for example a woman will have shorter vocal cords than a man, while the tension and distance apart of the vocal cords is under the control of muscles.

When a person speaks, his vocal cords vibrate and the resulting sounds, which are rich in harmonics, but of almost constant pitch, are carried to the cavities in the mouth, throat and nose. Here the sounds are given some of the characteristics of the desired speech by the emphasizing of some of the harmonics contained in the sound waveforms and the suppression of others. Sounds produced in this way are the vowels, a, e, i, o and u, and contain a relatively large amount of sound energy. Consonants are made with the lips, tongue and teeth and contain much smaller amounts of energy and often include some relatively high frequencies.

The sounds produced in speech contain frequencies which lie within the frequency band 100–10,000 Hz. The pitch of the voice is determined by the fundamental frequency of the vocal cords and is about 200–1,000 Hz for women and about 100–500 Hz for men.

The power content of speech is small, a good average being of the the order of 10–20 microwatt. However, this power is not evenly distributed over the speech frequency range, most of the power being contained at frequencies in the region of 500 Hz for men and 800 Hz for women.

MUSIC

The notes produced by musical instruments occupy a much larger frequency band than that occupied by speech. Some instruments, such as the organ and the drum, have a fundamental frequency of 50 Hz or less while many other instruments, for example the violin and the clarinet, can produce notes having a harmonic content in excess of 15,000 Hz. The power content of music can be quite large. A large orchestra may generate a peak power somewhere in the region of 90–100 watts while a bass drum well thumped may produce a peak power of about 24 watts.

HEARING

When sound waves are incident upon the ear they cause the ear drum to vibrate. Coupled to the ear drum are three small bones which transfer the vibration to a fluid contained within a part of the inner ear known as the cochlea. Inside the cochlea are a number of hair cells and the nerve fibres of these are activated by vibration of the fluid. Activation of these nerve fibres causes them to send signals, in the form of minute electric currents, to the brain where they are interpreted as sound.

The ear can only hear sounds whose intensity lies within certain limits; if a sound is too quiet it is not heard and, conversely, if a sound is too loud it is felt rather than heard and causes discomfort or even pain. The minimum sound intensity, measured in N/m^2, that can be detected by the ear is known as the "threshold of hearing or audibility" and the sound intensity that just produces a feeling of discomfort is known as the "threshold of feeling". The ear is not, however, equally sensitive at all frequencies, as shown in Fig. 1.6.

In this diagram curves have been plotted showing how the thresholds of audibility and feeling vary with frequency for an average person.

It can be seen that the frequency range over which the average human ear is capable of responding is approximately 30–16,500 Hz, but this range varies considerably with the individual. It is also to be noted that the ear is most sensitive in the region of 1,000 to 2,000 Hz and becomes rapidly less sensitive as the upper and lower limits of audibility are approached. The limits of audibility are clearly determined not only by the frequency of the sound but also by its intensity. See, for example, the increase in the audible frequency

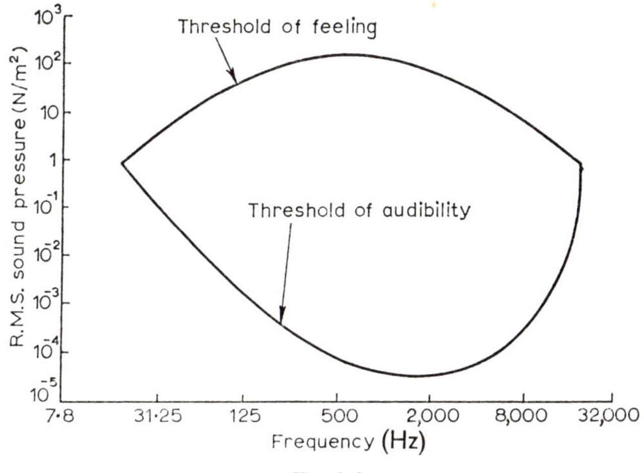

Fig. 1.6

The thresholds of audibility and feeling

range when the sound intensity is increased from, say, 1×10^{-3} N/m² to 1×10^{-2} N/m². At the upper and lower limits of audibility the thresholds of audibility and feeling coincide and it becomes difficult for an observer to distinguish between hearing and feeling a sound.

THE TRANSMISSION OF SOUND OVER A SIMPLE TELEPHONE CIRCUIT
The intensity of a sound wave rapidly diminishes as it travels away from the source producing it and if conversation over a long distance is desired a telephone circuit must be employed.

The arrangement of a simple, unidirectional telephone circuit is shown in Fig. 1.7.

Details of the construction and operation of transmitters (or microphones) and receivers are given in Chapter 4. For the moment it is sufficient to consider the microphone as a device whose electrical resistance varies in accordance with the waveform of the sound incident upon it, and to consider the receiver as a device which vibrates when a varying current is passed through it and in so doing produces sound waves having the same waveform as the current.

When no sound is incident upon the microphone the resistance of the microphone is constant and a steady current flows into the line because of the presence of the battery. When a person speaks into

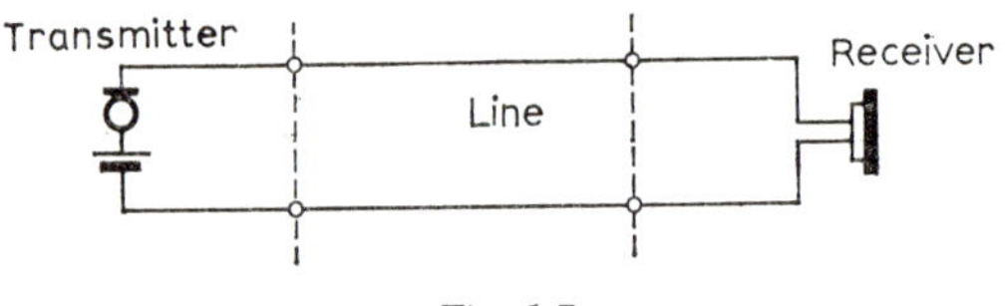

Fig. 1.7

A simple unidirectional speech circuit

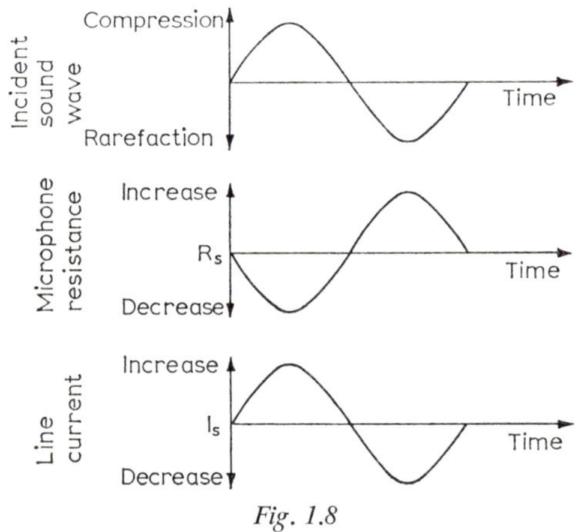

Fig. 1.8

Showing the relationship between the sound incident on a microphone, the microphone resistance and the current flowing to line

R_s, microphone resistance when no sound is incident upon it
I_s, steady current to line when no sound is incident on the microphone

the microphone its resistance varies in the same way as the speech waveform and hence so does the current flowing to line. For example, during one half-cycle of the speech waveform the resistance of the microphone is decreased and so the line current is increased, while during the next half-cycle the microphone resistance is increased and the line current decreased. Thus the line current is continuously varying about its steady value, Fig. 1.8. The varying line current passes through the receiver at the distant end of the line and causes the receiver to vibrate and reproduce the original speech.

For a two-way conversation this simple arrangement would have to be duplicated, the second circuit having the positions of the microphone and battery and the receiver reversed. Such an arrangement would be uneconomic since it would require two pairs in a telephone cable and telephone cables are very expensive. A further disadvantage of the simple circuit is that as the length of line is increased the variation in the resistance of the microphone becomes an increasingly smaller fraction of the line resistance. This means that the magnitude of the changes in line current decreases with increase in line length until the current changes can no longer operate the receiver. A telephone circuit which overcomes these disadvantages is shown in Fig. 1.9. In this circuit, each microphone is connected

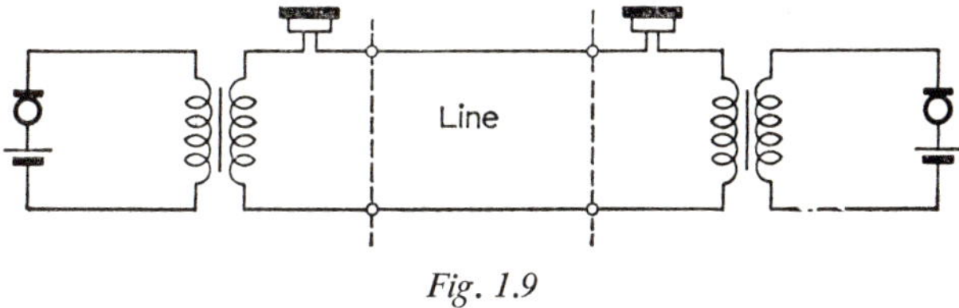

Fig. 1.9

A simple bothway speech circuit

to the line via a transformer but the two receivers are directly connected to line. When a person speaks into either of the two microphones the resulting changes in microphone resistance cause a relatively large varying current to flow in the local microphone circuit. In passing through the transformer primary winding this varying current induces an e.m.f. in the secondary winding and the induced e.m.f. drives a varying current, having the same waveform, to line. At the distant end of the line the varying current flows in the receiver and causes it to vibrate and so produce sound waves which are similar to the original speech.

Range of Frequencies Transmitted by Communication Systems

It has been assumed so far that all the frequencies present in a speech or music waveform are converted into electrical signals, transmitted over the communication system, and then reproduced as sound at the distant end. In practice, this is rarely the case for two reasons. Firstly, for economic reasons the devices employed in circuits that carry speech and music signals have a limited bandwidth; secondly, particularly for the longer-distance routes, a number of circuits are often transmitted over a single telecommunication system and this practice provides a further limitation of bandwidth. It is thus desirable to have some idea of the effect on the ear when it is responding to a sound waveform, the frequency components of which have amplitude relationships differing from those existing in the original sound.

By international agreement the frequency band for a "commercial quality" speech circuit is restricted to 300–3,400 Hz; this means that both the lower and upper frequencies contained in the average speech waveform are not transmitted. To the ear the pitch of a complex, repetitive, sound waveform is the pitch corresponding to the frequency difference between the harmonics contained in the waveform, i.e. the pitch is that of the fundamental frequency. Hence, even though the fundamental frequency itself may have been suppressed, the pitch of the sound heard by the listener is the same as the pitch of the original sound. However, much of the power contained in the original sound is lost. Suppression of all frequencies above 3,400 Hz reduces the quality of the sound but does not affect its intelligibility. Since the function of a telephone system is to transmit intelligible speech the loss of quality can be tolerated, sufficient quality remaining to allow a speaker's voice to be recognized.

Where the transmission of music is concerned the enjoyment of the listener must also be considered. The enjoyment of a person listening to music transmitted over a communication system may be considerably impaired if too many of the higher harmonics have been suppressed, since the loss of these harmonics could well result in it becoming difficult, if not impossible, to distinguish between the various kinds of musical instruments. For music circuits routed over line communication systems a wider bandwidth must be allowed than is allocated to commercial speech circuits. A typical bandwidth employed in practice is 30 to 10,000 Hz, excellent quality reproduction then being possible. Land-line music circuits of this kind are employed to connect together two B.B.C. studios or a studio and a transmitting station. When music is broadcast in the long and medium wavebands a bandwidth as wide as 30–10,000 Hz cannot be achieved because the wavebands are shared by so many different broadcasting stations. By international agreement broadcasting stations in Europe are spaced approximately 9,000 Hz apart in the frequency spectrum, and this means that to make it possible for any particular station to be selected by a radio receiver, without undue interference from adjacent (in frequency) stations, the output sound bandwidth of the receiver cannot be much greater than 4,500 Hz. Thus the effective bandwidth of a broadcast transmission, be it music or speech, is of the order of 50–4,500 Hz. For various reasons the same selectivity is not demanded of very high frequency (v.h.f.) sound broadcast and television receivers (405 line), or of ultra high frequency (u.h.f.) television receivers (625 line), and consequently the sound bandwidths handled by these receivers are of the order of 12,000 Hz or more.

Exercises

1. Speech is to be transmitted from one point to another by electrical means and without the intervention of an exchange. Draw a circuit diagram of a suitable arrangement. State the functions and the general principle of operation of each of the devices used in the arrangement you adopt.

   State, in general terms, the effect of any line resistance on the performance of your arrangement. (E 1962)

2. Give a brief description of the means whereby sound waves are converted into electric currents suitable for transmission along a pair of wires and reconverted at the receiving end into sound waves. (P1 1959)

2 Modulation

The bandwidth required for the transmission of commercial-quality speech is 300–3,400 Hz and as long as a physical pair of wires, or two pairs of wires, are provided to connect the two parties to a conversation no problem exists. Telephone cable is, however, very expensive and, together with the associated duct work and manholes, comprises the major part of the cost of linking two points. It is desirable, therefore, to be able to transmit more than one conversation over a given link and thus economize in telephone cable. If a number of telephone conversations were merely transmitted into one end of a line it would not be possible to separate them at the distant end of the line since each conversation would be occupying the same part of the frequency spectrum. How then can many different conversations be transmitted over a single circuit and yet be separable at the distant end? Two main methods exist: in one, known as Frequency-division Multiplex, each conversation is shifted, or translated, to a different part of the frequency spectrum, in the other, known as Time-division Multiplex, the conversations each occupy the same frequency band but are applied in rotation to the common line.

FREQUENCY-DIVISION MULTIPLEX

To illustrate the principle of a frequency-division multiplex (f.d.m.) system, consider the simple case where it is required to transmit three telephone channels, of bandwidth 300 to 3,400 Hz, over a common line. The first of these channels can be transmitted directly over the common line and it will then occupy the band 300 to 3,400 Hz. The second and third channels cannot also be transmitted directly over the line because they would be inseparable from the first channel and from each other. Suppose, therefore, that instead these two channels are each passed into a device which frequency translates, or shifts, them to the frequency bands 4,300 to 7,400 Hz and 8,300 to 11,400 Hz respectively, before transmission to line. The three channels can now all be transmitted over the common line but since there is a frequency gap of 900 Hz between them no inter-channel interference will occur. At the receiving end of the line, devices known as filters must be employed to separate the three channels and further devices introduced to restore the second and third channels to their original frequency bands. The bandwidth which must be provided by the common circuit is 300–11,400 Hz.

The frequency translation of a channel to a position higher in the frequency spectrum is known as modulation and the device which achieves it as a modulator; the restoration of a channel to its original position in the frequency spectrum is known as demodulation, the device employed being called a demodulator. Modulation may be defined as the process by which one of the characteristics of a carrier wave is modified in accordance with the characteristics of a modulating signal.

A block schematic diagram of the equipment required for one direction of transmission in the three-channel f.d.m. system just

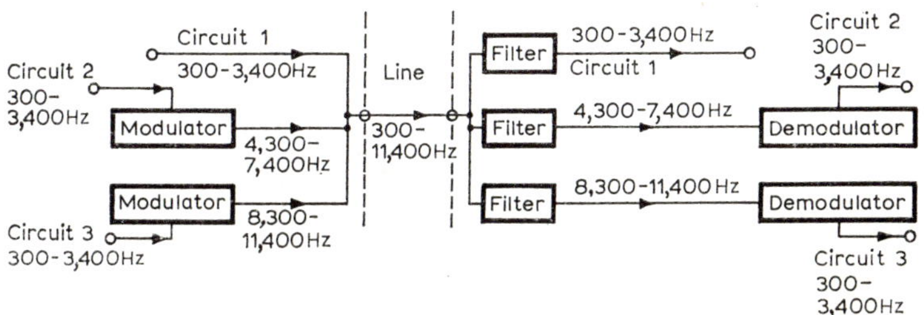

Fig. 2.1

A simple 3-channel f.d.m. system

described is shown in Fig. 2.1. It should be noted that since the equipment employed is unidirectional, it needs to be duplicated for transmission in both directions. Furthermore, the line may be a telephone cable or may be a v.h.f. or u.h.f. radio link.

Frequency translation is also employed for radio and television broadcasting. It is well known that such broadcasts are transmitted by the broadcasting authority and received in the home by means of aerials, but no type of aerial is able to operate at audio frequencies. It is therefore necessary to shift each programme originally produced in the audio-frequency band to a point higher in the frequency spectrum where aerials can operate with reasonable efficiency. Since there is usually a large number of broadcasting stations within a given geographical area it is necessary to arrange that each station, as far as possible, occupies a different part of the usable frequency spectrum. Hence the programmes radiated by different broadcasting stations are frequency translated to their own, internationally agreed, fixed frequency bands. For example, consider the medium-wave Light and Third programmes of the B.B.C. The former is shifted to frequency bands centred on 200 kHz and 1,214 kHz and the latter to frequency bands centred on 647 kHz and 1,546 kHz.

TIME-DIVISION MULTIPLEX

With time-division multiplex (t.d.m.) a number of different channels can be transmitted over a common circuit by allocating the common circuit to each channel in turn for a given period of time, i.e. at any particular instant only one channel is connected to the common circuit. The principle of a t.d.m. system is illustrated by Fig. 2.2 which shows the basic arrangement of a two-channel t.d.m. system.

The two channels which are to share the common circuit are each connected to it via a channel gate. The channel gates are electronic switches which only permit the signal present on a channel to pass when opened by the application of a controlling pulse. Hence,

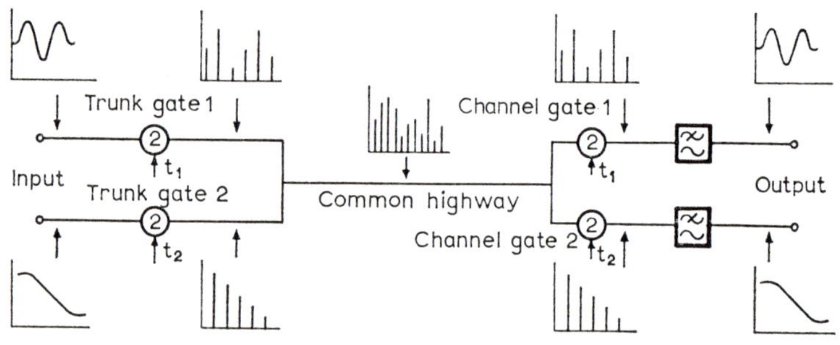

Fig. 2.2

A simple 2-channel t.d.m. system

t_1, a series of pulses occurring at fixed intervals
t_2, a series of pulses occurring at the same periodicity as t_1 but commencing later
   by an amount equal to half the time interval

(From the *Post Office Electrical Engineers' Journal*)

if the controlling pulse is applied to gate 1 at time t_1 and not to gate 2, gate 1 will open for a time equal to the duration of the pulse but gate 2 will remain closed. During this time, therefore, a pulse, or sample, of the amplitude of the signal waveform on channel 1 will be transmitted to line. At the end of the pulse both gates are closed and no signal is transmitted to line. If now the controlling pulse is applied to gate 2, at a later time t_2, gate 2 will open and a sample of the signal waveform on channel 2 will be transmitted. Thus if the pulses applied to control the opening and shutting of gates 1 and 2 are repeated at regular intervals, a series of samples of the signal waveforms existing on the two channels will be transmitted.

At the receiving end of the system gates 1 and 2 are opened, by the application of control pulses, at those instants when the incoming waveform samples appropriate to their channel are being received.

This requirement demands accurate synchronization between the controlling pulses applied to the gates 1, and also between the controlling pulses applied to the gates 2. If the time taken for signals to travel over the common circuit was zero then the system would require controlling pulses as shown, but since, in practice, the transmission time is not zero, the controlling pulses applied at the receiving end of the system must occur slightly later than the corresponding controlling pulses at the sending end. If the pulse synchronization is correct the waveform samples are directed to the correct channels at the receiving end. The received samples must then be converted back to the original waveform, i.e. demodulated. Provided the sampling rate, i.e. the number of controlling pulses per second, is at least equal to twice the highest frequency contained in the original waveform, correct demodulation can be achieved by merely passing the samples through a filter network that passes freely all frequencies lower than the sampling frequency.

In the foregoing description of a simple t.d.m. system it has been assumed that the samples of the information signal are directly transmitted to line. Very often, however, it is desirable for the transmitted pulses to occupy an entirely different part of the frequency spectrum. For example radio-frequency pulses may be employed so that the samples may be efficiently radiated by an aerial. When this is the case the equipment required at each end of a t.d.m. system is somewhat more complex than that shown in the diagram.

Types of Modulation

For a signal to be frequency translated to another part of the frequency spectrum it is necessary for the signal to vary one of the characteristics of a sinusoidal wave—usually known as the carrier wave—whose frequency occupies the required part of the spectrum. The process by which one of the characteristics of a carrier wave is modified in accordance with the characteristics of a modulating signal is known as modulation.

The general expression for a sinusoidal carrier wave is

$$v = V \sin (\omega t + \theta) \tag{2.1}$$

where $v =$ instantaneous voltage of the wave,
    $V =$ peak value or amplitude of the wave,
    $\omega =$ angular velocity of the wave in radians/second; ω is related to the frequency of the wave by the expression $\omega = 2\pi f$ where f is the frequency in hertz,
    $\theta =$ the phase of the wave at the instant when $t = 0$.

For modulation it is necessary to cause one of the characteristics of the wave to be varied in accordance with the waveform of the modulating signal. Three possibilities exist—

(*a*) the amplitude V of the wave may be varied to give amplitude modulation,

(*b*) the frequency $\omega/2\pi$ may be modified to give frequency modulation,

(*c*) the phase θ may be the variable, giving phase modulation.

To illustrate, graphically, the difference between these three types of modulation consider the case in which the modulating signal is a positive-going pulse as in Fig. 2.3*a*.

In the case of the amplitude-modulated wave, the frequency of the carrier wave remains constant but its amplitude is suddenly increased when the leading edge of the modulating pulse occurs and then suddenly reduced to its original value when the pulse ends, Fig. 2.3*c*. The phase-modulated carrier wave has a constant amplitude and frequency, but experiences an abrupt change of phase at the instants corresponding to the beginning and end of the modulating signal pulse, Fig. 2.3*d*. Finally, the frequency-modulated carrier wave also has a constant amplitude, but the frequency of the wave is abruptly increased at the beginning of the modulating pulse and maintained at the new frequency until the pulse ends, when the frequency is suddenly decreased to its original value, Fig. 2.3*e*.

Amplitude modulation is the most commonly used method in Great Britain. It is employed for f.d.m. telephony and telegraphy systems, radio broadcasting in the long and medium wavebands, the picture signals of both 405 line and 625 line television broadcasting, the sound signals of 405 line television broadcasting and for various base-to-mobile radio-telephony systems such as police and taxi-cab communication.

Frequency modulation is employed for v.h.f. radio broadcasting and for the sound signal of 625 line television broadcasting, as well as for some base-to-mobile radio-telephony systems. Phase modulation is rarely employed in its own right but it is often used as a stage in the production of a frequency-modulated signal. Frequency modulation has the advantage over amplitude modulation in that its performance in the presence of interfering signals and noise is much better. It suffers from the disadvantages of added complexity and cost and, perhaps most important, requiring a larger bandwidth. Because of the large frequency band required for a frequency-modulated system the use of frequency modulation is restricted to very high frequencies where the bandwidth required can be more readily provided.

Systems employing the principle of time-division multiplex are generally based on the sampling of the amplitude of the information signal at regular intervals, and the subsequent transmission of a

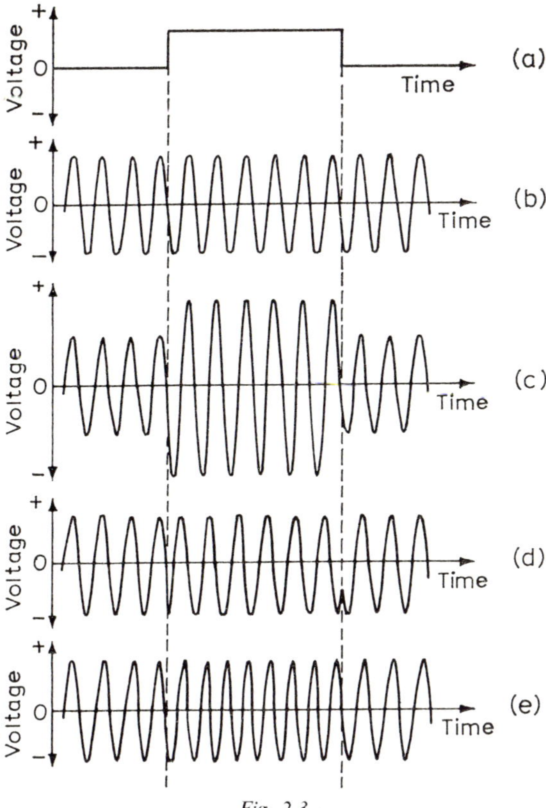

Fig. 2.3

Illustrating the difference between amplitude, frequency and phase modulation

    (*a*) Modulating signal
    (*b*) Unmodulated carrier wave
    (*c*) Amplitude-modulated wave
    (*d*) Phase-modulated wave
    (*e*) Frequency-modulated wave

short pulse of radio-frequency carrier for each sample. For the intelligence contained in the information signal to be transmitted, the pulse characteristics must, in some way, be varied in accordance with the amplitude of the signal at the instant of sampling. Four main types of pulse modulation exist, namely pulse-amplitude modulation (p.a.m.), pulse-duration modulation (p.d.m.), pulse-position modulation (p.p.m.) and pulse-code modulation (p.c.m.). The principle of each of the first three types of pulse modulation is shown in Fig. 2.4.

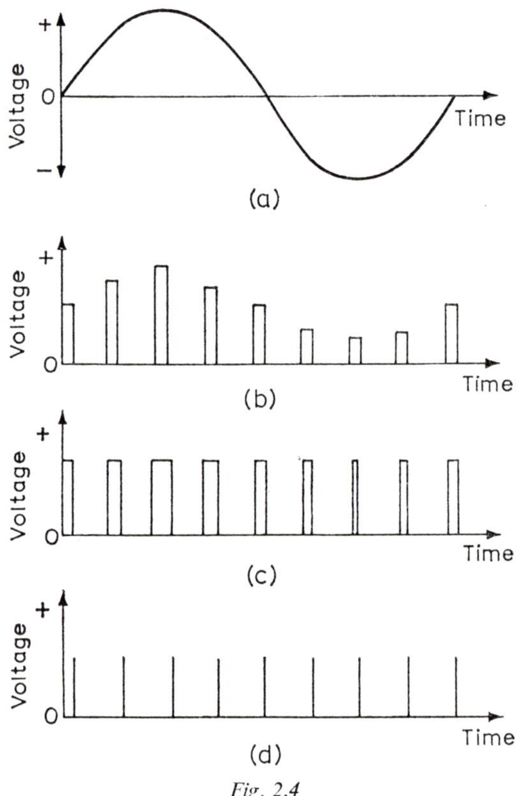

Fig. 2.4

Illustrating the difference between p.a.m., p.d.m. and p.p.m.

    (*a*) Modulating signal
    (*b*) Pulse-amplitude modulation
    (*c*) Pulse-duration modulation
    (*d*) Pulse-position modulation

With p.a.m., pulses of radio-frequency carrier of equal width and common spacing are employed and have their amplitudes varied in proportion to the instantaneous amplitude of the modulating signal. P.D.M. employs pulses of constant amplitude and common spacing between their leading edges, but whose width, or duration, is varied in accordance with the waveform of the modulating signal. Finally p.p.m. employs constant amplitude and width pulses whose position (in time) is dependent upon the instantaneous amplitude of the modulating signal.

Each of these three methods of pulse modulation suffers from the disadvantage that distortion and noise is cumulative along the system—although the p.p.m. system is much less affected than the two others in this way. The fourth type of pulse modulation, i.e. pulse code modulation (p.c.m.), is a more sophisticated method which largely overcomes the effects of distortion and noise, but at the expense of added complexity and a wider bandwidth. With a p.c.m. system, samples of the information signal are taken, as for the other systems, and the sampled amplitude information is translated into a suitable code. The code is transmitted to line as a series of pulses and equipment at the receiving end of the system is able to interpret this code and reproduce the original information signal. A p.c.m. system is relatively free from the effects of distortion and noise because it is merely necessary for the receiving equipment to be able to distinguish between the presence or absence of a pulse, the pulse shape being unimportant.

Pulse modulation systems have not so far been employed to any great extent in Great Britain but increasing attention is now being given to the possibility of employing p.c.m. systems in the telephone network.

Only amplitude modulation will receive further attention in this chapter.

Amplitude Modulation

If a carrier wave is amplitude modulated the amplitude of the carrier is caused to vary in accordance with the instantaneous value of the modulating signal. For example, consider the most simple case when the modulating signal is itself sinusoidal. The amplitude of the modulated wave must then vary in a sinusoidal manner as shown in Fig. 2.5. The difference in frequency between the carrier wave and the modulating signal will normally be much higher than is indicated by the diagram. In practice, the frequency difference would often be of the order of thousands of hertz.

The amplitude of the modulated carrier wave can be clearly seen to vary between a maximum value, which is greater than the amplitude of the unmodulated wave, and a minimum value, which is less than the amplitude of the unmodulated wave. The outline of the modulated carrier waveform is known as the modulation envelope.

To convey intelligence a signal containing at least two components at different frequencies is required and hence, in practice, a sinusoidal modulating signal is rarely employed. However, whatever the waveform of the modulating signal the same principle still holds good: the envelope of the modulated wave must be the same as the waveform of the modulating signal. For example, Fig. 2.6 shows the

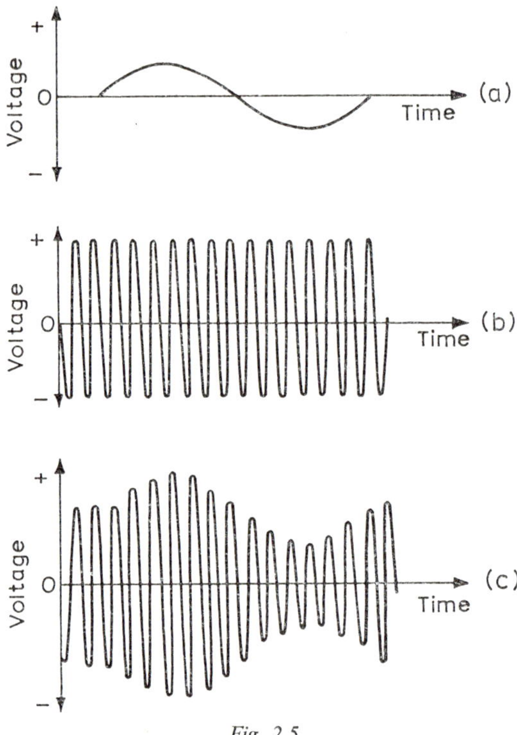

Fig. 2.5

A carrier wave amplitude modulated by a sinusoidal signal

(*a*) Modulating signal
(*b*) Unmodulated carrier wave
(*c*) Amplitude-modulated carrier wave

envelopes of a carrier wave modulated by (*a*) a signal consisting of components at a fundamental frequency and its third harmonic, both components being in phase at time $t = 0$, (*b*) note C played on a cello organ pipe and (*c*) a telegraphy signal produced by a teleprinter.

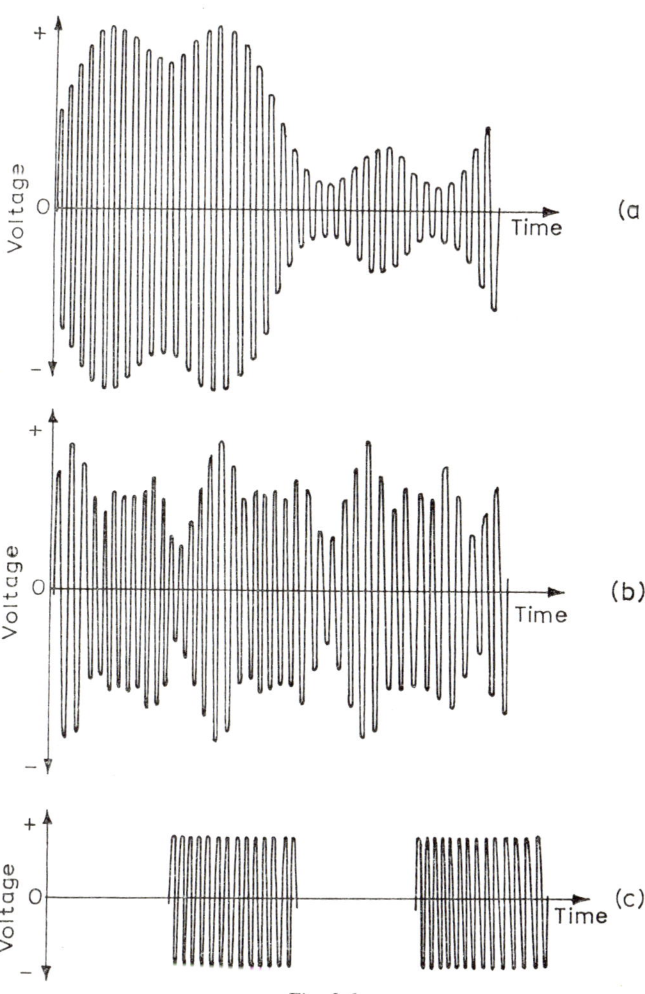

Fig. 2.6

Waveforms of a carrier wave modulated by (*a*) a fundamental plus third harmonic, (*b*) a cello organ pipe playing C, and (*c*) a telegraphy signal

THE FREQUENCIES CONTAINED IN AN AMPLITUDE-MODULATED WAVE

It is possible to show, with the aid of mathematics beyond the scope of this book, that when a sinusoidal carrier wave is amplitude modulated each frequency component in the modulating signal gives rise to two frequencies in the modulated wave, one below the carrier frequency and one above. When, for example, the modulating signal is a sinusoidal wave of frequency f_m the modulated carrier wave contains three frequencies; these are (*a*) the carrier frequency f_c, (*b*) the "lower side frequency" $(f_c - f_m)$, and (*c*) the "upper side frequency" $(f_c + f_m)$. The two new frequencies are the upper and lower side frequencies $(f_c + f_m)$ and $(f_c - f_m)$ respectively, and these are equally spaced either side of the carrier frequency by an amount equal to the modulating signal frequency f_m. The frequency f_m of the modulating signal itself is *not* present.

The bandwidth required to transmit a modulated carrier wave, or any other waveform for that matter, is equal to the difference between the highest frequency to be transmitted and the lowest. In the case of sinusoidal modulation and with $f_c > f_m$, the bandwidth required is given by

$$B = (f_c + f_m) - (f_c - f_m) = 2f_m$$

i.e. the required bandwidth is equal to twice the frequency of the modulating signal.

Example 2.1

A 100 kHz carrier wave is amplitude modulated by a sinusoidal tone of frequency 3,000 Hz. Determine the frequencies contained in the amplitude-modulated wave and the bandwidth required for its transmission.

Solution. The frequencies contained in the modulated wave are

(1) The carrier frequency $f_c = 100,000$ Hz   *Ans.*
(2) The lower side frequency $(f_c - f_m) = 100,000 - 3,000 = 97,000$ Hz *Ans.*
(3) The upper side frequency $(f_c + f_m) = 100,000 + 3,000 = 103,000$ Hz *Ans.*
   The bandwidth required $= 103,000 - 97,000 = 6,000$ Hz *Ans.*

The required bandwidth is obviously equal to twice the frequency f_m of the modulating signal. This relationship does not hold, however, if $f_m > f_c$; for example, if $f_c = 100$ kHz, as before, and $f_m = 103$ kHz the lower side frequency is 3 kHz and the upper side frequency is 203 kHz. The required bandwidth is 200 kHz and this is not equal to $2f_m$.

If the modulating signal is non-sinusoidal it will contain components at a number of different frequencies; suppose that the highest frequency contained in the modulating signal is f_2 and the lowest frequency is f_1. Then the frequency f_2 will produce an upper side frequency component $(f_c + f_2)$ and a lower side frequency component $(f_c - f_2)$, while frequency f_1 will produce upper and lower side frequencies $(f_c \pm f_1)$. Thus the modulating signal will produce a number of lower side frequency components lying in the range $(f_c - f_2)$ to $(f_c - f_1)$ and a number of upper side frequency components in the range $(f_c + f_1)$ to $(f_c + f_2)$. The band of frequencies below the carrier frequency, i.e. $(f_c - f_2)$ to $(f_c - f_1)$, is known as the lower sideband, while the band of frequencies above the carrier frequency is known as the upper sideband. When the carrier frequency is higher than the modulating signal frequency the sidebands are symmetrically situated, with respect to frequency, on either side of the carrier frequency. The lower sideband is said to be inverted because the highest frequency in it $(f_c - f_1)$ corresponds to the lowest frequency f_1 in the modulating signal and vice versa. Similarly, the upper sideband is said to be erect because the lowest frequency in it $(f_c + f_1)$ corresponds to the lowest frequency f_1 in the modulating signal.

Example 2.2
A 108 kHz carrier wave is amplitude modulated by a band of frequencies, 300–3,400 Hz. What frequencies are contained in the upper and lower sidebands of the a.m. wave and what is the bandwidth required to transmit the wave?

Solution. The lower sideband will contain frequencies in the band 108,000 − 3,400 Hz to 108,000 − 300 Hz, or 104,600 to 107,700 Hz   *Ans.*

The upper sideband will contain frequencies between 108,000 + 300 Hz to 108,000 + 3,400 Hz or 108,300 to 111,400 Hz   *Ans.*

The required bandwidth B is equal to the maximum frequency contained in the modulated wave minus the minimum frequency:

$$B = 111,400 - 104,600$$
$$= 6,800 \text{ Hz} \quad \textit{Ans.}$$

Note that the bandwidth is equal to twice the highest frequency contained in the modulating signal.

It is possible to confirm graphically the presence of a number of frequencies in an amplitude-modulated wave. Consider, for example, the case of a 10,000 Hz carrier wave and a 2,000 Hz modulating signal. The upper side frequency will be 12,000 Hz and the lower side frequency will be 8,000 Hz. The waveforms of the carrier, the upper side frequency and the lower side frequency components are shown in Figs. 2.7a, b and c respectively. The carrier component can be seen to complete 10 cycles, the upper side frequency component 12 cycles and the lower side frequency component 8 cycles in the time of one millisecond. To obtain the waveform of the amplitude-modulated wave, which is composed of these three components, it is necessary to sum the instantaneous values of the three waves. It can be seen that the envelope of the modulated wave shows two complete variations in the time of one millisecond, i.e. it varies at the modulating frequency of 2,000 Hz.

There are two ways in which the frequency spectrum occupied by a modulated carrier wave may be shown. Each component may be represented by an arrow drawn perpendicular to the frequency axis

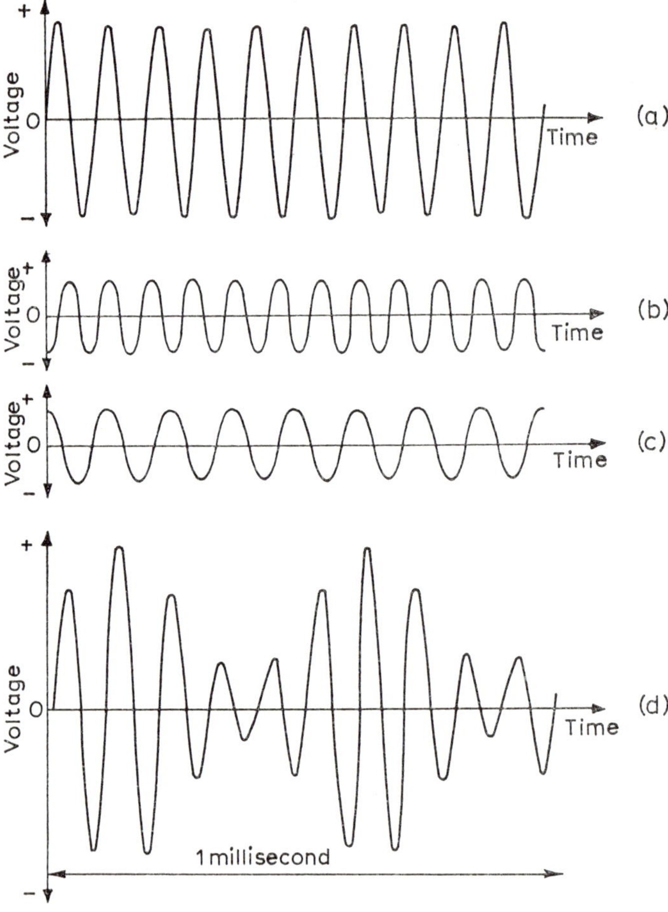

Fig. 2.7

Showing the formation of an amplitude-modulated wave by adding the components at the carrier and upper and lower side frequencies

(*a*) Carrier wave
(*b*) Upper side frequency
(*c*) Lower side frequency
(*d*) Amplitude-modulated carrier wave

as shown in Fig. 2.8, where $f_c, f_1,$ and f_2 have the same meanings as before. The lengths of the arrows are made proportional to the amplitude of the component they each represent. If many of the frequencies between f_1 and f_2 were present in the modulating signal a large number of arrows would be required and the diagram would become impracticable. It is usual, particularly in connection with multi-channel carrier telephony systems, to represent the sidebands produced by a complex modulating signal by truncated triangles, in which the vertical ordinates are made proportional to the modulating frequency and no account is taken of amplitude—see Fig. 2.9.

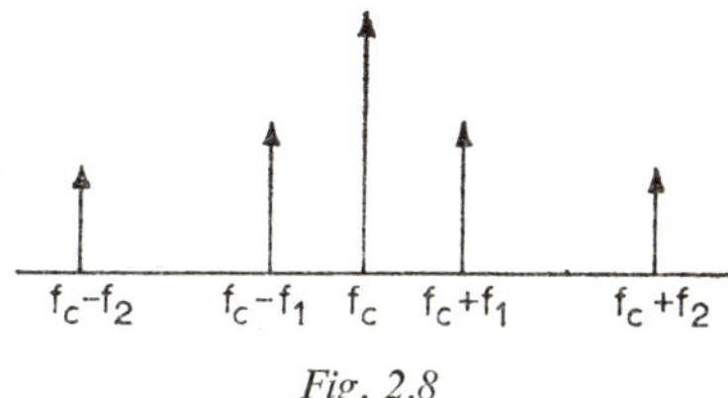

Fig. 2.8

Frequency spectrum of an amplitude-modulated carrier wave

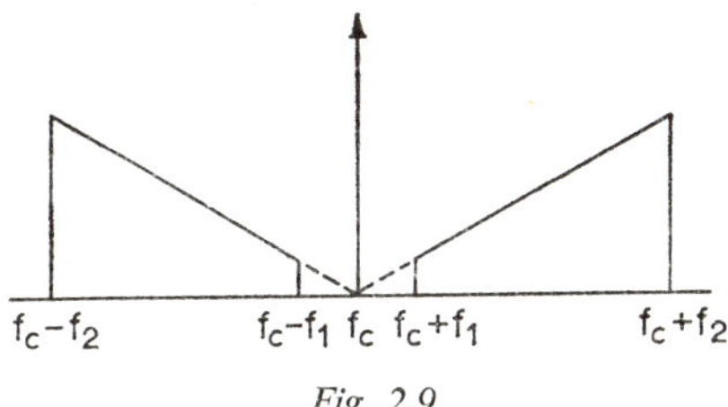

Fig. 2.9

A method of representing the sidebands of amplitude modulation

This method of representing sidebands gives an immediate indication of which sideband is erect and which is inverted. This is useful when considering systems employing more than one stage of modulation when the inverted sideband is not necessarily the lower sideband.

The products of the amplitude modulation of a carrier wave are two sidebands, each of which contains all the intelligence contained in the modulating signal, i.e. every frequency in the modulating signal has a corresponding frequency in each sideband. To transmit intelligence, therefore, it is only necessary to transmit *one* sideband; the other sideband, together with the carrier component, may be suppressed. Single-sideband (s.s.b.) working, as it is termed, has several advantages over the transmission of both sidebands and the carrier (double-sideband working or d.s.b.), the most important, perhaps, of which is that a considerable economy in bandwidth is obtained. The major disadvantage is the need for more complex receiving equipment and this has restricted the use of s.s.b. to f.d.m. line telephony systems and point-to-point radio telephony links.

MODULATION DEPTH

The envelope of an amplitude-modulated carrier wave varies in accordance with the waveform of the modulating signal and hence there must be a relationship between the maximum and minimum values of the modulated wave and the amplitude of the modulating

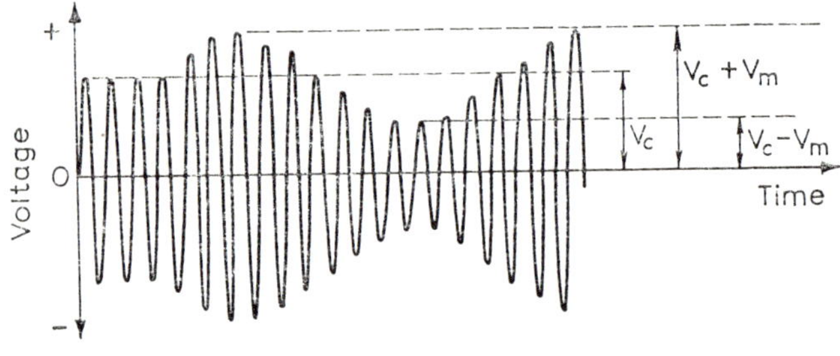

Fig. 2.10

Characteristics of an a.m. wave required to calculate modulation depth

signal. This relationship is expressed in terms of the modulation factor of the modulated wave.

The modulation factor m of an amplitude-modulated wave is defined by the expression

$$m = \frac{\text{maximum amplitude} - \text{minimum amplitude}}{\text{maximum amplitude} + \text{minimum amplitude}} \qquad (2.2)$$

When expressed as a percentage m is known as the modulation depth, or the depth of modulation or the percentage modulation.

Consider a sinusoidally modulated wave such as the wave shown in Fig. 2.10. Since the envelope of the modulated carrier wave must vary in accordance with the modulating signal its maximum amplitude must be equal to the amplitude of the carrier wave plus the amplitude of the modulating signal, i.e. $(V_c + V_m)$ Similarly, the minimum amplitude of the modulated wave must be equal to $(V_c - V_m)$, where V_c is the amplitude of the carrier wave and V_m is the amplitude of the modulating signal.

For *sinusoidal modulation*, therefore, the modulation factor m becomes

$$m = \frac{(V_c + V_m) - (V_c - V_m)}{(V_c + V_m) + (V_c - V_m)} = \frac{V_m}{V_c} \qquad (2.3)$$

i.e. the modulation factor is equal to the ratio of the amplitude of the modulating signal to the amplitude of the carrier wave.

Example 2.3
Draw the waveform of an amplitude-modulated carrier wave that is sinusoidally modulated to a depth of 25%.

Solution. For a modulation depth of 25%, $m = 0.25$, and since the modulating signal is sinusoidal, $m = V_m/V_c$ or $V_m = 0.25V_c$. Thus the maximum amplitude of the modulated wave is $V_c + 0.25V_c = 1.25V_c$ and the minimum amplitude is $V_c - 0.25V_c = 0.75V_c$. The required waveform is shown in Fig. 2.11.

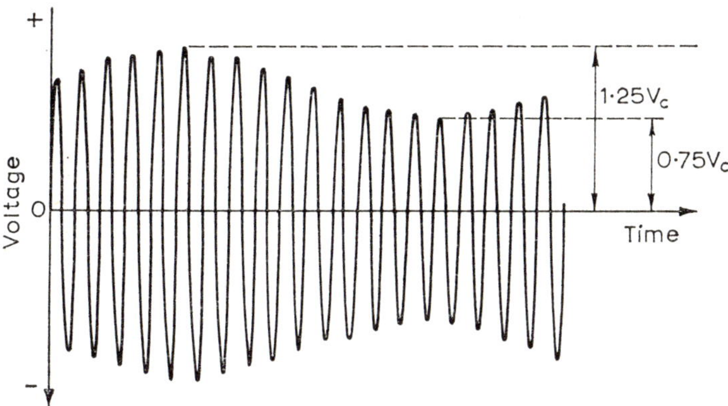

Fig. 2.11

Amplitude-modulated wave of modulation depth 25%

Example 2.4
Determine the depth of modulation of the amplitude-modulated wave shown in Fig. 2.12.

Solution. The maximum voltage of the wave is 10 V and the minimum voltage is 5 V; hence the depth of modulation is

$$\frac{10 - 5}{10 + 5} \times 100\% = 33.3\% \qquad Ans.$$

It can be shown, mathematically, that for a sinusoidally modulated carrier wave the amplitudes of the two side frequencies are the same and equal to $\frac{1}{2}m$ times the amplitude of the carrier wave.

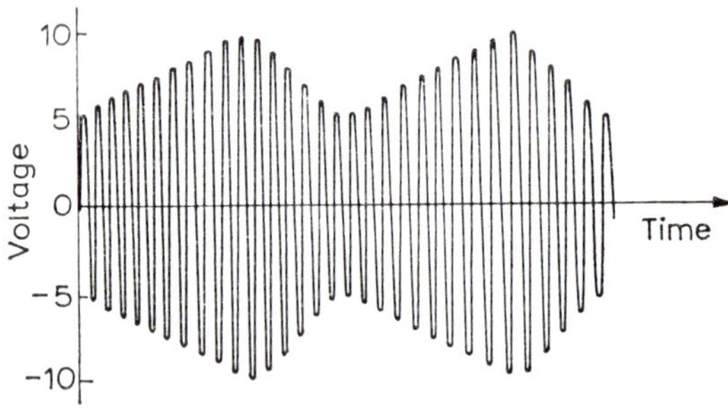

Fig. 2.12

Carrier wave amplitude modulated by a saw-tooth signal

Example 2.5

The envelope of a sinusoidally modulated carrier wave varies between a maximum value of 8 V and a minimum value of 2 V. Find (*a*) the amplitude of the carrier frequency component, (*b*) the amplitude of the modulating signal and (*c*) the amplitude of the two side frequencies.

Solution.

(*a*)
$$\text{Maximum value } V_c + V_m = 8 \qquad (2.4)$$
$$\text{Minimum value } V_c - V_m = 2 \qquad (2.5)$$

Adding equations (2.4) and (2.5)

$$2V_c = 10 \text{ or } V_c = 5 \text{ V} \qquad Ans.$$

(*b*) Subtracting equation (2.5) from equation (2.4)

$$2V_m = 6 \text{ or } V_m = 3 \text{ V} \qquad Ans.$$

(*c*) The amplitude of each of the two side frequencies is given by

$$\tfrac{1}{2}mV_c, \text{ i.e. } \frac{V_m}{V_c} \times \frac{V_c}{2} \text{ or } \tfrac{1}{2}V_m$$

$\therefore$ Amplitude of side frequencies = 3/2 or 1·5 V   *Ans.*

Carrier Frequencies Employed for Different Services

The various methods of modulation are employed in telecommunication systems to enable the best use to be made of the frequency spectrum available in a given medium. An outline of these various methods has been given and it now remains to discuss briefly the typical carrier frequencies and bandwidths employed for the most common systems, both line and radio.

The principles of operation of the various types of telecommunication system will be the subject of the final chapter of this book. At the moment the intention is merely to give some idea of the frequencies commonly employed. Only f.d.m. systems and broadcasting will be considered, since t.d.m. systems are, in general, still in the experimental or field trial stages.

Line Communication Systems

(1) TELEPHONY

Telephone cables are capable of transmitting a band of frequencies well in excess of the normal speech frequency range and can therefore be employed to carry single-sideband amplitude-modulated f.d.m. telephony systems. A number of channels are made available by such a system, each channel being allocated a different carrier frequency. The range of carrier frequencies employed depends upon the number of channels provided by the system and the part of the frequency spectrum which the system is to occupy. In Great Britain the vast majority of multi-channel telephony systems are either 24 circuit links or coaxial systems, but usually the systems consist basically of a suitable combination of the 12 channel carrier system which forms the basic "building block". Single sideband working is employed and the lower sidebands are selected.

The 12 carrier frequencies of the basic 12 channel carrier system are spaced at 4 kHz intervals in the frequency band 60–108 kHz. When, rarely nowadays, the system is directly transmitted to line it is usual to frequency translate it, by modulation of a 120 kHz carrier wave, to the band 12–60 kHz.

To form a 24 circuit link two such 12 channel systems are combined, one system being transmitted in the band 12–60 kHz and the other in the band 60–108 kHz. Thus a 24 circuit link occupies the frequency band of 12–108 kHz.

Several different types of coaxial system are in current use but their principle of operation is similar. One modern system can carry 2,700 telephone circuits, or 1,200 telephone circuits plus a 5 MHz bandwidth television channel, in the frequency band 300–12,435 kHz, while a more recent transistorized system, which is designed for use over relatively short distances, has a capacity for 300 telephone circuits in the frequency band 60–1,300 kHz.

(2) TELEGRAPHY

Multi-channel f.d.m. telegraphy systems employing both amplitude modulation and frequency modulation are in use, the former being by far the most common. Both types of system are capable of providing up to 24 channels within the speech frequency band of 300–3,400 Hz. The carrier frequencies employed start at 420 Hz and increase at 120 Hz intervals to the maximum carrier frequency of 3,180 Hz. Double-sideband amplitude modulation is employed.

Facsimile, or picture, telegraphy is employed for the transmission of photographs, drawings and similar information over both line and radio links. The carrier frequency employed and the range of transmitted frequencies depend upon a number of factors and varies according to the particular system considered. A typical carrier frequency is 1,800 Hz with the transmitted frequencies lying in the band 800–2,800 Hz.

Radiocommunication Systems

(1) GENERAL

The radiocommunication systems in use today may be divided into one of two main classes: firstly, broadcasting—both radio and

Table 2.1

FREQUENCY BAND	CLASSIFICATION	ABBREVIA-TION	PROPAGATION CHARACTERISTICS
10–30 kHz	very-low frequencies	v.l.f.	World-wide transmission by ground wave and reflection from ionosphere.
30–300 kHz	low frequencies	l.f.	Transmission over hundreds of miles by ground wave. Reflection by ionosphere, at night only, gives even greater distance.
300–3,000 kHz	medium frequencies	m.f.	Range of ground wave somewhat less. Reflection from ionosphere at night only.
3–30 MHz	high frequencies	h.f.	Ground wave transmission only tens of miles. Transmission by multiple reflections between earth and ionosphere for thousands of miles.
30–300 MHz	very-high frequencies	v.h.f.	Transmission by "line of sight" waves only for distances slightly greater than the optical horizon.
300–3,000 MHz	ultra-high frequencies	u.h.f.	
3–30 GHz	super-high frequencies	s.h.f.	
30–300 GHz	extra-high frequencies	e.h.f.	

television—and, secondly, radio links for providing point-to-point telephonic communication.

Before discussing briefly the carrier frequencies employed for each type of system, it will be advantageous to note the nomenclature commonly employed for identifying the various frequency bands which may be employed and, very briefly, mentioning their propagation characteristics.

Table 2.1 gives the classification of the various frequency bands together with a brief note on their propagation characteristics.

Note 1. 1 kHz = 1,000 Hz, 1 MHz = 1×10^6 Hz, 1 GHz = 1×10^9 Hz.
Note 2. Frequencies above about 1 GHz are also known as microwave frequencies.
Note 3. Radio waves are propagated through the atmosphere in three ways, generally known as ground waves, sky waves and space waves.

    The ground wave is a wave which is supported at its lower edge by the surface of the earth and which therefore follows the curvature of the earth as it travels, Fig. 2.13*a*. The sky wave is directed upwards from

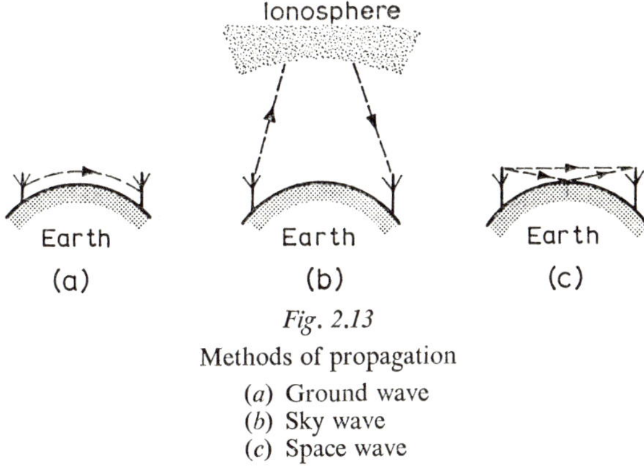

Fig. 2.13

Methods of propagation

(*a*) Ground wave
(*b*) Sky wave
(*c*) Space wave

the earth into the atmosphere where, under some conditions, it may be returned to the earth, Fig. 2.13*b*. Finally, the space wave, shown in Fig. 2.13*c*, is one which may be considered to consist of two rays, one travelling in a straight line between two points and the other travelling between the same two points by means of a single reflection from the earth.

(2) BROADCASTING

The B.B.C. broadcast their Home, Light and Third programmes at a number of different frequencies in the medium and v.h.f. bands and at one frequency in the low-frequency band. In the medium-frequency band amplitude-modulated d.s.b. transmissions are employed using carrier frequencies in the range 647–1,546 kHz and having a bandwidth of approximately 9,000 Hz. By international agreement carrier frequencies in the medium-frequency band are spaced at 9,000 Hz intervals and, consequently, the highest audio frequency transmitted is in the region of 4,500 Hz, i.e. a bandwidth of 9,000 Hz. If the bandwidth of a receiver is in excess of 9,000 Hz, sidebands from adjacent (in frequency) radio stations will be simultaneously received giving rise to interference, while if the bandwidth of a receiver is less than 9,000 Hz the quality of the reproduction will suffer. In the v.h.f. band, frequency modulation is employed with carrier frequencies lying between 88·1–96·8 MHz. In this band the spacing of carrier frequencies is much greater, approximately 200 kHz, and audio frequencies up to about 15 kHz can be transmitted. Hence better quality reproduction is possible with v.h.f. broadcasts. The Light programme is also broadcast on 200 kHz in the low-frequency band using d.s.b. amplitude modulation and having a 9,000 Hz bandwidth.

At the time of writing three television services are available to the British viewing public, two provided by the B.B.C. and the third provided by the various commercial television companies but transmitted by the Independent Television Authority (I.T.A.). One of the B.B.C. services, B.B.C.1, and the I.T.A. programmes are transmitted in the v.h.f. band. B.B.C.1 is transmitted on five different carrier frequencies, channels 1–5, in the range 41·5–66·75 MHz, and the I.T.A. programmes are transmitted on channels 6–12 in the band 176·25–209·75 MHz. Both services employ vestigial-sideband amplitude modulation for the picture signal, and d.s.b. amplitude modulation for the sound signal (vestigial sideband is a form of amplitude modulation in which the lower sideband plus a *part only* of the upper sideband is transmitted). B.B.C.2 television is broadcast in the u.h.f. band in the frequency range 470–585 MHz and employs vestigial-sideband amplitude modulation for the picture signal and frequency modulation for the sound signal.

(3) POINT-TO-POINT

Post Office radio links are employed for the routing of telephone calls to overseas destinations where line communication links via submarine cable are not available or have inadequate capacity. Different carrier frequencies are employed for each direction of transmission; and receiving stations are geographically situated well apart to minimize interference between signals passing in different directions. Radio links may be employed to carry ordinary telephone conversations or to carry facsimile telegraphy or Press broadcasts. The majority of radio links employ carrier frequencies in the high-frequency band and generally use a form of s.s.b. amplitude modulation. Radio-telegraphy services, such as the Press Broadcast Service, which is provided for the use of news agencies, are also generally provided in the high-frequency band.

Telephonic services to ships at sea are provided in different bands according to the position of the ship concerned. For example, transmissions to and from ships in seas near to the British Isles are in the medium-frequency band of 415–525 kHz, and include bothway telephone and telegraph circuits, weather forecasts and distress signals (500 and 2,182 kHz). For communication between a ship and a harbour authority the v.h.f. band is employed, 156·8 MHz, while for long-distance services the following frequencies are used, 4, 6, 8, 13, 17 and 22 MHz.

A mobile radio telephone service which provides telephonic communication between mobile radio telephones in motor cars and the public telephone network exists in both South Lancashire and London. The system employs frequency modulation in the v.h.f. band (160 MHz).

Many organizations, e.g. police, taxi-cab firms and ambulance services, employ base-to-mobile systems to provide telephonic communication between a headquarters and a mobile unit. Systems of this kind are operated in the v.h.f. band, 71·5 to 88 MHz or 156 to 174 MHz and may use amplitude modulation, frequency modulation or, in some cases, phase modulation.

(4) MULTI-CHANNEL TELEPHONY/TELEVISION LINKS

In the u.h.f. and s.h.f. bands multi-channel f.d.m. telephony systems, known as radio relay systems, provide an alternative to line systems for the provision of large numbers of circuits between two points and are increasingly employed. In Great Britain microwave point-to-point links will be provided in one of the following frequency bands—

> 1,700–1,900 MHz, 1,900–2,300 MHz (the 2,000 MHz band)
> 3,790–4,200 MHz (the 4,000 MHz band)
> 5,925–6,425 MHz (the lower 6,000 MHz band)
> 6,430–7,110 MHz (the upper 6,000 MHz band)
> 10,700–11,700 MHz (the 11,000 MHz band).

At the time of writing the two highest bands are not in use but it is expected that they will be used in the same way as the other bands. The 2,000 MHz, 4,000 MHz, and lower 6,000 MHz bands may each be split into a number of broadband channels, each of which is capable of handling either 1,800 telephony channels or a television link. However, equipment that is able to handle this number of channels is, at present, only available in the lower 6,000 MHz band. The present capacity of the equipment employed in the other frequency bands is 960 channels. Systems provided in these three bands provide inter-city links, or links between a television studio and a television transmitter. The 1,700–1,900 MHz band is split into broadband channels that have a capacity for 120 telephone channels and is employed for the provision of spur links within a city.

Phase Velocity, Frequency and Wavelength

The propagation of a voltage or current wave along a telephone line, or of a radio wave through the atmosphere, is not an instantaneous process but occupies a definite interval of time. There is thus a time lag between the application of a voltage at one end of a line and the detection of a voltage change at a distant point along the line. This means that there will be a phase difference between the a.c. voltages existing at two points along a line at a given instant, since the instantaneous voltage at each point is continuously changing. For example, suppose that for a given line the voltage at a point x miles from the sending end lags behind the sending-end voltage by $90°$. Then at a distance of $2x$ miles from the sending end the voltage will be lagging by $2 \times 90°$ or $180°$ on the sending-end voltage. The voltage $4x$ miles from the sending end will be in phase with the sending-end voltage, i.e. a complete cycle of instantaneous values has been completed. The distance $4x$ is known as "one wavelength," symbol λ, see Fig. 1.3.

The voltage of a signal travelling along a line, or the field strength of a signal propagated through the atmosphere, varies continuously with both distance and time in a similar manner to that discussed in Chapter 1 with regard to the propagation of sound waves. Consequently, the same relationship between the frequency f, the wavelength λ, and the phase velocity of propagation v_p, of a sinusoidal wave applies, i.e.

$$v_p = f\lambda \tag{2.6}$$

Although the frequency of a given wave remains constant as the wave travels its velocity is a variable quantity, being dependent upon the characteristics of the medium in which the wave is travelling. For a wave propagating in the atmosphere very little error is involved in taking the phase velocity to be constant at the speed of light $c = 3 \times 10^8$ m/s.

Example 2.6

A 5 MHz radio transmitter is connected to an aerial by a length of transmission line. If the wavelength of a signal on the transmission line is 0.8 times the wavelength of a signal in the atmosphere, calculate the velocity with which a signal is propagated along the cable.

Solution. For the wave in the atmosphere,

$$\lambda_{air} = \frac{c}{f}$$
$$= \frac{3 \times 10^{8}}{5 \times 10^{6}} = 60 \text{ m}$$

Hence, the wavelength in the cable is

$$\lambda_{cab} = 0 \cdot 8 \times 60 = 48 \text{ m}$$

and therefore the velocity of propagation in the transmission line is

$$v_{p} = \lambda_{cab} f$$
$$= 48 \times 5 \times 10^{6} = 2 \cdot 40 \times 10^{8} \text{ m/s} \quad Ans.$$

Exercises

1. With reference to an amplitude-modulated wave what is meant by the terms side frequency, sideband and depth of modulation?

   Sketch the waveform of an r.f. wave, amplitude-modulated by a sine-wave tone to a depth of 75 per cent.

   If a radio-frequency carrier wave is amplitude-modulated by a band of frequencies, 300–3,400 Hz, what will be the bandwidth of the transmission and what frequencies will be present in the transmitted wave if the carrier frequency is 104 kHz? (A 1963)

2. With reference to an amplitude-modulated wave, what is meant by the term sidebands?

   Why do medium-wave broadcast receivers, used for the reception of amplitude-modulated signals, require a total bandwidth of about 9 kHz?

   Briefly describe the effects on the reception of a medium-wave broadcast signal if the total bandwidth of the receiver were made (*a*) much less than, (*b*) much greater than, 9 kHz. (A 1962)

3. Sketch the waveforms of a radio-frequency carrier wave, amplitude-modulated by a sine-wave tone, when the depth of modulation is (*a*) 100 per cent, and (*b*) 25 per cent. If a radio-frequency carrier wave is amplitude-modulated by a band of speech frequencies, 50 Hz–4,500 Hz, what will be the bandwidth of the transmission and what frequencies will be present in the transmitted wave, if the carrier frequency is 506 kHz? (A 1959)

4. What is an amplitude-modulated wave?

   Why do broadcast receivers used for the reception of amplitude-modulated signals have a total bandwidth of about 9 kHz? (1 1957)

5. An audio-frequency signal of 5,000 Hz is used to amplitude-modulate a radio-frequency signal of 50,000 Hz. Sketch the resulting waveform.

   If, instead of modulating the radio-frequency carrier wave, the audio-frequency signal is simply added to it, what would be the resulting waveform?

   State the frequencies of the signal components present in each case. (1 1956)

6. Sketch the waveform of a radio-frequency carrier wave amplitude-modulated by a sinusoidal tone when the depth of modulation is (*a*) 50% and (*b*) 100%. Label the axes clearly to show the quantities concerned.

   If the amplitude of the unmodulated carrier is 1 volt (r.m.s.) and its frequency is 1,000 kHz, what are the amplitudes and frequencies of the side-frequency components in cases (*a*) and (*b*) when the modulation frequency is 1,000 Hz? (1 1955)

7. State the approximate values of carrier frequencies you would expect to find used for the following applications—

   (*a*) a high-quality sound broadcast service to serve a relatively small area,
   (*b*) a group of 12 telephone channels to be transmitted over a pair-type cable,
   (*c*) a point-to-point television radio-relay system.

   A radio wave is propagated through a block of polythene in which its velocity of propagation is 2×10^8 m/s. If its wavelength is found to be 4 cm, what is the frequency of the wave? (A 1960)

8. If the maximum frequency tolerance for fixed service stations operating in the frequency band 4·0 to 27·5 MHz is ± 30 parts in 10^6 ($\pm 0·003\%$) and for television stations operating in the frequency band 40·0 to 70·0 MHz is ± 10 parts in 10^6 ($\pm 0·001\%$), to what frequency variations in hertz do these tolerances correspond for

   (*a*) a fixed service station on 12·3 metres,
   (*b*) a television station on 5·5 metres?

   At what wavelength does a fixed-service station, having a variation of ± 300 Hz, become outside the permitted tolerance? (A 1959)

9. State for what types of radio transmission service the following frequencies are used: 16 kHz, 1 MHz and 45 MHz.

   Indicate, very approximately, the distance over which useful signals are obtained in each case.

   A half-wave dipole aerial is to be used for the reception of signals at a frequency of 45 MHz. What will be the length of the aerial? (1 1958)

10. Indicate the frequency bands in common use for present-day radio communication.

    Outline the general uses of the medium-frequency and high-frequency bands. (1 1957)

11. Explain briefly the relationship between the frequency and wavelength of a radio wave. Broadcasting stations work at a constant frequency separation of about 10 kHz. What difference in wavelength separates stations when their frequencies are 1,000 kHz and 1,010 kHz? (1 1956)

12. State the relationship between the velocity, frequency and wavelength of radio waves, indicating the units employed. What wavelength corresponds to the frequencies: (*a*) 600 kHz, (*b*) 9 MHz and (*c*) 50 MHz?

    State a suitable application for each frequency. (1 1955)

13. With reference to amplitude modulation explain the terms "modulation envelope" and "depth of modulation" and distinguish between "side frequencies" and "sidebands."

    The amplitude of a 310 kHz wave is modulated sinusoidally at a frequency of 5 kHz between 0·9 V and 1·5 V. Determine the amplitude of the unmodulated carrier, the depth of modulation and the frequency components present in the modulated wave. (A 1965)

3 The Construction of Resistors, Capacitors, Inductors and Transformers

Components such as resistors, capacitors, inductors and transformers are widely employed in all kinds of telecommunication and electronic equipment and it is essential that a student of radio and/or line transmission should possess an adequate knowledge of the constructional features and relative merits of the different types. Many different types of each kind of component are in common use, however, and it is impracticable in a book of this kind even to attempt to mention most of them. Consequently, this chapter will restrict itself to a brief discussion of a representative sample of each kind of component.

Resistors

The function of a resistor is to introduce a definite amount of resistance into an electric circuit. The resistance thus introduced may be required for any one of a large number of reasons and the characteristics of a suitable resistor, such as its power rating, may often depend upon its intended use. The power rating of a resistor is the maximum power, in watts, which can be dissipated by the resistor without raising its temperature above a certain critical value at which the resistor would be damaged. A factor of safety is usually allowed.

In general, resistors can be placed into one of two categories, that is, high stability or general purpose. Stability is not the same thing as accuracy: stability is the change in resistance of a resistor during its life; accuracy is the tolerance to which the resistor was originally made or selected.

WIRE-WOUND RESISTORS

Wire-wound general-purpose resistors are normally employed in situations where a relatively large amount of power must be dissipated but a high degree of resistance accuracy is not demanded. This type of resistor consists of a length of resistance wire wound on a ceramic rod or tube. The resistance wire employed must satisfy a number of requirements; for example, it must have a high resistivity, it must have a small temperature coefficient of resistance, it must be ductile and resistant to corrosion, it must have uniform and permanent characteristics and a high melting point, and it must be easy to solder. The materials most commonly employed that satisfy these requirements are (*a*) alloys of nickel and chromium and (*b*) alloys of nickel and copper.

The construction of one type of general-purpose wire-wound resistor is shown in Fig. 3.1. The resistance wire is wound on a ceramic tube and is welded to end connections which are made from a similar material. The end connections are firmly held in grooves at each end of the ceramic tube. Copper wire terminals are soldered

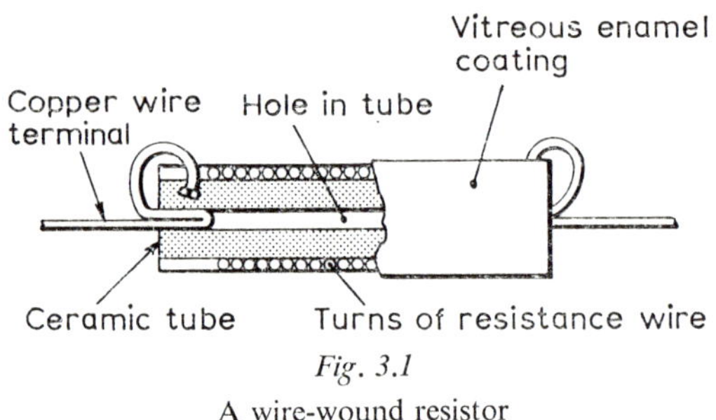

Fig. 3.1

A wire-wound resistor

to the end connections and folded before being forced into a central hole in the tube. This method of connecting the leads ensures that mechanical stresses are not transmitted to the resistance wire. The vitreous enamel covering provides both chemical and mechanical protection for the resistance wire.

Resistors of this type are normally supplied to a tolerance of $\pm 5\%$ and a stability of $\pm 1\%$. The relatively high power rating is achieved by the ease with which heat is removed from the resistor.

At radio frequencies the self-inductance and self-capacitance of a wire-wound resistor have a noticeable effect, and it is generally necessary to employ special winding techniques. Three methods of winding a radio-frequency wire-wound resistor are commonly employed, they are Bifilar, Ayrton–Perry and Fish-line. Consider briefly the cause of the self-inductance and self-capacitance of an ordinary winding. If a length of wire is wound in a number of turns a magnetic field is created in and around those turns whenever a current flows in them. When the current changes the resulting change in magnetic flux induces an e.m.f. in the winding. The induced e.m.f. has such a polarity that the original change in current is opposed (Lenz's Law), and the winding is then said to possess self-inductance. The adjacent turns in a winding and the insulation separating them provide a very small capacitance and the total effect of many such capacitances is as though a capacitor were connected directly across the winding. An ideal resistor for use at radio frequencies should have the same resistance at all frequencies, and this requires the self-inductance and self-capacitance to be as low as possible. The self-inductance can be minimized by arranging that the magnetic field set up by one turn is cancelled by the field set up by an adjacent turn. The self-capacitance can be minimized by reducing the potential difference between adjacent turns and separating the turns as widely as possible.

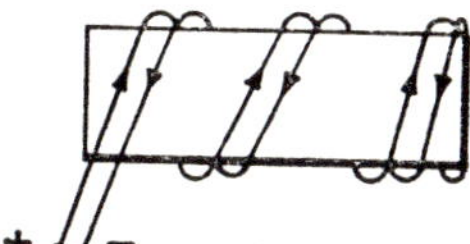

Fig. 3.2

Bifilar winding of a resistor

(*a*) *Bifilar.* With this method of winding, the wire is bent back on itself and the doubled wire is wound on a former as shown in Fig. 3.2. The resulting winding is almost completely non-inductive, because the adjacent turns carry current in opposite directions, but its self-capacitance is fairly high because the potential difference between adjacent turns is relatively great.

(*b*) *Ayrton–Perry.* The principle of an Ayrton–Perry winding is shown in Fig. 3.3. A single wire is wound on a thin Bakelite card so that the spacing between adjacent turns is slightly greater than the diameter of the wire. Another wire is then wound in the opposite direction in the space left between the turns of the first winding. The

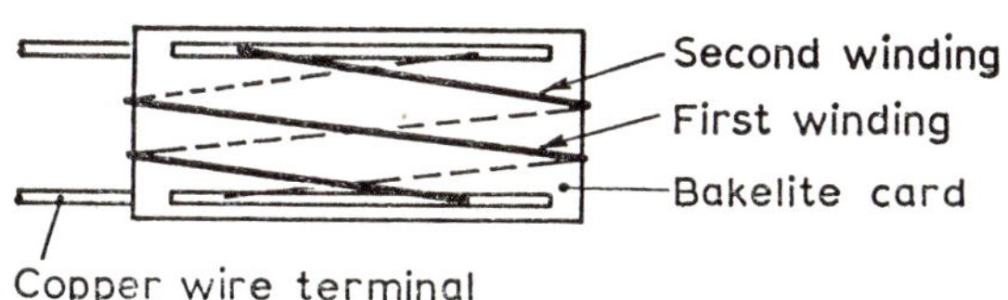

Fig. 3.3

Ayrton–Perry winding of a resistor

self-inductance of this type of winding is very small because almost complete cancellation of the magnetic fields is achieved. The self-capacitance is also very small since (*a*) adjacent turns are at very nearly the same potential and (*b*) the permittivity of the Bakelite card is low.

(*c*) *Fish-line.* The wire is wound in a single layer with a relatively large inter-turn spacing around a silk cord. The coil thus formed is then wound in a helix in grooves on a cylindrical former and this results in almost complete cancellation of the magnetic field. The self-capacitance is also small because the potential difference between adjacent turns is small and the adjacent turns are spaced relatively far apart.

CARBON RESISTORS

Carbon resistors are by far the most common type of general-purpose resistor in use today and Figs. 3.4*a* and *b* show the construction of two

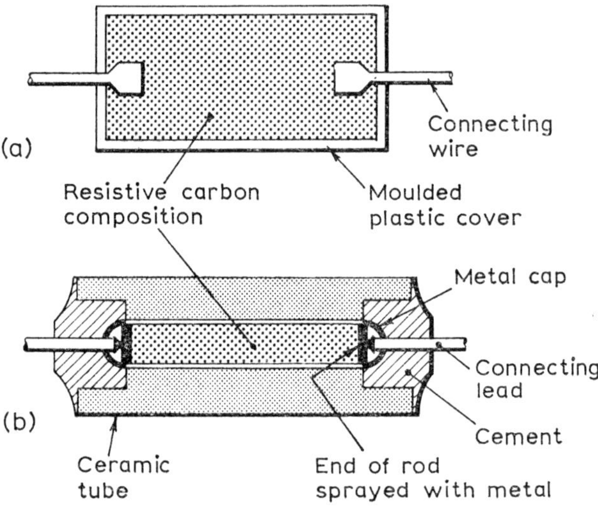

Fig. 3.4

Two types of carbon resistor

different types. The resistive element consists of a mixture of carbon black, resin and refractory filling that is mixed in the correct proportions and then sifted to produce a black powder. This powder is then compressed into the required shape by a die press, placed in an oven and heated until it is solid. Connection is made to the resistive element either by moulding the enlarged ends of the connecting wires directly into the carbon, Fig. 3.4*a*, or by spraying metal over the ends of the carbon rod and then pressing metal over the sprayed ends, Fig. 3.4*b*.

Carbon resistors have a low power rating because (*a*) their small physical size limits access to the surrounding air, and the insulating case prevents air coming into contact with the resistive element, and (*b*) there is only a small mass of good thermal conductivity material available to conduct heat away. Carbon resistors also tend to be noisy but have the advantages of small size, low cost, a large range of values and negligible self-inductance. This type of resistor is normally supplied to a stability of $\pm 15\%$ and a tolerance of $\pm 5\%$, 10% or 20%.

CRACKED-CARBON RESISTORS

High-stability composition resistors can be obtained by using the "cracked-carbon" process, in which a suitable hydrocarbon vapour is decomposed onto a ceramic rod to produce a layer of carbon that has a particular value of resistance per unit length. The desired resistance value is then obtained by spiralling the layer to give the appropriate length. The construction of a typical cracked-carbon resistor can be seen in Fig. 3.5. Cracked-carbon resistors are supplied

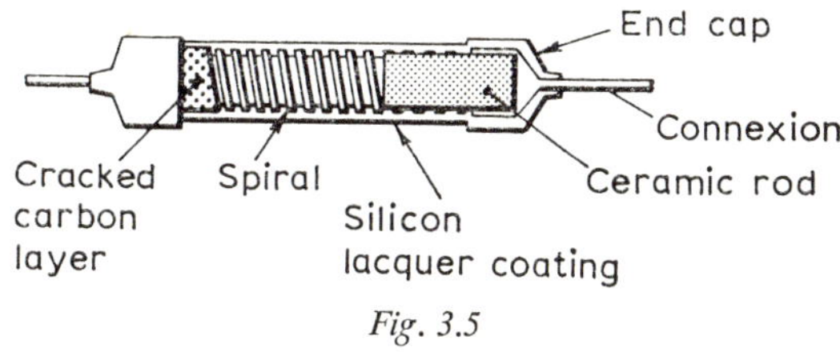

Fig. 3.5

A cracked-carbon resistor

to $\pm 2\%$ stability and $\pm 1\%$ tolerance and have the same sort of advantages as the carbon resistor, the main disadvantage being their low power rating.

METAL-FILM RESISTORS

A metal-film resistor is made by coating a sheet of clean glass with a solution of gold and platinum in oil and then heating it in an oven.

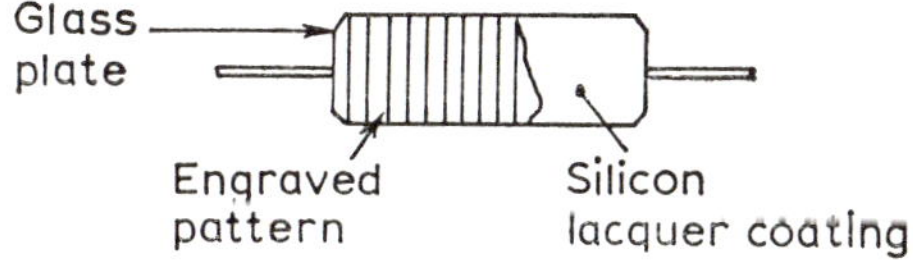

Fig. 3.6

A metal-film resistor

The metal solution is converted into a thin metal film that can be engraved to increase the resistance to the required value. The glass sheet is then replaced in the oven and heated to a greater temperature than before and this makes the metal film adhere to the glass. Connections are then soldered to the sheet and a protective coating of silicon lacquer is applied.

A metal-film resistor is extremely stable and can be made to an accuracy of approximately $\pm 1\%$. The construction of a metal film resistor is shown in Fig. 3.6.

METAL-OXIDE RESISTORS

A metal-oxide resistor consists of a metal-oxide film, of constant resistivity, applied to the surface of a glass rod. The glass is of optical

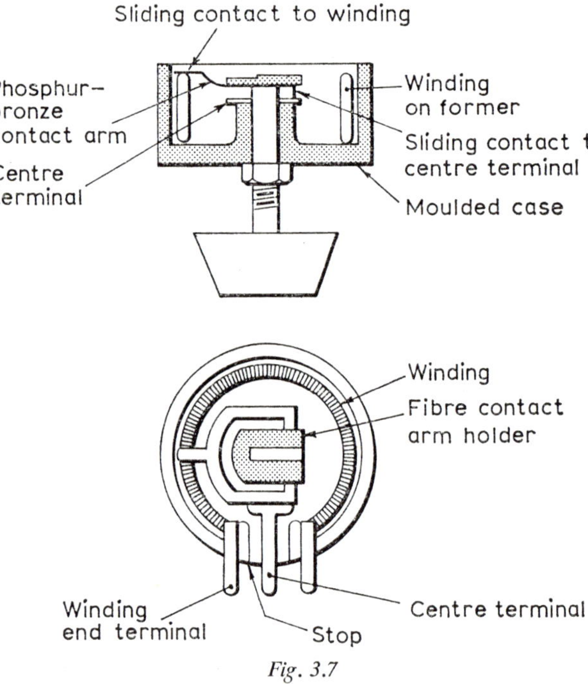

Fig. 3.7

A variable wire-wound resistor

(From the *Post Office Electrical Engineers' Journal*)

quality and is alkali-free and the oxide is often tin oxide. The film is impervious to moisture, is not affected by the heat produced by soldering the resistor into a circuit, is very resistant to abrasion and expands only very little. The resistive element is coated with a material that is both an electrical and a heat insulant. This type of resistor is very reliable, and is supplied to a stability of $\pm 5\%$ and a tolerance of $\pm 5\%$.

VARIABLE RESISTORS

In general, variable resistors can be divided into two classes, (*a*) wire-wound and (*b*) carbon. The construction of a variable wire-wound resistor is shown in Fig. 3.7. A flat strip winding, which may be made of nickel-chromium or nickel-copper and be bare or insulated, is wound on a flat former and bent into an arc. If bare wires are used they are spaced well apart and are tightly wound, but

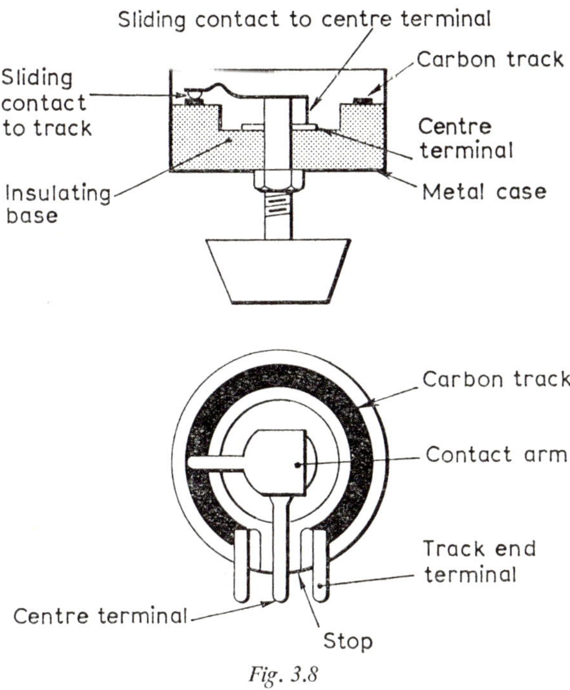

Fig. 3.8

A variable carbon resistor

(From the *Post Office Electrical Engineers' Journal*)

insulated wire is wound touching. The metal sliding contact consists of a bent phosphor-bronze spring that has a split tip to allow it to be tensioned and to give a good contact area. Phosphor-bronze is chosen because this material is resistant to wear and corrosion. The slider also carries a contact which makes on the bronze contact ring that is mounted around the centre terminal. The slider is carried by a fibre holder attached to the top of the brass spindle but insulated from it.

Fig. 3.8 shows the construction of a variable carbon resistor. The resistive element is made in a similar way to that described previously for fixed carbon resistors and is in the shape of a ring held within an insulated holder. A carbon slider is employed and is mounted on a bent phosphor-bronze contact arm that, in turn, is mounted on, but insulated from, the spindle. A sliding contact that makes with the phosphor-bronze contact ring of the centre terminal is also carried on the contact arm. The complete track assembly is enclosed in a metal case.

For the same physical dimensions a variable carbon resistor can have a greater resistance, but its tolerance and stability figures are of the order of $\pm 25\%$ as opposed to a tolerance of $\pm 10\%$ and a stability of $\pm 2\%$ for a wire-wound resistor. Also, the power rating of a wire-wound resistor is greater.

RESISTOR COLOUR CODE

With physically small resistors it is inconvenient to mark the resistance value and tolerance in figures, and an easily identifiable method of representing resistance values by a colour code has been developed. A resistor is marked by having up to four rings around its body as in Fig. 3.9. The resistance value is indicated by the colours of the rings, according to Table 3.1.

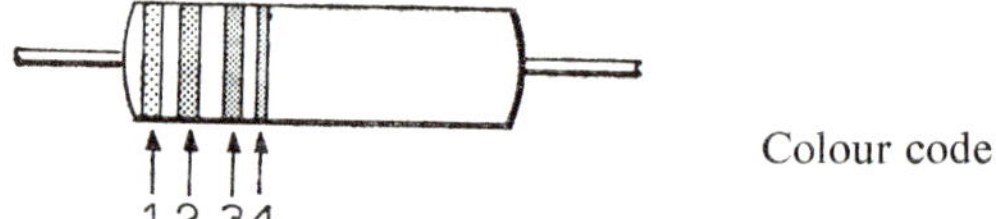

Fig. 3.9

Colour code marking of a resistor

Table 3.1

COLOUR	RING 1 (gives value of first digit)	RING 2 (gives value of second digit)	RING 3 (gives the number of noughts)	RING 4 (gives the tolerance)
black	0	0	0	—
brown	1	1	1	—
red	2	2	2	—
orange	3	3	3	—
yellow	4	4	4	—
green	5	5	5	—
blue	6	6	6	—
violet	7	7	7	—
grey	8	8	8	—
white	9	9	9	—
gold	—	—	—	5%
silver	—	—	—	10%
none	—	—	—	20%

In an alternative method, which uses the same colour code, the body colour determines the first digit, the tip colour the second digit, and a coloured dot the number of noughts. The tolerance is shown by the appropriate colouring of the other tip.

Capacitors

A capacitor basically consists of two conductors, which are known as plates, separated by an insulating material, such as air or paper, that is known as the dielectric. When a voltage is applied across a capacitor it will become charged, that is, it will store a quantity of electricity in the electric field in its dielectric. The charge of a capacitor is directly proportional to the voltage applied across the capacitor, the constant of proportionality being known as its capacitance. Thus, for any capacitor

$$Q = VC \qquad (3.1)$$

where Q = charge stored, in coulombs; V = voltage across the capacitor, in volts; C = capacitance, in farads.

The unit of capacitance, the farad, is inconveniently large and the microfarad $(1\ \mu\text{F} = 10^{-6}\ \text{F})$ and the picofarad $(1\ \text{pF} = 10^{-12}\ \text{F})$ are commonly used.

The capacitance of a capacitor is directly proportional to the total area of its plates and to the relative permittivity of its dielectric, but is inversely proportional to the distance between the plates. For a high capacitance, therefore, a large plate area and a thin dielectric are necessary. The relative permittivity (symbol ϵ_r) of a material is the ratio of the capacitance of a capacitor having the material as its dielectric to the capacitance of a similarly dimensioned capacitor having air as its dielectric.

The stability of a capacitor is its ability to maintain its characteristics during its life.

IMPREGNATED-PAPER CAPACITORS

Fig. 3.10 shows the construction of an impregnated-paper capacitor. The capacitor consist of two strips of aluminium foil interleaved with four strips of impregnated paper and then rolled up tightly. A special type of paper is employed and is carefully dried before it is impregnated with, for example, paraffin wax, mineral oil or petroleum jelly. Impregnation of the paper has the effect of increasing its permittivity and increasing the maximum voltage it can withstand. Connection to the metal foils is made by means of two copper strips and the assembly is placed inside an aluminium tube. A capacitor of this type has a capacitance in the region of 0·01 to 12 μF, a tolerance of $\pm 20\%$ and a stability of $\pm 2\%$.

METALLIZED-PAPER CAPACITORS

To obtain a larger capacitance within a smaller volume than is possible with an impregnated-paper capacitor, the form of construction shown in Fig. 3.11 may be employed. The dielectric consists of a lacquered paper on which aluminium has been evaporated

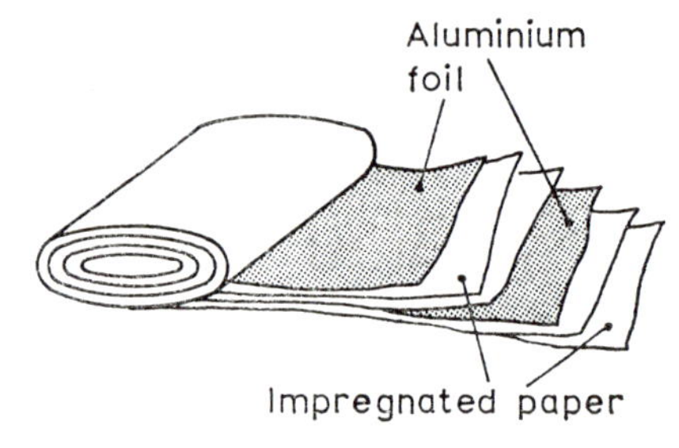

Fig. 3.10

An impregnated-paper capacitor

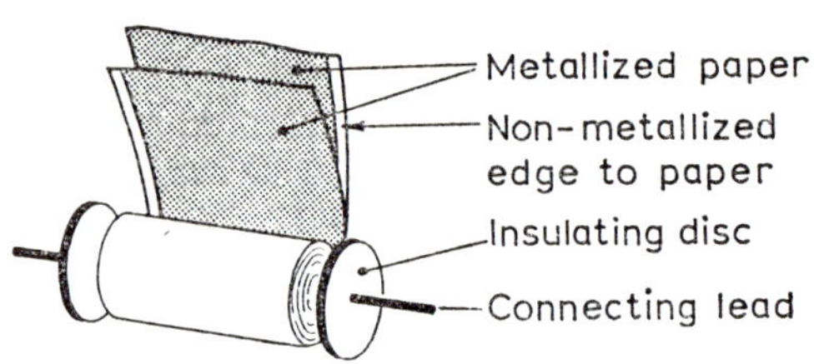

Fig. 3.11

A metallized-paper capacitor

in a vacuum. Apart from a narrow strip at one edge the entire width of one side of the paper is metallized. Two papers are superimposed so that the metallized edge of one paper lies over the unmetallized edge of the other and the papers are then rolled up. Since the unmetallized part of each paper serves as the dielectric no further paper is required and thus a very thin dielectric is obtained. Each end of the roll is sprayed with copper to connect the edges of the aluminium coating together and then wires are attached. A wrapping of insulated paper is then wound around the rolled metallized paper and the complete assembly is placed inside a metal case.

If an excess voltage is applied to this type of capacitor the paper is punctured and the metal around the hole evaporates and prevents a short-circuit. This self-restoring action, which only applies at voltages up to about one-and-half times the normal working voltage, and the small size are the advantages of this type of capacitor. The main disadvantage is its relatively low insulation resistance.

A capacitance range of approximately 0·0001 to 40 μF is available with a tolerance of $\pm 20\%$ and stability $\pm 10\%$. Paper capacitors are employed as interstage coupling capacitors in audio-frequency valve amplifiers.

MICA CAPACITORS

Two main types of mica capacitor are to be found, these are (*a*) stacked-mica capacitors and (*b*) silvered-mica capacitors. A stacked mica capacitor is constructed by interleaving a number of sheets of mica with thin sheets of tin or copper foil, clamping them together

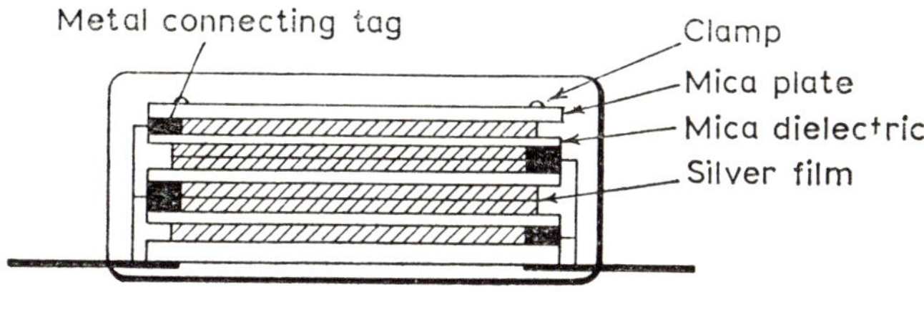

Fig. 3.12

A silvered-mica capacitor

and then encasing the assembly in a plastic case. For increased stability and reduced size the silvered-mica capacitor is employed (see Fig. 3.12). Mica sheets are cut to size and a solution of silver oxide powder in lavender oil is coated onto the surface. The sheets are then placed in an oven and heated, volatizing the oil and converting the oxide to silver. The sheets are silvered on both sides and stacked as shown. Metal tags are interleaved on alternate sides of the capacitor to make contact with the silver films, and two, somewhat thicker, mica plates are fitted at each end before the whole assembly is clamped, impregnated with wax, and finally given a wax coating to prevent the entry of moisture.

Silvered-mica capacitors are made to a stability of $\pm 1\%$ and a tolerance of $\pm 10\%$ and have values in the range 10 to 1,100 pF. This type of capacitor has low losses and can be operated at high voltages, it is, however, rather costly. Silvered-mica capacitors are employed as fixed tuning capacitors in tuned circuits.

ELECTROLYTIC CAPACITORS

Electrolytic capacitors have the advantage of providing a large capacitance in a small volume, particularly if the working voltage is low. The large capacitance/volume ratio arises because a very thin dielectric film is employed. Electrolytic capacitors are of two main types, (*a*) aluminium and (*b*) tantalum, the construction of the former type being shown in Fig. 3.13.

A thin film of aluminium oxide is coated on to both sides of a strip of aluminium foil by an electrolytic forming process. The formed aluminium foil and a strip of plain aluminium foil are then interleaved with two layers of porous paper that have previously

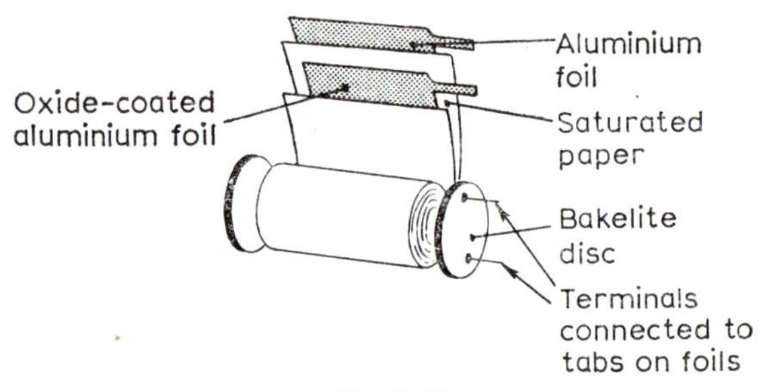

Fig. 3.13

An aluminium electrolytic capacitor

been soaked with electrolyte; the electrolyte is usually a paste of glycol and ammonium tetraborate. The assembly is rolled up, its ends are sealed with wax, and then it is sealed into an aluminium container. To increase the capacitance the oxide-coated foil can be etched to increase its surface area before the electrolytic forming process is begun. An effective increase in capacitance of about seven times can be achieved. This type of capacitor has relatively high losses, a low insulation resistance, can only be made to wide tolerances, such as -20% to $+50\%$, and must be operated with the oxide-coated foil positive with respect to the plain foil. The great advantages of an electrolytic capacitor are its high capacitance/volume ratio and its low cost.

The construction of a tantalum electrolytic capacitor is very similar to the construction of the aluminium electrolytic capacitor just described. The main differences are (*a*) the use of tantalum foil instead of aluminium foil and (*b*) only the middle paper strip is saturated with electrolyte.

Tantalum electrolytic capacitors have a number of advantages over the aluminium type; they are smaller, have lower losses and a higher insulation resistance and do not require periodic reforming when in storage. Their main disadvantage is their extra cost.

Both types of capacitor are supplied in values in the approximate range of 1 to 10,000 μF with a tolerance of $\pm 20\%$ for tantalum and -20% to $+50\%$ for aluminium. Electrolytic capacitors are employed as h.t. smoothing capacitors, interstage coupling capacitors in audio-frequency transistor amplifiers, and as emitter-resistor decoupling capacitors.

VARIABLE CAPACITORS

The change in the capacitance of an air-dielectric variable capacitor, with angular rotation of its spindle, is determined by the shape of its plates. By suitable choice of the shape of the plates any one of the following four characteristics may be obtained: (*a*) linear

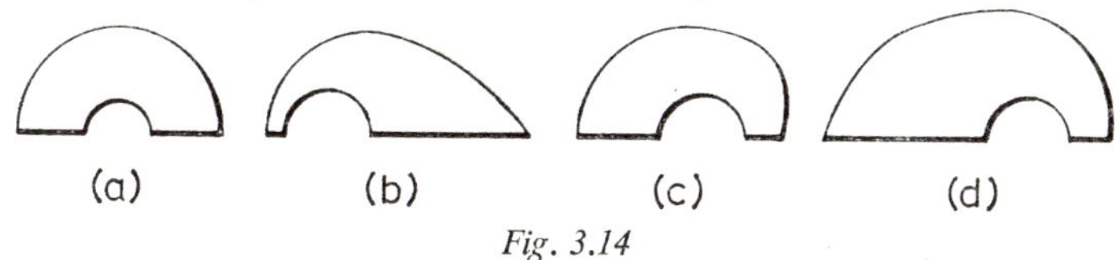

Fig. 3.14

The shapes of plate employed in variable air capacitors

capacitance change, (*b*) linear frequency change, (*c*) linear wavelength change, and (*d*) logarithmic capacitance change. The plate shapes giving these characteristics are shown in Figs. 3.14*a*, *b*, *c* and *d* respectively.

The shape of plate chosen for a particular capacitor depends upon the type of scale to be used in conjunction with the capacitor. For example, if plates of the shape to be seen in Fig. 3.14*c* are employed a linear wavelength scale will be obtained. Such a scale is desirable for a radio receiver when the stations to which the receiver is most likely to be tuned are evenly distributed over the tuning range of the receiver. In the case of the B.B.C. medium-wave programmes, however, most stations operate on frequencies above 1 MHz and a linear wavelength scale would make them appear very close together, making tuning difficult. Thus, a modified form of Fig. 3.14*c* is generally used which has the effect of spreading the B.B.C. stations more evenly over the tuning range. A similar procedure has been adopted for some modern transistorized receivers to enable "pop" stations to be more easily obtained.

The construction of a two-gang variable air capacitor is shown in Fig. 3.15. The term "two-gang" means that two separate capacitors are mounted on a single spindle. Ganged capacitors are employed in superheterodyne radio receivers, where it is required to be able to

tune two, or more, tuned circuits by the rotation of a single knob.
The framework of the capacitor consists of two end-plates that are
connected by pillars. The tuning control of the receiver is mounted
on the end of the brass spindle; also mounted on this spindle are two
sets of movable plates (only one of which is shown), the individual

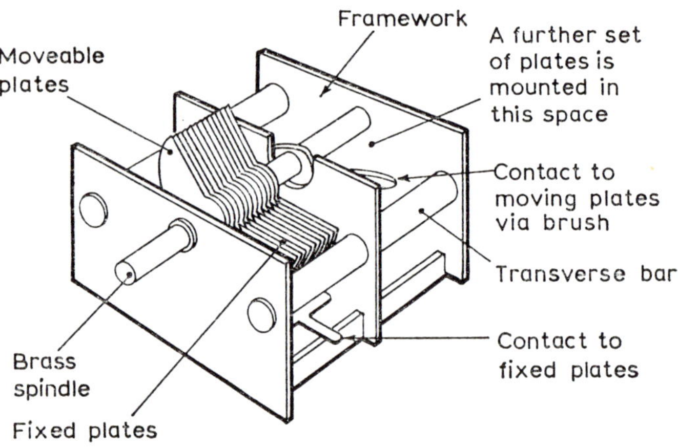

Fig. 3.15

An air-dielectric variable capacitor

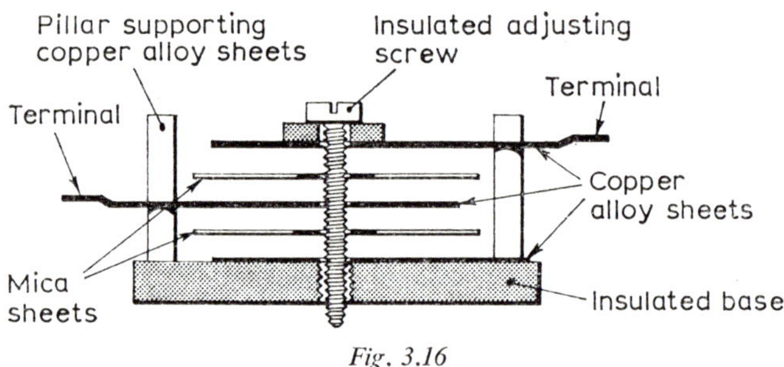

Fig. 3.16

A compression-type mica-dielectric variable capacitor

plates being spaced apart by washers. The fixed plates are mounted
on ceramic supports that are fixed to two transverse bars attached to
the framework. Both sets of plates are made of aluminium or brass,
and are rigidly constructed and mounted to prevent microphonic
noise and capacitance instability. Rotation of the tuning control
rotates the moving plates and interleaves them with the fixed plates.
Maximum capacitance is obtained when the plates are fully
enmeshed and minimum capacitance when the plates are fully

withdrawn; the minimum capacitance is generally of the order of 10–20 pF.

Finally, consider the construction of a mica compression-type variable capacitor, Fig. 3.16. The capacitor consists of a number of thin sheets of mica and copper alloy interleaved as shown. By turning the adjustable screw the spacing between the sheets can be altered to give any desired capacitance value within the range of the device. Typical capacitance ranges available are approximately 5–60 pF for the smaller types and 1,400–2,000 pF for the larger types.

Inductors and Transformers

An inductor is a component that has been designed to have a definite amount of inductance and, ideally, should possess neither resistance nor self-capacitance. Whenever a current flows in an inductor it produces a magnetic flux that links with the current; the inductance of an inductor is defined on this basis, i.e.

$$\text{Inductance } L = \frac{\text{Flux linkages}}{\text{Current producing the flux in amps}} \text{ henrys} \quad (3.2)$$

(*Note.* One flux linkage corresponds to one flux line linking the current once.)

When the current flowing in an inductor changes, the flux changes also and in so doing induces an e.m.f. in the inductor. The magnitude of this induced e.m.f. provides an alternative method of defining inductance, i.e.

Induced e.m.f.

$$= -L \times \text{(rate of change of current in amps/sec) volts} \quad (3.3)$$

The minus sign denotes that the polarity of the induced e.m.f. is opposite to the direction of the change in current producing it.

Basically, an inductor consists of a coil of wire wound on a former. The differences between types of inductor are mainly a question of the method of winding, the presence or absence of a magnetic core and the core material when one is employed.

A transformer consists of two coils of wire which are inductively coupled together, the degree of coupling being dependent upon the proximity of the windings, and upon the core material. When a changing current flows in one winding of a transformer the changing flux links the turns of the other winding and induces an e.m.f. in it. The magnitude of the e.m.f. induced in the winding is equal to the product of the e.m.f. induced per turn and the number of turns. By suitable choice of the number of turns in the two windings a voltage (or current) step-up or step-down action can be obtained. One of the windings, the one to which the input signal is generally applied, is known as the primary winding, the other, from which the output signal is generally taken, is called the secondary winding.

Unfortunately, the changing magnetic flux not only links the windings of an inductor or transformer, but also links the core around which the windings are wound, and therefore an e.m.f. is induced in the core. This induced e.m.f. causes eddy currents to flow in the core where they produce an I^2R power loss. To reduce the eddy-current loss the type of core employed in an inductor or transformer must be carefully chosen.

At audio frequencies the cores of inductors and transformers consist of thin strips of magnetic material, called laminations, which

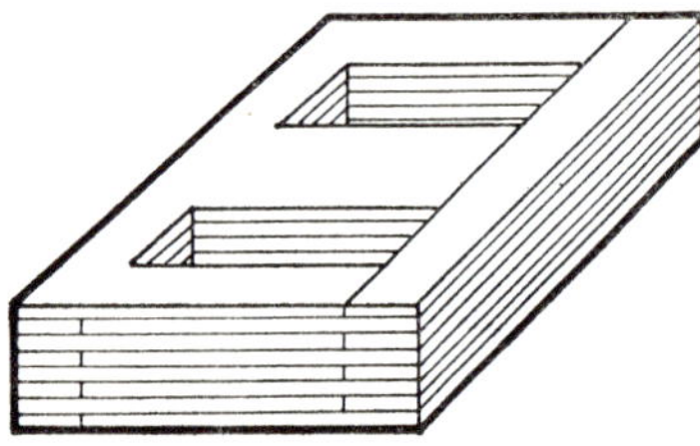

Fig. 3.17

A laminated core

are insulated from one another, see Fig. 3.17. A laminated core reduces the flow of eddy currents because the currents cannot flow around the complete core; each eddy current is restricted to the relatively high-resistance path offered by an individual lamination. The windings, one for an inductor, two or more for a transformer, are wound on a former around the centre limb of the core. The only requirements for the former are low cost and adequate physical strength, and suitable materials are Paxolin and resin-bonded hard-compressed paper. If an inductor, or transformer, is to be employed in a situation where it will have to pass a direct current an air gap is generally included in the centre limb to prevent the core becoming magnetically saturated. Laminated cores are made from such materials as silicon-steel and nickel-iron alloy. Enamelled or wax-impregnated cotton-covered copper wire is used for the windings. For a transformer the secondary winding may be wound directly over the primary winding or, alternatively, a part of the secondary winding may be wound first, then the primary winding, and finally, the remainder of the secondary winding wound on top. Layers are separated with impregnated paper and the complete winding is impregnated with either mineral wax or varnish to keep out moisture.

The power loss caused by the flow of eddy currents in a core is proportional to the square of the frequency of operation and hence increasingly thin laminations are required as the operating frequency is raised. At radio frequencies, therefore, the use of laminations becomes impracticable and dust and ferrite cores are used.

A dust core consists of a magnetic material reduced to powder form and then mixed with an insulating binding agent before being compressed into the required core shape. The eddy-current losses of a dust core are low because each eddy current is restricted to the high-resistance path presented by a particle of the magnetic material. The disadvantage of a dust core is that its effective permeability is rather low. To overcome the disadvantages of laminated and dust cores at radio frequencies a ferrite core may be used; ferrite is a

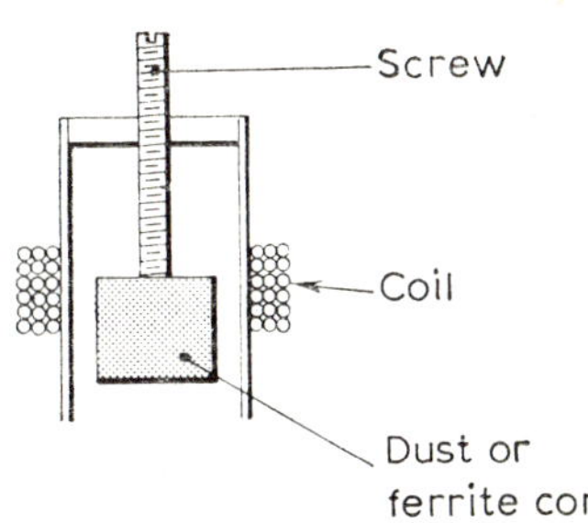

Fig. 3.18

Permeability tuning of an inductor
or transformer winding

manufactured material that has a high permeability and, since it has the insulating properties of ceramic, very low eddy-current losses.

With both dust and ferrite cores the coil is often wound on an insulating former around the core itself. To reduce dielectric losses, low-permittivity materials, such as polythene and steatite, are used for formers. Both types of core are often fitted on to a screw so that they can be screwed into, and out of, the coil to vary its inductance. The method, known as permeability tuning, is to be seen in Fig. 3.18.

At radio frequencies the self-capacitance of the windings may have a significant effect and it is usual to wind a radio-frequency winding in such a way as to minimize its self-capacitance. As in the case of a wire-wound resistor the self-capacitance of an inductor, or transformer, can be minimized by preventing turns having a relatively large potential difference between them from being in close proximity to one another. If the turns are wound in a single layer all that is necessary is to space the turns well apart. Very often more than one layer is necessary and then the turns must be arranged in a particular manner. Several methods of winding a coil to minimize self-capacitance exist but here only the "pie-type" construction will be mentioned, see Fig. 3.19. The coil winding is split into three "pies" each of which is wound so that the turns of one layer are separated by air and the turns of adjacent layers cross one another at an angle, i.e. the winding is zig-zag.

At frequencies in excess of about 2 MHz air-cored inductors and transformers begin to be employed. A typical construction for a transformer is shown in Fig. 3.20 in which the coils are wound on a tube or rod of an insulating material having a low dielectric loss.

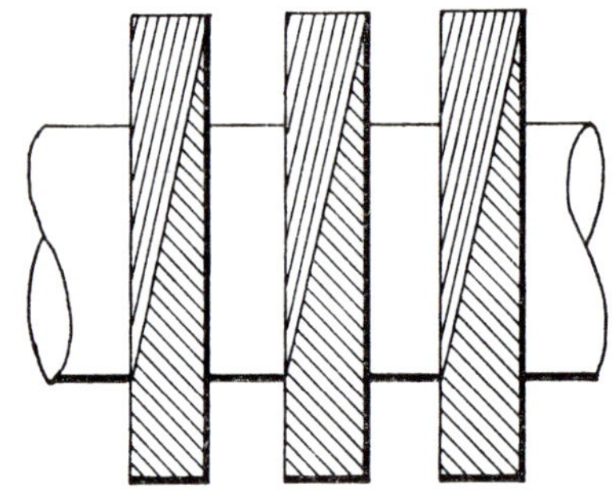

Fig. 3.19

"Pie-type" winding of an inductor or transformer

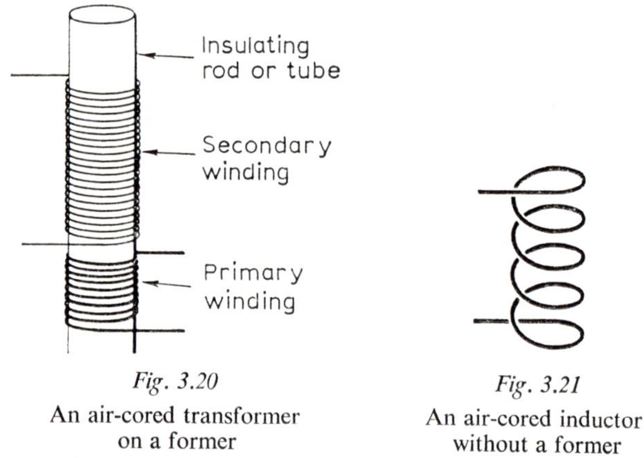

Fig. 3.20	*Fig. 3.21*
An air-cored transformer on a former	An air-cored inductor without a former

Finally, at frequencies in the v.h.f. band an inductor or transformer may consist of a few turns of wire wound around an insulating former, or, when sufficiently rigid wire is employed, merely consist of a few turns of wire alone, as in Fig. 3.21.

THE USE OF A TRANSFORMER FOR IMPEDANCE MATCHING

In a transformer having a laminated core almost all the flux set up by the current in the primary winding links with the turns of the secondary winding. For such a transformer, it can be shown that the ratio of the voltage appearing across the terminals of the secondary winding to the voltage applied across the terminals of the primary winding is very nearly equal to the ratio of the number of secondary turns to the number of primary turns. Thus, referring to Fig. 3.22,

$$\frac{V_s}{V_p} = \frac{N_s}{N_p} \qquad (3.4)$$

where V_s, V_p are respectively the voltages across the secondary and primary windings and N_s, N_p are respectively the number of turns in the secondary and primary windings.

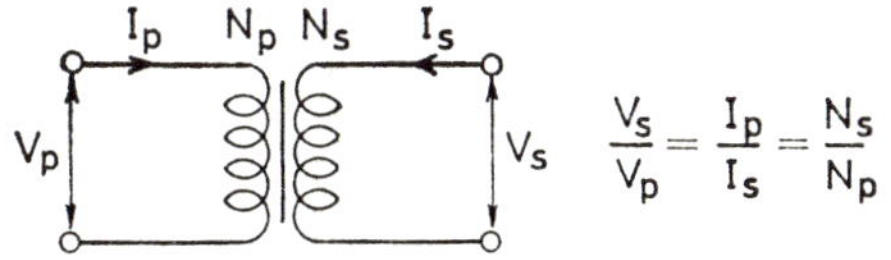

Fig. 3.22

Representation of a transformer

It can further be shown that the current transformation ratio is the inverse of the voltage transformation ratio, i.e.

$$\frac{I_s}{I_p} = \frac{N_p}{N_s} \qquad (3.5)$$

Now $I_p = V_p/Z_p$ and $I_s = V_s/Z_s$ where Z_p and Z_s are the impedances of the primary and secondary circuits respectively; hence, substituting in equation (3.5),

$$\frac{V_s}{Z_s} \bigg/ \frac{V_p}{Z_p} = \frac{N_p}{N_s}$$

$$\frac{V_s Z_p}{V_p Z_s} = \frac{N_p}{N_s}$$

But, from equation (3.4),

$$\frac{V_s}{V_p} = \frac{N_s}{N_p}$$

$$\therefore \quad \frac{Z_p}{Z_s} = \left(\frac{N_p}{N_s}\right)^2$$

or $$\frac{N_p}{N_s} = \sqrt{\left(\frac{Z_p}{Z_s}\right)} \qquad (3.6)$$

This relationship can be employed to effect a desired impedance transformation by suitable choice of the numbers of primary and secondary turns.

Example 3.1

Determine the turns ratio of a transformer which is to be employed to match a 50 Ω load to a 1,000 Ω source. (The term "match" means to make the impedance effectively connected across the source equal to the source impedance, see Fig. 3.23.)

Solution.
$$\frac{N_p}{N_s} = \sqrt{\left(\frac{1,000}{50}\right)} = \sqrt{20} = 4.47$$

Thus the primary winding must have 4·47 times as many turns as the secondary winding.

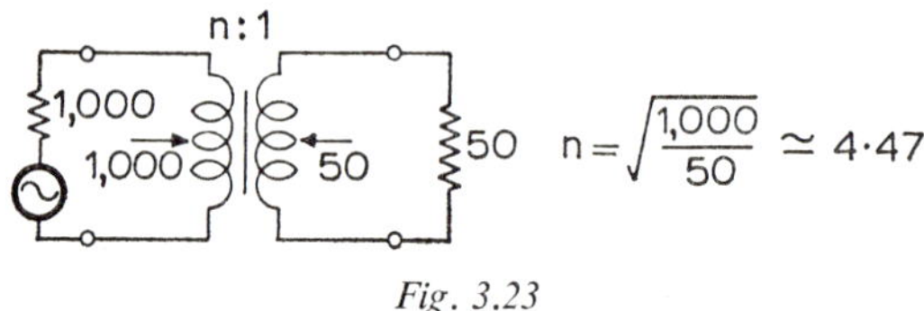

Fig. 3.23

Impedance matching with a transformer

Exercises

1. What do you understand by the following terms with respect to resistors: (*a*) Bifilar, (*b*) Ayrton–Perry, (*c*) Fish-line?

   Explain, in some detail, how each arrangement achieves its purpose and comment on the relative efficiencies of the three arrangements.   (E 1964)

2. A wire-wound resistor for high-frequency application is required to have minimum self-inductance and minimum self-capacitance. Sketch and describe the construction of a suitable type, explaining how the self-inductance and self-capacitance are minimized.

   Under what circumstances can carbon resistors be used for such applications? Explain why carbon resistors are not always used, the wire-wound resistor being preferred.   (E 1963)

3. Explain why it is sometimes necessary, in radio-frequency applications, to minimize the self-capacitance of (*a*) an inductor, (*b*) a resistor.

   Describe and explain the principle of any constructional features which may be adopted to achieve this purpose in each case.   (E 1962)

4. Explain why it is sometimes necessary in radio-frequency applications to minimize (*a*) the resistance of an inductor, (*b*) the inductance of a resistor.

   Describe, and explain the principle of, any constructional features which may be adopted to achieve the two purposes (*a*) and (*b*).   (E 1961)

5. Describe the general form of construction of a wire-wound non-inductive resistor and of a composition-type resistor for use at radio frequencies.

   List the advantages and disadvantages of the composition type relative to the wire-wound type.   (A 1961)

6. Explain why resistors are sometimes included in electrical circuits.

   Sketch and describe the construction of two resistors, one having a relatively high, and the other a relatively low, power rating. Point out the features in the two constructions which contribute to the respective power ratings.   (E 1960)

7. Describe, with the aid of sketches, the construction of any types of adjustable resistor of small physical size.

   State, briefly and in general terms, any advantages and disadvantages of the types you describe.   (E 1959)

8. Describe the construction of (*a*) a preset trimming capacitor as used in a tuned radio circuit and (*b*) an electrolytic capacitor.

   Quote typical capacitance ranges for both types and mention two uses of an electrolytic capacitor. (A 1962)

9. Explain why capacitors are sometimes included in electrical circuits.

   Describe, with the aid of a sketch, the construction of a variable capacitor. Give reasons for the choice of materials used. State the factors on which the value of the capacitance depends. Explain how the value is varied in the capacitor you describe.

   Would the capacitor you describe be of high or low stability? Give reasons. (E 1962)

10. With the aid of sketches describe the construction, stating the materials used, of two of the following types of fixed capacitors—

    (*a*) a silvered-mica capacitor having a capacitance of about 500 pF,

    (*b*) a paper capacitor having a capacitance of about 0·5 μF,

    (*c*) an electrolytic capacitor having a capacitance of about 16 μF.

    State one application of each in a receiver. (A 1960)

11. Sketch the construction of a variable tuning capacitor of the type used in a medium-wave broadcast receiver.

    How, and in what circumstances, are two-gang, and sometimes three-gang, capacitors used in broadcast receivers? (1 1957)

12. Describe, with the aid of sketches, the construction of an output transformer for use in a high-fidelity audio-frequency amplifier, discussing the material used. (A 1959)

13. Describe with the aid of sketches, the constructional features of an audio-frequency output transformer.

    What factors determine the gauge of wire used for the primary and secondary windings? (1 1958)

14. Discuss briefly the factors which determine the choice of dielectric in the electrolytic, fixed, and variable types of capacitor used in a radio receiver.

    Quote typical capacitance values for (*a*) a preset trimming capacitor as used in a tuned radio circuit, and (*b*) an electrolytic capacitor. Mention two uses of an electrolytic capacitor.

    What determines the shape of the plates required in a variable tuning capacitor? (A 1965)

4 *The Principles and Construction of Microphones, Loudspeakers and Receivers*

The basic principle of a telecommunication system employed for the transmission of sound intelligence is that the intelligence is converted into an electrical signal, and the signal is transmitted over the system and re-converted to the form of sound intelligence at the receiving end. The conversion of sound intelligence to electrical signal is the function of a microphone (sometimes called a transmitter), and the re-conversion of electrical signal to sound intelligence is the function of a receiver. The requirements for either of these devices depend upon the use to which it is to be put. For example, the microphones and receivers used in the subscribers' instruments in a telephone network must be cheap and mechanically robust, because there are so very many of them and their treatment may not be gentle. Further, their performance should be stable over a fairly long period of time and be adequate for the transmission and repro-duction of reasonable quality speech. On the other hand, micro-phones destined for use in broadcasting studios are relatively few and their cost is not a primary consideration, a good performance over a wide frequency range being much more important.

Microphones

A number of different types of microphone are used today and they can be roughly divided into one of two main groups, (*a*) microphones whose electrical output depends upon the magnitude of the changes in air pressure on the microphone and (*b*) microphones whose electri-cal output is a function of the difference between the air pressures at two closely situated points. In the first group are to be found the carbon microphone, the crystal microphone and the capacitor micro-phone while the second group contains the moving-coil microphone and the ribbon microphone. The crystal and capacitor microphones are not very common and will not be described in this chapter. Perhaps the major problem with all types of microphone is to obtain an electrical output that is an exact replica of the incident sound wave, because of mechanical vibrations in the microphone itself which make its responses at different frequencies unequal.

THE CARBON MICROPHONE

Carbon is a material of low conductivity, it is cheap and does not oxidize readily, and, most important, if two pieces of carbon are in contact they have a contact resistance that decreases with increase in pressure. The principle of operation of a carbon microphone is to be seen in Fig. 4.1. In Fig. 4.1*a* the air pressure on the diaphragm of the microphone is such that the movable electrode is in the position shown and the carbon granules are just touching. The resistance between the two electrodes is then R ohms and a current $I = E/R$ flows. If the air pressure on the diaphragm is increased the movable electrode is moved to the right and the carbon granules are compressed. This increases the area of contact between neighbouring

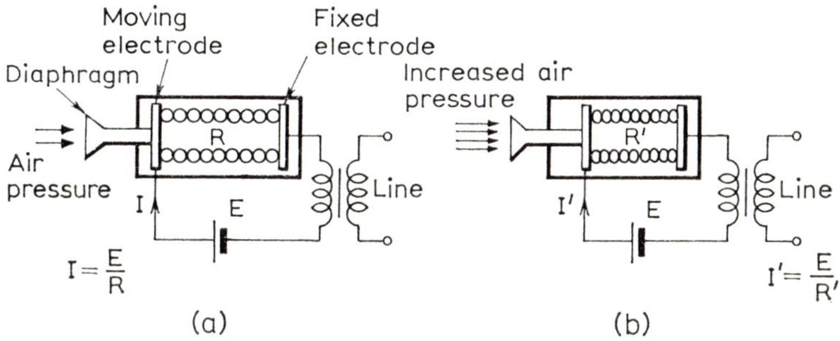

Fig. 4.1

The principle of the carbon microphone

granules and so the resistance between the electrodes falls to a new value R'. The current flowing in the primary winding of the transformer increases to a new value, $I' = E/R'$, and the change in current induces an e.m.f. in the secondary winding of the transformer. If the air pressure on the diaphragm continually changes, an alternating e.m.f. will appear across the secondary winding.

The constructional details of a typical carbon microphone are given in Fig. 4.2. The front of the microphone is circular and has a diameter of just over 2 inches. The conical aluminium diaphragm is rigidly clamped around its edge but is free to vibrate in sympathy with changes in air pressure. Fixed to the centre of the diaphragm is a hollow aluminium cylinder and fixed to the other end of this is the front carbon electrode. This electrode is free to move in the direction of the axis of the cylinder under the influence of any movement of the diaphragm. The rear carbon electrode is fixed to, but insulated from, the microphone case. The faces of the two electrodes are parallel to one another and the remaining space within the chamber is filled with carbon granules. To ensure that the current flowing from one electrode to the other passes only via the carbon granules, the inner surfaces of the chamber are insulated by spraying them with varnish. Equalization of the air pressure at both sides of the diaphragm is necessary to prevent damping its movement and it is achieved by the provision of the small breathing hole; the felt pad allows the passage of air but prevents the carbon granules escaping. The granules are prevented from escaping around the moving electrode by the mica and silk washers that are held firmly in position by a spring ring. Electrical connection to the rear electrode is made by a pin protruding into the insulated connection

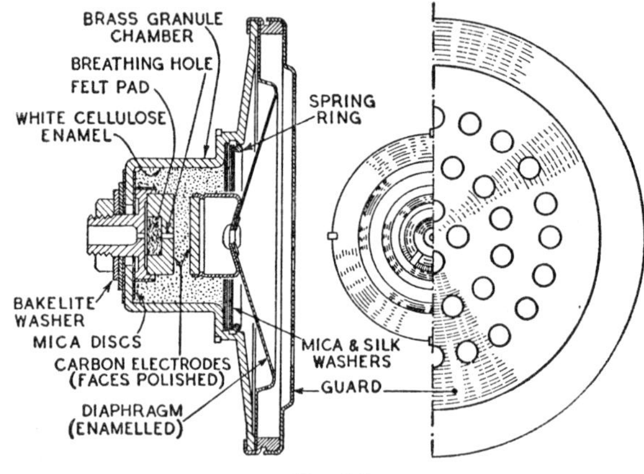

Fig. 4.2

A common type of carbon microphone

(From the *Post Office Electrical Engineers' Journal*)

socket and electrical connection to the front electrode is via springs making contact with the metal case. The diaphragm is protected against mechanical damage by a metal cover which has a number of small holes in it to allow the sound waves to pass. To protect against corrosion the diaphragm is coated with enamel. Sound waves incident upon the diaphragm set it in vibration and the cylinder transfers these vibrations to the movable front electrode. The pressure exerted on the carbon granules varies and a variation in the electrical resistance of the microphone is achieved. A mathematical analysis of the operation of this type of microphone shows that the changes in electrical output are not identical with the changes in diaphragm position causing them; this means that some distortion is introduced. Further distortion is introduced because the electrical output, for a given amplitude of incident sound wave, varies with change in the frequency of that sound wave.

This type of carbon microphone is widely employed in telephone networks in spite of the distortion it introduces because it possesses two great advantages. The first is that it is cheaper than any other type of microphone and the second is that it is extremely sensitive. For the average level of speech the power output of this type of transmitter is of the order of half a milliwatt; this might seem small but it is more than sufficient to operate a telephone receiver.

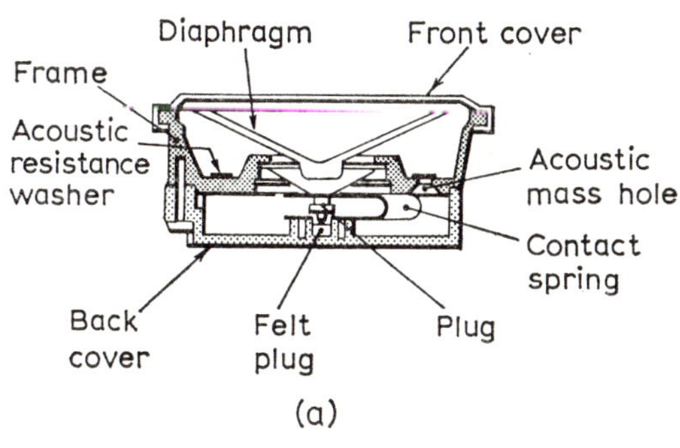

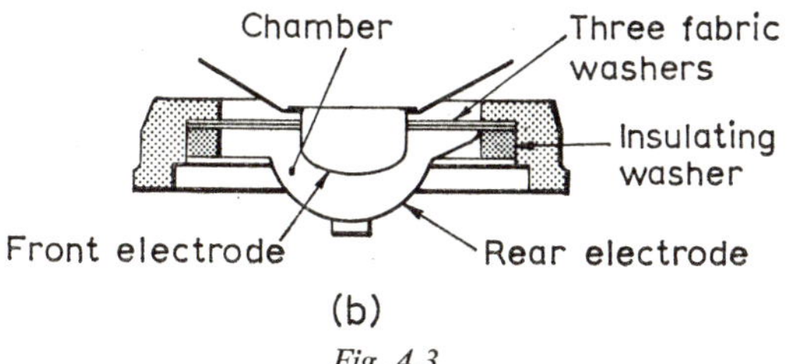

Fig. 4.3

A more modern carbon microphone

(From the *Post Office Electrical Engineers' Journal*)

Recently a new type of carbon microphone has been introduced into the British telephone network that is slightly more sensitive than the old and introduces less distortion. The operation of this microphone is similar to the operation of the older type, but its construction is different and is shown in Fig. 4.3*a*. Fig. 4.3*b* shows an enlarged view of the carbon granule chamber. The aluminium-magnesium alloy diaphragm is clamped around its edge to the die-cast aluminium frame and the edge sealed with a moisture-resistance compound. Fixed to the diaphragm is the hemispherical front electrode and this is able to move when the diaphragm vibrates. The front and rear electrodes are both made of thin carbonized nickel and are shaped so that they can, in conjunction with the three fabric washers, form the carbon granule chamber. To protect the diaphragm against corrosion it is enamelled. Air pressure equalization is provided by three holes in the frame over which is stuck a washer made of an acoustic-resistance material.

THE MOVING-COIL MICROPHONE

A moving-coil microphone consists, basically, of a conductor that is caused to move within a magnetic field by a sound wave incident upon a diaphragm attached to the conductor. Any conductor moving in a magnetic field has an e.m.f. induced in it according to the expression

$$e = Blv \qquad (4.1)$$

where e = induced e.m.f. in volts; B = magnetic flux density in webers/metre^2; l = length of the conductor in metres; v = the velocity with which the conductor moves in the field in metres/sec.

The moving coil thus has an e.m.f. induced in it which, in the absence of distortion, has the same waveform as the incident sound wave.

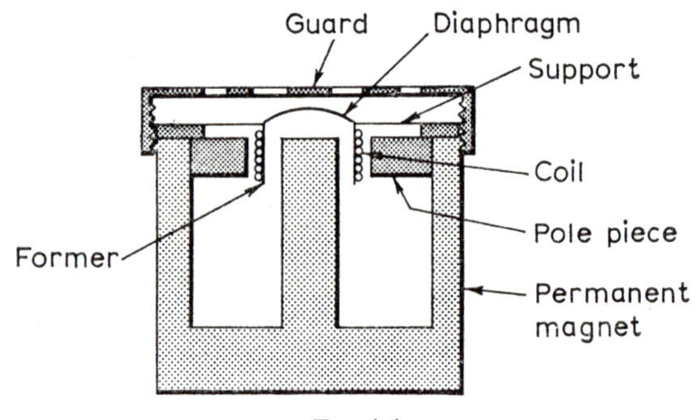

Fig. 4.4

A moving-coil microphone

The construction of a typical moving-coil microphone is shown in Fig. 4.4. The circular diaphragm is made of aluminium and is supported around its edge. Mounted on the diaphragm is the aluminium coil former, the former being situated in the radial magnetic field between the circular pole pieces. The aluminium former has a slit cut in it to minimize the flow of eddy currents and around the former is wound the coil. When a sound wave is incident on the diaphragm the coil is caused to move within the magnetic field and an e.m.f. is induced in it. For the microphone to have the same sensitivity at all frequencies the velocity of the coil for a given sound amplitude should be independent of the frequency of the sound. In practice, this is not quite so and some distortion occurs.

A moving-coil microphone is fairly robust and reliable but is more expensive than the carbon microphone. Because of its good frequency characteristic the moving-coil microphone is often used, for example, in public address systems and in broadcasting.

THE RIBBON MICROPHONE

Essentially, a ribbon microphone is a moving-coil microphone having but a single turn, a typical construction being shown in Fig. 4.5. The microphone is activated by the difference between the sound pressures at the front and back of the ribbon. When a

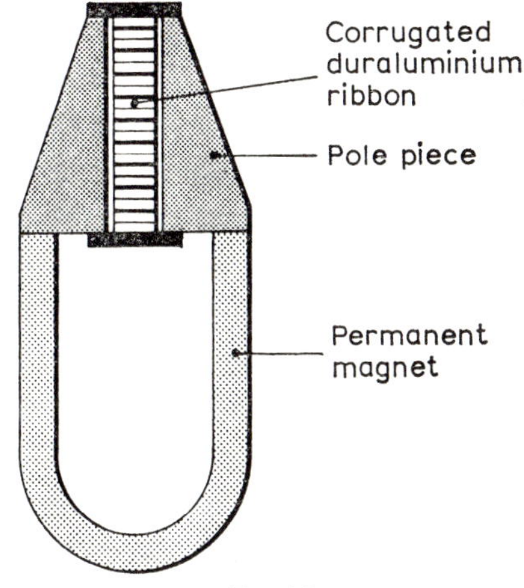

Fig. 4.5

A ribbon microphone

sound wave is incident on the microphone the ribbon vibrates between the pole pieces and an e.m.f. is induced in it. The sensitivity of a ribbon microphone is low and its electrical output requires amplification, but its output/frequency characteristic is extremely uniform over a wide bandwidth. This type of microphone responds much better to sound waves coming from some directions than from others, i.e. it is directional.

The relative advantages and disadvantages of the microphones mentioned are listed in Table 4.1.

Receivers

For many years the receivers used in telephone instruments have been of the moving-iron type in which a magnetic diaphragm is attracted, to a greater or lesser extent, by a pair of pole pieces. In the last ten years or so a new type of telephone receiver has been progressively introduced into the telephone network. The new receiver, known as the "rocking armature" receiver, has an increased sensitivity and a more uniform output/frequency characteristic.

THE MOVING-IRON RECEIVER

The construction of a moving-iron receiver is shown in Fig. 4.6. Two coils of wire are wound on a pair of nickel-iron pole pieces that are connected to a nickel-aluminium-iron permanent magnet. The pole pieces attract, but do not quite touch, the light and high-permeability cobalt-iron diaphragm. When no current is flowing in the

Table 4.1

MICROPHONE	ADVANTAGES	DISADVANTAGES
Carbon	Robust, reliable, cheap, very sensitive.	Poor output/frequency characteristic. Produces background noise. Requires a source of steady current. Granules are liable to pack.
Moving-coil	Robust, reliable, quiet, fairly good output/ frequency characteristic.	Diaphragm may vibrate in the wind. More expensive than carbon microphone, insensitive.
Ribbon	Robust, reliable, good frequency response, quiet, directional.	Insensitive, diaphragm very liable to be vibrated by the wind.

coils the steady magnetic flux of the permanent magnet attracts the diaphragm towards the pole pieces. When an alternating current is passed through the coils the alternating magnetic flux produced alternately adds to, and subtracts from, the steady flux. When the flux is increased the pull on the diaphragm is increased and the diaphragm moves nearer to the pole pieces; when the flux is reduced the pull on the diaphragm is less and the natural springiness of the diaphragm causes it to move further from the pole pieces. Hence, if a speech current is passed through the coils of the receiver the diaphragm will vibrate and sound waves will be produced.

Since the only source of power energizing a receiver of this type is the energy contained in the currents flowing in the coils, it might be thought that the permanent magnet was unnecessary. However, it can readily be demonstrated that if the permanent magnet were not present the sensitivity of the receiver would be considerably less and the output speech would be an octave too high (because the diaphragm would make one complete vibration for each half-cycle of current).

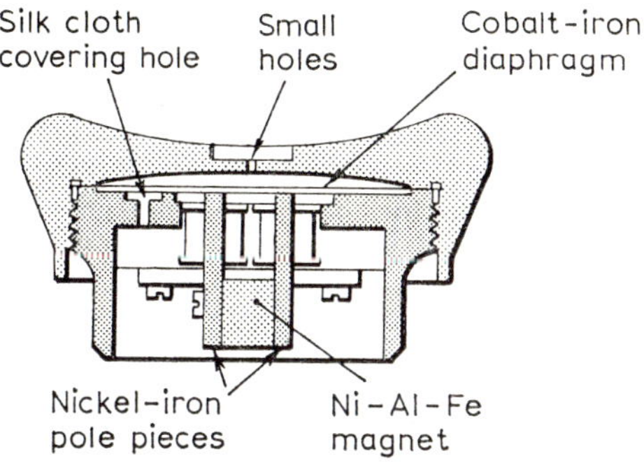

Fig. 4.6

A moving-iron receiver

(From the *Post Office Electrical Engineers' Journal*)

THE ROCKING-ARMATURE RECEIVER

The basic construction of a rocking armature receiver is shown in Fig. 4.7. With no current flowing in the coils and the armature in the equilibrium position shown the flux set up by the permanent

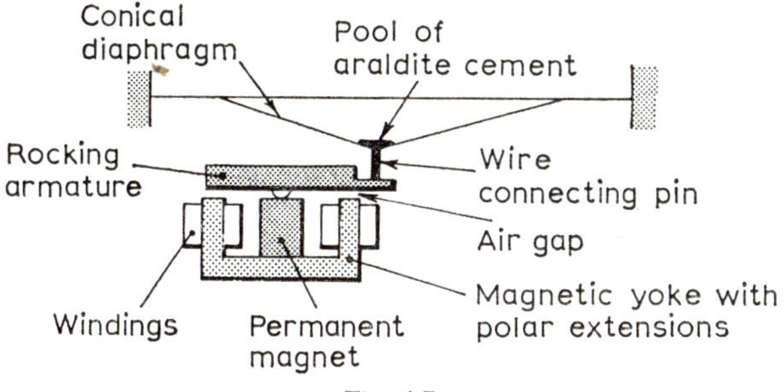

Fig. 4.7

A rocking-armature receiver

(From the *Post Office Electrical Engineers' Journal*)

magnet has two identical parallel paths to follow, each path consisting of one arm of the yoke and a half of the armature. Thus the armature is attracted equally to each of the pole pieces and remains still. Should the armature be moved off balance towards either of the pole pieces the resulting magnetic unbalance would tend to deflect the armature still further. To prevent this happening, and so give static stability, a restoring force must be applied to the armature. This force is provided by two torsion members (not shown) which are attached to the armature. The two coils are connected in series and wound in opposite directions and so the flux created when a current flows in the coils assists the permanent flux at one pole piece and opposes it at the other. As a result the pull on the armature is increased at one end and decreased at the other, and the armature rocks in the appropriate direction. Since the diaphragm is connected to one end of the armature by a connecting pin, any movement of the armature is transmitted to the diaphragm. If an alternating current is passed through the coils the armature rocks about its pivot and the diaphragm vibrates and produces sound waves. The rocking-armature receiver is more sensitive than the moving-iron receiver, because in the former the functions of the diaphragm and the magnetic circuit are carried by two mechanically separate items allowing each to be designed for optimum performance.

The receiver is supplied in the form of a sealed capsule unit having exterior terminals for connection. The capsule is placed into the receiver end of the telephone handset, and an earpiece, of similar shape to the one shown in Fig. 4.6, is screwed on to completely enclose it.

THE MOVING-COIL RECEIVER
The construction of a moving-coil receiver is identical with that of a moving-coil microphone, see Fig. 4.4; indeed the one device can be used for either purpose. When used as a receiver the operation is similar to that of the moving-coil loudspeaker to be described. A moving-coil receiver produces less distortion than either of the types previously mentioned but is more expensive and less robust.

Loudspeakers

A loudspeaker is required to handle a relatively large acoustic output power, but a good sensitivity is not demanded since adequate electrical power is generally supplied. Since a loudspeaker is normally required to produce music as well as speech it should have a fairly uniform output/frequency characteristic over as wide a band as possible. Typically, a cheap loudspeaker might have a fairly uniform response over the band 100–10,000 Hz. and an expensive loudspeaker might have a uniform response over the band 50–15,000 Hz.

Although several different types of loudspeaker exist the vast majority in use today are of the moving-coil type. A moving-coil loudspeaker is constructed in basically the same way as a moving-coil microphone or receiver, differing mainly in its larger size and the use of a baffle board or cabinet. Fig. 4.8 shows the construction of a typical moving-coil loudspeaker; the materials employed for the components of the magnetic circuit have not been shown since they vary considerably according to the manufacturer. A few turns of non-magnetic wire are wound on a light cylindrical former that is

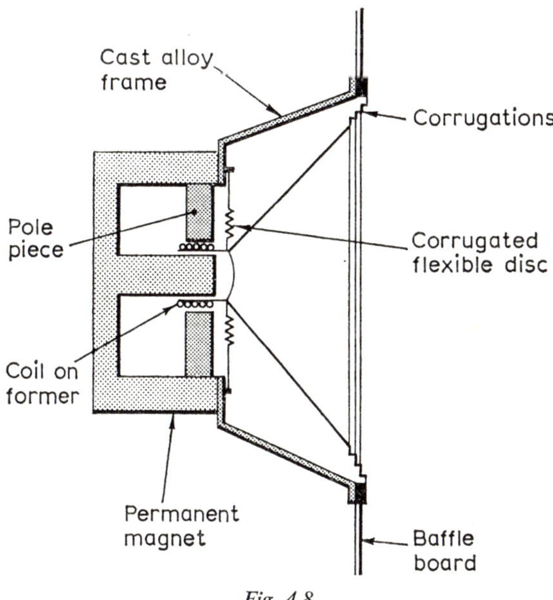

Fig. 4.8

A moving-coil loudspeaker

mounted so that it can move backwards and forwards but not sidewards. The coil and former are situated in the radial magnetic field produced by the pole pieces of the permanent magnet and hence the field is always at right angles to the coil. When a current is passed through the coil a force, whose direction is determined by Fleming's left-hand rule, is exerted on the former, which therefore moves. If the current is alternating at speech frequencies the cone of the loudspeaker, which is cemented to the former, will vibrate and a sound wave will be produced. The cone is generally circular in shape and some 3 to 15 inches in diameter, but other shapes, such as elliptical, are also in common use. The corrugations around the edge of the cone affect the output/frequency characteristic of the loudspeaker and some control can be achieved by their careful design. The cone is prevented from moving sideways by the flexible corrugated disc connecting the apex of the cone to the frame of the loudspeaker.

If the maximum sound energy is to be radiated the air compressed at the front of the loudspeaker by movement of the cone must be prevented from passing to the rear of the cone where, at the same time, the air is rarefied. This may be achieved either by the provision of a baffle board made of a material such as plywood, or by enclosing the speaker within a suitable cabinet.

Generally, a loudspeaker has an input impedance of some 3 to 15 ohms and, since such values are far too low to provide a suitable load for a valve or transistor, a moving-coil loudspeaker is usually fed via a matching transformer (see Example 3.1). However, some small modern loudspeakers, which have been designed for use in transistorized radio receivers, have a sufficiently high input impedance to obviate the need for a matching transformer.

Exercises

1. Describe, with the aid of sketches, the construction and operation of:
   (*a*) a moving-coil telephone receiver and
   (*b*) a moving-iron telephone receiver.
   Briefly explain the effect on the operation of each type if the permanent magnet is removed. (A 1964)

2. With the aid of a sketch, describe a moving-coil loudspeaker and explain its operation. Why are baffle boards normally used with such loudspeakers?
   An output tetrode valve is designed to work into an anode load of 5,000 Ω. What turns ratio is required in order to match a 3 Ω moving-coil cone loudspeaker? (A 1962)

3. Explain, by reference to the operation of a simple two-way telephone circuit, the function of:
   (*a*) the permanent magnet in the telephone receiver,
   (*b*) the direct current in the carbon microphone. (A 1961)

4. Describe, with the aid of a sketch, the constructional features and the principle of operation of the carbon-granule microphone.
   Explain clearly, using a diagram, how the output alternating energy is derived. (A 1960)

5. Describe, with the aid of a sketch, the construction and operation of a typical telephone receiver.
   Explain the need for the permanent magnet in such a receiver. (A 1959)

6. Explain with the aid of a sketch the principle of operation of a moving-coil loudspeaker. What is the function of a baffle board?
   The optimum load for a particular output valve is 4,900 Ω. What turns ratio is required in the output transformer to match the valve to a 4 Ω moving-coil loudspeaker? (A 1965)

5 *The Decibel and the Neper*

In telecommunication engineering an engineer is concerned with the transmission of intelligence from one point to another, the intelligence being transmitted in the form of electrical signals. A telecommunication system employed to carry such signals may consist of a number of links in tandem and, certainly, each link will consist of a number of different items, such as transmission lines and amplifiers, also connected in tandem. Each item will introduce a certain loss, or gain, of power into the system and the overall ratio (output power/input power), which is a measure of the efficiency of the system, is equal to the product of the power ratios of the individual items. Consider, as a simple example, the arrangement of Fig. 5.1.

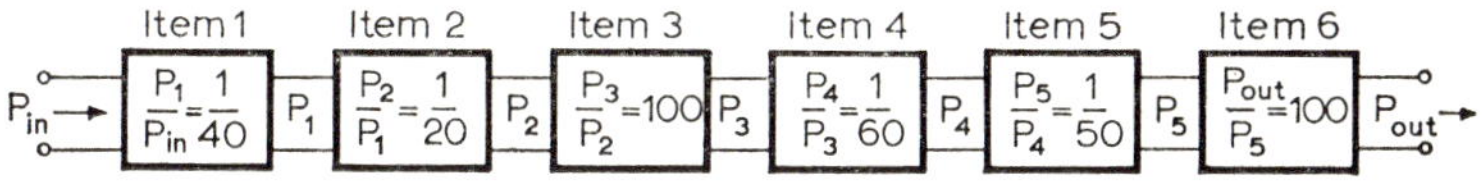

Fig. 5.1

Block schematic of a generalized telecommunication system

If it is assumed that the input and output terminals of each item of equipment are matched* then the input power to item 2 is equal to $P_{IN} \times 1/40$, the input power to item 3 is equal to $P_{IN} \times 1/40 \times 1/20$ and so on. Hence the output power P_{OUT} of the circuit is

$$P_{OUT} = P_{IN} \times \frac{1}{40} \times \frac{1}{20} \times 100 \times \frac{1}{60} \times \frac{1}{50} \times 100$$

or $\qquad \dfrac{P_{OUT}}{P_{IN}} = \dfrac{1}{240} = 0{\cdot}00417$

In this example simple figures have been selected and the arithmetic is easy. However, it should be appreciated that in practice

---

* The term "matched" means that the input terminals are closed with an impedance equal to the impedance seen looking into the terminals. For example, if the input impedance of item 1 is R ohms then the impedance of the source of the input power, P_{IN}, must also be R ohms. Similarly the output impedance of item 3 must be equal to the input impedance of item 4 and so on. If this is not so and a pair of terminals are "mismatched" some of the power arriving at those terminals is reflected back whence it came and the calculation of the overall loss, or gain, is much more difficult and beyond the scope of this book.

the power ratios of the various items comprising a link would almost certainly involve somewhat less convenient numbers, and recourse would probably be had to logarithmic tables, or a slide rule, to aid the computation. Should logarithms be employed the method of attack, again considering the simple example, would be to look up in tables the logarithm of each of the power ratios, add the values thus found, and then look up the antilogarithm of the result. Normally, logarithms to the base 10 are employed and assuming this here gives

$$\log_{10} \frac{P_{OUT}}{P_{IN}}$$

$$= \log_{10} \frac{1}{40} + \log_{10} \frac{1}{20} + \log_{10} 100 + \log_{10} \frac{1}{60} + \log_{10} \frac{1}{50} + \log_{10} 100$$

$$= \log_{10} 1 - \log_{10} 40 + \log_{10} 1 - \log_{10} 20 + \log_{10} 100$$
$$+ \log_{10} 1 - \log_{10} 60 + \log_{10} 1 - \log_{10} 50 + \log_{10} 100$$

$$\left(\text{since } \log \frac{a}{b} = \log a - \log b\right)$$

Now $\log_{10} 1 = 0$,

$$\therefore \quad \log_{10} \frac{P_{OUT}}{P_{IN}} = -\log_{10} 40 - \log_{10} 20 + \log_{10} 100 - \log_{10} 60$$
$$- \log_{10} 50 + \log_{10} 100$$

$$= -1 \cdot 6021 - 1 \cdot 3010 + 2 - 1 \cdot 7782 - 1 \cdot 6990 + 2$$

$$= -2 \cdot 3803$$

$$= -3 + 0 \cdot 6197$$

$$\therefore \quad \frac{P_{OUT}}{P_{IN}} = \text{antilog}_{10} \bar{3} \cdot 6197 = 0 \cdot 00417 \text{ as before}$$

Very often in practical calculations the power ratios involved are enormous and the direct use of power ratios would involve inconveniently large, or small, numbers. This suggests the possibility of directly quoting the power losses, or gains, of the items of equipment in a logarithmic form, so that the overall loss, or gain, also quoted in logarithmic form, can be obtained by the mere algebraic addition of the individual losses, or gains. This, in fact, is the system used in practice, the logarithmic unit employed being known as the decibel.

The Decibel

The decibel can be defined in the following way:

> If the ratio of two powers P_1 and P_2 is to be expressed in decibels, the number of decibels, x, is given by

$$x = 10 \log_{10} \frac{P_1}{P_2} \qquad (5.1)$$

As an illustration consider again the system shown in Fig. 5.1.

Power ratio of item 1 $= 10 \times -1{\cdot}6021 = -16{\cdot}021$ dB
$\simeq -16{\cdot}0$ dB*
Power ratio of item 2 $= 10 \times -1{\cdot}3010 = -13$ dB
Power ratio of item 3 $= 10 \times 2 = 20$ dB
Power ratio of item 4 $= 10 \times -1{\cdot}7782 = -17{\cdot}8$ dB
Power ratio of item 5 $= 10 \times -1{\cdot}6990 = -17{\cdot}0$ dB
Power ratio of item 6 $= 10 \times 2 = 20$ dB

The overall power ratio is equal to the algebraic sum of these ratios, i.e. $-23{\cdot}8$ dB.

$$\therefore \qquad 10 \log_{10} \frac{P_{OUT}}{P_{IN}} = -23{\cdot}8 \text{ dB}$$

The negative sign means that P_{OUT} is less than P_{IN}, i.e. there is a loss of $23{\cdot}8$ dB. (Note that to speak of a *loss* of $-23{\cdot}8$ dB would mean a *gain* of $23{\cdot}8$ dB.)

Example 5.1
Convert the following power ratios into decibels.

(a) $\dfrac{P_1}{P_2} = 2$; (b) $\dfrac{P_1}{P_2} = 1{,}000$; (c) $\dfrac{P_1}{P_2} = 2{,}000$; (d) $\dfrac{P_1}{P_2} = \frac{1}{2}$; (e) $\dfrac{P_1}{P_2} = \frac{3}{10}$.

Solution

(a) $\dfrac{P_1}{P_2} = 2$, or in dB,

$$\frac{P_1}{P_2} = 10 \log_{10} 2 = 3 \text{ dB} \qquad \textit{Ans.}$$

(b) $\dfrac{P_1}{P_2} = 1{,}000$, or in dB,

$$\frac{P_1}{P_2} = 10 \log_{10} 1{,}000 = 30 \text{ dB} \qquad \textit{Ans.}$$

---

* Decibel values need only be quoted to one place of decimals since it is impracticable to measure to a greater accuracy.

(c) $\dfrac{P_1}{P_2} = 2{,}000$, or in dB,

$$\frac{P_1}{P_2} = 10 \log_{10} 2{,}000 = 33 \text{ dB} \qquad Ans.$$

(d) $\dfrac{P_1}{P_2} = \frac{1}{2}$, or in dB,

$$\frac{P_1}{P_2} = 10 \log_{10} \tfrac{1}{2}$$

$$= 10 \log_{10} 1 - 10 \log_{10} 2 = 10 \times -0{\cdot}3 = -3 \text{ dB} \qquad Ans.$$

(e) $\dfrac{P_1}{P_2} = \dfrac{3}{10}$, or in dB,

$$\frac{P_1}{P_2} = 10 \log_{10} \frac{3}{10}$$

$$= 10 \log_{10} 3 - 10 \log_{10} 10 = 10\,(0{\cdot}4771 - 1) = -5{\cdot}2 \text{ dB} \qquad Ans.$$

Two things should be noted from Example 5.1. Firstly, a doubling, or halving, of power is equivalent to an increase, or decrease, of 3 dB. Thus, if a particular power ratio P_r/P_s is equivalent to 60 dB, then the ratio $2P_r/P_s$ is equivalent to 63 dB and the ratio $P_r/2P_s$ corresponds to 57 dB. Secondly, where a power ratio of less than unity is concerned the method of calculation shown is straightforward and easy if the ratio is quoted as a fraction. Often, however, a power ratio of less than unity is quoted as a decimal when calculation will involve the use of bar numbers.

Now, it is true that $\log_{10} a/b = \log_{10} a - \log_{10} b$

$$= -(\log_{10} b - \log_{10} a)$$

$$= -\log_{10} b/a$$

and this relationship shows that the number of decibels corresponding to a particular power ratio can be calculated by *always* making the larger power the numerator in equation (5.1), and quoting the result as a gain if the output is the larger and as a loss if the input power is the larger. Bar numbers are avoided and the calculation is simplified.

Example 5.2
Calculate the overall loss, or gain, in decibels of the arrangement shown in Fig. 5.2. If the input power is 10 mW calculate the output power.

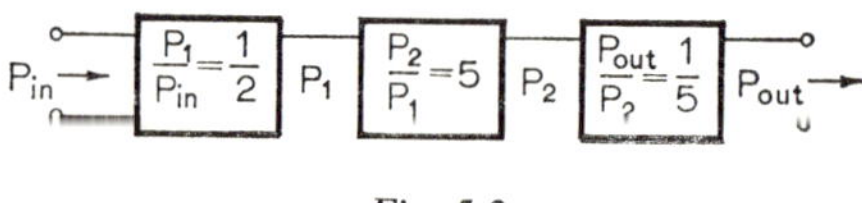

Fig. 5.2

Solution

$$\text{\textit{Loss} of item 1} = 10 \log_{10} \frac{P_{IN}}{P_1} = 10 \log_{10} 2 = 3 \text{ dB}$$

$$\text{\textit{Gain} of item 2} = 10 \log_{10} \frac{P_2}{P_1} = 10 \log_{10} 5 = 7 \text{ dB}$$

$$\text{\textit{Loss} of item 3} = 10 \log_{10} \frac{P_2}{P_{OUT}} = 10 \log_{10} 5 = 7 \text{ dB}$$

Overall loss $= (7 + 3) - 7$, or 3 dB   *Ans.*

$$\therefore \qquad 10 \log_{10} \frac{P_{IN}}{P_{OUT}} = 3$$

$$\frac{P_{IN}}{P_{OUT}} = \text{antilog}_{10} \, 0{\cdot}3 \simeq 2$$

and so the output power P_{OUT} is equal to $\frac{1}{2}P_{IN}$, or $P_{OUT} = 5 \text{ mW}$ *Ans.*

VOLTAGE AND CURRENT RATIOS

A power ratio of x decibels is defined as

$$x = \log_{10} \frac{P_1}{P_2} \tag{5.1}$$

The power P dissipated in a resistance R may be written $P = I^2 R$ or $P = V^2/R$, where I is the current flowing in the resistance and V is the voltage developed across the resistance. Hence equation (5.1) may be rewritten as either

$$x = 10 \log_{10} \frac{I_1^2 R_1}{I_2^2 R_2} \qquad \text{or} \qquad x = 10 \log_{10} \frac{V_1^2/R_1}{V_2^2/R_2}$$

If, *and only if*, $R_1 = R_2$ the resistances cancel and the equations become

$$x = 10 \log_{10} \frac{I_1^2}{I_2^2} \qquad \text{or} \qquad x = 10 \log_{10} \frac{V_1^2}{V_2^2}$$

$$\therefore \qquad x = 20 \log_{10} \frac{I_1}{I_2} \tag{5.2}$$

or $$x = 20 \log_{10} \frac{V_1}{V_2} \tag{5.3}$$

Equations (5.2) and (5.3) may *only* be employed when the resistances in which the currents I_1 and I_2 flow, or across which the voltages V_1 and V_2 are developed, are identical. When unequal resistances are involved the power dissipated in each should be calculated and equation (5.1) employed.

A change in current, or voltage, at a point can always be quoted in dB, using either equation (5.2) or equation (5.3), since the same resistance is involved (provided, of course, that the resistance is not changed in value by the change in current, or voltage).

Example 5.3
An item of telecommunication equipment has an input resistance of 600 Ω and its output terminals are correctly terminated in a 600 Ω resistor. When a voltage of 1·5 V is applied across the input terminals a current of 15 mA flows in the load resistor. Calculate the loss, or gain, of the equipment.

Solution. Three methods of attack are possible: (*a*) calculate the input and output powers and use equation (5.1), (*b*) calculate the input current and use equation (5.2), and (*c*) calculate the output voltage and use equation (5.3). Methods (*b*) and (*c*) can be employed because the input and output powers are dissipated in equal resistances of 600 Ω.
Employing method (*c*),

$$\text{Output voltage} = 15 \times 10^{-3} \times 600 = 9 \text{ V}$$

$$\therefore \quad \text{Gain of equipment} = 20 \log_{10} 9/1 \cdot 5 = 15 \cdot 6 \text{ dB} \quad \textit{Ans.}$$

Example 5.4
An amplifier has a gain of 60 dB. If the input resistance of the amplifier is 75 Ω and its output terminals feed a matched load of 140 Ω, calculate the current flowing in the load when a voltage of 100 μV r.m.s. is applied to the input terminals.

Solution. The resistances in which the input and output powers are dissipated are unequal and hence equation (5.1) must be employed.

$$\text{Input power to the amplifier} = \frac{(100 \times 10^{-6})^2}{75}$$

$$= \frac{1 \times 10^{-8}}{75} \text{ watts}$$

$$\therefore \quad 60 = 10 \log_{10} \frac{P_{OUT}}{(1 \times 10^{-8})/75}$$

$$\text{antilog}_{10} 6 = 75 P_{OUT} \times 10^{8}$$

$$1 \times 10^{6} = 75 P_{OUT} \times 10^{8}$$

$$P_{OUT} = \frac{1}{7,500} = I_{OUT}^{2} \times 140$$

$$\therefore \quad I_{OUT} = \sqrt{\left(\frac{1}{7,500 \times 140}\right)}$$

$$= 0 \cdot 976 \text{ mA} \quad \textit{Ans.}$$

REFERENCE LEVELS: THE DBM, DBR AND DBW
The decibel is not an absolute unit but is only a measure of a power *ratio*. It is meaningless to say, for example, that an amplifier has an output of 60 dB unless a reference level is quoted or is clearly understood. For example, a 60 dB increase on 1 microwatt gives a power level of 1 watt and a 60 dB increase on 1 watt gives a power

level of 1 megawatt: here the same 60 dB difference expresses power differences of less than 1 watt in one case and nearly one million watts in the other. It is therefore customary in telecommunication engineering to express power levels as so many decibels above, or below, a clearly understood reference power level. This practice makes the decibel a more significant unit and allows it to be used for absolute measurements. The reference level most commonly used is 1 milliwatt and a larger power, P_1 watts, is said to have a level of $+x$ dBm, where $x = 10 \log_{10} (P_1/1 \times 10^{-3})$, and a smaller power, P_2 watts, is said to have a level of $-y$ dBm, where $y = 10 \log_{10} (1 \times 10^{-3}/P_2)$.

Example 5.5
Express in dBm the following power levels, (*a*) 1 watt, (*b*) 1 milliwatt and (*c*) 1 microwatt.

Solution.

$$(a) \qquad 1 \text{ watt} = 10 \log_{10} \frac{1}{1 \times 10^{-3}} = 10 \times 3 = 30 \text{ dBm} \qquad Ans.$$

$$(b) \qquad 1 \text{ milliwatt} = 10 \log_{10} \frac{1 \times 10^{-3}}{1 \times 10^{-3}} = 10 \times 0 = 0 \text{ dBm} \qquad Ans.$$

$$(c) \qquad 1 \text{ microwatt} = -10 \log_{10} \frac{1 \times 10^{-3}}{1 \times 10^{-6}} = -10 \times 3 = -30 \text{ dBm} \qquad Ans.$$

In microwave radio-relay telephony/television systems a reference level of 1 watt is employed and power levels expressed in decibels relative to this level are quoted in dBW. A power level of 1 milliwatt is equal to $-10 \log_{10} (1/1 \times 10^{-3}) = -30$ dBW.

A further unit that is employed, particularly in connection with multi-channel carrier telephony systems, is the dBr. This unit expresses in decibels the power level at a point, relative to the power level at some reference point. Normally the reference point is taken as the two-wire origin of a circuit.

Example 5.6
Fig. 5.3 represents, in extremely simplified form, a four-wire telephony circuit that is routed over a multi-channel carrier-telephony system. It may be assumed that the loss through the terminating sets (these are devices for converting a two-wire line into a four-wire line) is 4 dB from terminals 1,1 to terminals 2,2; 4 dB from terminals 3,3 to terminals 1,1; and infinite from terminals 3,3 to terminals 2,2. The gain of the carrier system is 5 dB from input terminals to output terminals in both directions of transmission. Take, as is usual, the two-wire origin (point A) of the circuit as the reference point.

(*a*) If the power at point A is 0·25 mW what are the power levels in dBr at the input and output terminals of the carrier system and at the point B?
(*b*) What are the levels in dBm at the same points?
(*c*) What is the output power?

Solution.

(*a*) Since the loss from terminals 1.1 to terminals 2.2 of a terminating set is 4 dB the power level at the GO input terminals of the carrier system is 4 dB below the power at the reference point A.

∴   Level at GO input terminals $= -4$ dBr   *Ans.*

The gain of the carrier system is 5 dB,

∴   Level at GO output terminals $= +1$ dBr   *Ans.*

The loss through a terminating set from terminals 3.3 to terminals 1.1 is 4dB,

∴   Level at point B $= -3$ dBr   *Ans.*

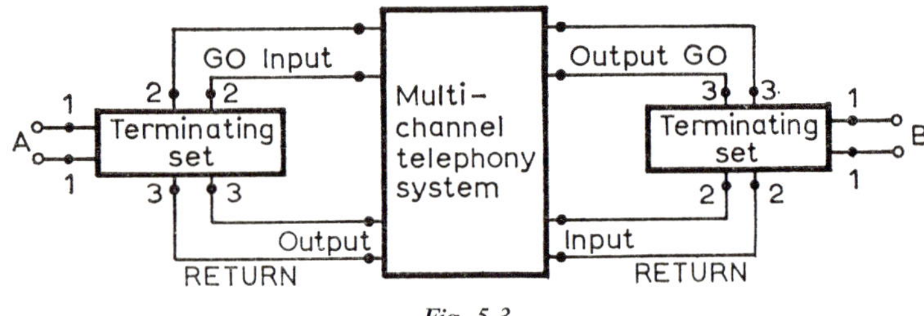

Fig. 5.3

(*b*) The input power level in dBm is equal to $-10 \log_{10} \dfrac{1 \times 10^{-3}}{0.25 \times 10^{-3}}$ or -6 dBm.

∴   The -4 dBr point has a level of $-6 - 4 = -10$ dBm   *Ans.*
         The $+1$ dBr point has a level of -5 dBm   *Ans.*
         The -3 dBr point has a level of -9 dBm   *Ans.*

(*c*) The output power level is -9 dBm.

∴
$$9 = 10 \log_{10} \frac{1 \times 10^{-3}}{P_{OUT}}$$

$$\text{antilog}_{10}\ 0.9 = \frac{1 \times 10^{-3}}{P_{OUT}}$$

∴
$$P_{OUT} = 0.125 \text{ mW}   \textit{Ans.}$$

In part (*c*) the same answer can be obtained by noting that the point B is a -3 dBr point and remembering that -3 dB corresponds to a power ratio of one-half. Thus the output power is equal to one-half the input power.

THE DECIBEL AND THE HUMAN EAR

The human ear is capable of responding to a wide range of sound intensities and has a sensitivity which varies with change in frequency in a logarithmic manner. This makes the decibel a convenient unit for use in conjunction with sound measurement and the measurement of sound equipment. Further, if the gain/frequency characteristic of an audio-frequency amplifier is to give a true indication of its aural effect it should consist of the gain expressed in decibels plotted to a base of frequency to a logarithmic scale.

The Neper

The neper is a logarithmic transmission unit which is widely used on the Continent and which expresses the ratio of two *currents*, or of two *voltages*, but *not* two powers as does the decibel.

A device is said to have a gain of x nepers if

$$x = \log_e \frac{I_{OUT}}{I_{IN}} \tag{5.4}$$

or

$$x = \log_e \frac{V_{OUT}}{V_{IN}} \tag{5.5}$$

where $I_{OUT} > I_{IN}$, $V_{OUT} > V_{IN}$,

or a loss of y nepers if

$$y = \log_e \frac{I_{IN}}{I_{OUT}} \tag{5.4a}$$

or

$$y = \log_e \frac{V_{IN}}{V_{OUT}} \tag{5.5a}$$

where $I_{IN} > I_{OUT}$, $V_{IN} > V_{OUT}$.

It should be carefully noted that the neper is based on logarithms to the base e ($= 2\cdot71828$) and *not* common (base 10) logarithms.

RELATIONSHIP BETWEEN THE DECIBEL AND THE NEPER

When a device, or system, having equal (matched) input and output impedances is concerned a simple relationship exists between the loss or gain of the device or system quoted in nepers and quoted in decibels. Suppose a device has a loss of x nepers, then

$$x = \log_e \frac{I_{IN}}{I_{OUT}}$$

$$\therefore \quad \frac{I_{IN}}{I_{OUT}} = e^x \quad \text{(from the definition of a logarithm)}$$

The loss y, in decibels, since equal impedances are concerned, is

$$y = 20 \log_{10} \frac{I_{IN}}{I_{OUT}}$$

$$= 20 \log_{10} e^x$$
$$= 20x \log_{10} e$$
$$= 20x \times 0\cdot4343$$
$$= 8\cdot686x$$

Thus, in the particular case of matched impedances, 1 neper is equal to $8\cdot686$ dB.

Exercises

1. Define the decibel and explain why it is a convenient unit for use in transmission problems.

   The input level to an amplifier is $+24$ dB relative to 1 μV and the amplifier has a gain of 30 dB. If the input and output impedances of the amplifier are equal and the output is matched to the load, calculate the input and output voltages.   (A 1964)

2. Define the decibel. Two amplifiers, having the gain/frequency responses given in the table below, are connected in tandem with a 10 dB resistive attenuator. Calculate and plot the overall gain of the combination, expressed in decibels, assuming that all input and output impedances are equal.

Frequency kHz	60	66	72	78	84	90	96	102	108
Voltage gain, first amplifier	29·8	34·5	38·0	38·9	38·9	38·5	38·0	34·7	29·0
Voltage gain, second amplifier	28·2	37·2	37·6	36·7	36·7	36·7	38·7	39·1	29·8

(B 1963)

3. Explain what is meant by a decibel.

   The gain/frequency characteristics of two group amplifiers are given in the table.

Frequency in kHz	60	76	92	108
Voltage gain of amplifier 1	310	330	340	390
Voltage gain of amplifier 2	330	345	380	325

   If the two amplifiers are connected in tandem, separated by an attenuator having a loss of 15 dB, plot the overall gain/frequency characteristic of the combination, expressing the gain in decibels. It may be assumed that the amplifiers and the attenuator have the same input and output impedances.

(B 1959)

4. Define the decibel and give three reasons why its use is convenient in transmission problems.

   The input signal to an amplifier varies between 23·5 mW and 1·25 W. Express each power in dB relative to 1 mW and state the fluctuation in the level of the signal in dB.   (A 1965)

6 *Simplified Semiconductor Theory*

A semiconductor is defined as a material whose resistivity is much less than the resistivity of an insulator yet is much greater than the resistivity of a conductor, *and* whose resistivity decreases with increase in temperature. For example, the resistivity of copper is 10^{-8} ohm-metre, of quartz is 10^{12} ohm-metres and of the semiconductor materials of interest in this chapter, that is silicon and germanium, is 0·6 ohm-metre and 1,500 ohm-metres respectively. Before an appreciation can be gained of the operation of semiconductors and semiconductor devices it is necessary to have some familiarity with the basic concepts of the atomic structure of matter.

A Simple Outline of Atomic Theory

All the substances which occur in nature consist of one or more basic elements, a substance containing more than one element being known as a compound. An element is a substance that can neither be decomposed (broken into a number of other substances) by ordinary chemical action, nor made by a chemical union of a number of other substances. A compound consists of two, or more, different elements in combination and has properties different from the properties of its constituent parts. Water, for example, is a compound of oxygen and hydrogen. A molecule is the smallest amount of a substance that can occur by itself and still retain the characteristic properties of that substance and may consist, for example, of two atoms of hydrogen and two atoms of oxygen for hydrogen peroxide, of one atom of oxygen and one atom of carbon for carbon monoxide, and of two atoms of oxygen and one atom of carbon for carbon dioxide. An atom is the smallest unit of which a chemical element is built. The atoms of any particular element all have the same average mass and this average mass differs from the average mass of the atoms of any other element.

The elements are grouped in an arrangement known as the Periodic Table of the Elements (Table 6.1) according to their chemical properties. Elements with similar properties are placed in the same vertical column. More than 100 elements are known to science today; some of them exist in large quantities and are commonly found throughout the world, e.g. oxygen, hydrogen and carbon, others such as gold, uranium, and radium are relatively rare, and some do not exist at all in Nature and are artificially created in equipment operated by atomic physicists.

An atom of any element consists of a complex pattern of electrons orbiting around a positively charged nucleus. The electrons each have a negative charge equal to $1{\cdot}602 \times 10^{-19}$ coulomb (known as the electronic charge e) and exist in just sufficient number to make the overall electrical charge of the atom equal to zero. The nucleus itself consists of a certain number A of particles known as nucleons. A is the mass number of the atom. There are two kinds of nucleons; protons, which each have a positive charge of e coulomb and neutrons, which have zero charge. The number of protons in a nucleus is known as the atomic number of the atom, symbol Z, and the number of neutrons the neutron number N. Hence $A = Z + N$.

The difference between the atoms of the various elements is in the number and arrangement of the electrons, protons and neutrons of which the atoms are composed. There is no difference between an electron in one element and another electron in any other element.

THE HYDROGEN ATOM

The simplest atom is that of the element hydrogen and it consists merely of a single proton in the nucleus and a single electron in orbit around it, Fig. 6.1a. The helium atom is the next simplest

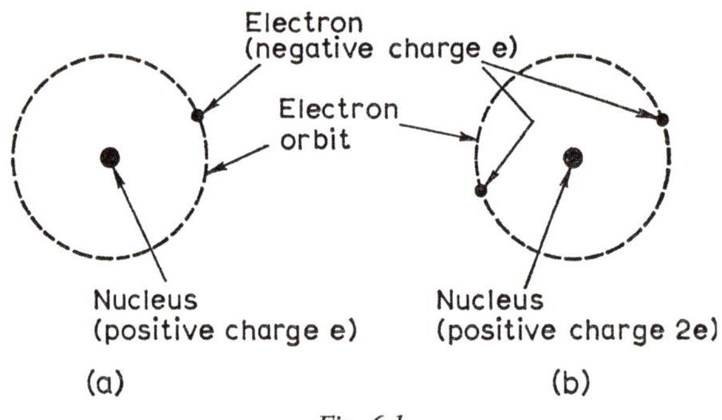

Fig. 6.1

The hydrogen and helium atoms

and can be seen from Fig. 6.1b to consist of a nucleus containing two protons and two neutrons, with two electrons orbiting around it.

For an electron to be able to move in a circular path around a nucleus, as shown in Fig. 6.1, it must have a force exerted on it pulling it towards the nucleus. This force is the electric attractive force exerted by the positive nucleus on the negative electron.

Work must be done in moving an electric charge through an electric field and so some work must have been done to move the electron from the nucleus to the orbit in which it is travelling. Thus an electron must possess a certain, discrete, quantity of energy before it can exist in an orbit around the nucleus. Analysis and hypothesis show that an electron can only exist in certain orbits of particular radii, and when in one of these orbits it must then possess the particular amount of energy that is associated with that orbit. An electron cannot occupy any orbit other than one of the orbits allowed as shown in Fig. 6.2. Normally the electron travels the

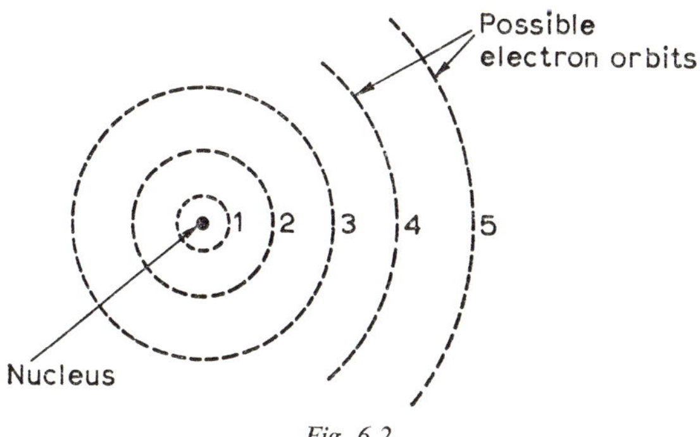

Fig. 6.2

Possible electron orbits in a hydrogen atom

innermost orbit since this is the orbit of least energy, but if it is given, in some way, extra energy, such as heat, it will move to another orbit. An electron can only absorb the exact amount of energy required to raise its total energy to the energy value associated with another orbit; if given this amount of energy the electron will move to its new orbit and remain there until it loses some of its energy. The electron can only lose the exact amount of energy that will allow it to fall into a lower-energy orbit. This means that an electron can only absorb or lose energy in discrete amounts.

The simplified picture of the hydrogen atom given so far is not adequate to account for all the observed phenomena, and it is necessary to extend the model somewhat by imagining that the electron can also move in elliptical orbits (Fig. 6.3). The number of elliptical orbits possible in equal to $n - 1$, where n is the number of the basic circular orbit. The innermost orbit, $n = 1$, has no elliptical paths associated with it, the next orbit, $n = 2$, has a single elliptical path and so on. The circular orbits, numbers 1, 2, 3, etc. are said to form the K, L, M, etc. shells and the elliptical orbits to form sub-shells within these shells.

OTHER ATOMS

The extra-nuclear make-up of the other, more complex, elements can be deduced with accuracy up to the element of atomic number 18 (argon), by adding one more electron for each element in turn, bearing in mind that the number, x, of electrons permitted in a particular shell is given by the expression $x = 2n^2$, where n is the order of the shell. The innermost shell can only contain 2×1^2 or 2 electrons, the next shell 2×2^2 or 8 electrons, the next 2×3^2 or 18 electrons, and so on. The electrons in a particular shell follow paths of different eccentricities. Above atomic number 18 some gaps appear in this system because some orbits in the N shell have lower energy than some orbits in the M shell and are filled first.

The Periodic Table of the Elements is given in Table 6.1.

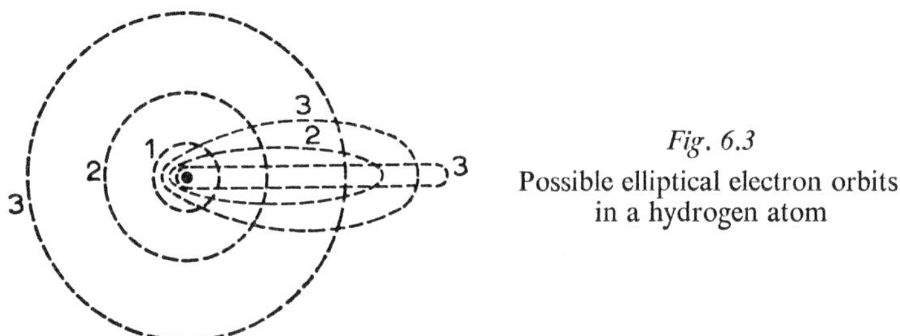

Fig. 6.3

Possible elliptical electron orbits
in a hydrogen atom

The atoms of all elements in group III have three electrons that are not part of a closed shell or sub-shell. Aluminium, for example, has 13 electrons, 10 of which completely fill the K and L shells; the M shell has only three electrons and is incomplete (since 18 electrons are necessary to fill it).

Indium is also in group III and has 49 electrons, 28 of which complete the K, L, and M shells. Of the remaining 21 electrons, 18 completely fill three of the four sub-shells of the N shell and the remaining three enter the O shell (leaving the fourth sub-shell of the N shell unfilled). For both aluminium and indium, therefore, the extra-nuclear structure consists of a number of tightly bound closed shells and sub-shells with three electrons outside and not so tightly bound to the nucleus.

The electronic structure of all the other atoms is also in the form of a number of closed shells and sub-shells with a number of electrons in orbit outside. The nucleus of an atom plus the closed shells and sub-shells of electrons can be considered to be a positively charged central core, the positive charge being equal to $e \times n$, where e is the electronic charge and n is the number of electrons outside the central core. The number of electrons outside this central core is equal to

I	II	III	IV	V	VI	VII	VIII	0
Hydrogen 1								Helium 2
Lithium 3	Beryllium 4	Boron 5	Carbon 6	Nitrogen 7	Oxygen 8	Fluorine 9		Neon 10
Sodium 11	Magnesium 12	Aluminium 13	Silicon 14	Phosphorus 15	Sulphur 16	Chlorine 17		Argon 18
Potassium 19	Calcium 20	Scandium 21	Titanium 22	Vanadium 23	Chromium 24	Manganese 25	Iron 26 Cobalt 27 Nickel 28	
Copper 29	Zinc 30	Gallium 31	Germanium 32	Arsenic 33	Selenium 34	Bromine 35		Krypton 36
Rubidium 37	Strontium 38	Yttrium 39	Zirconium 40	Niobium 41	Molybdenum 42	Technetium 43	Ruthenium 44 Rhodium 45 Palladium 46	
Silver 47	Cadmium 48	Indium 49	Tin 50	Antimony 51	Tellurium 52	Iron 53		Xenon 54
Caesium 55	Barium 56	Rare Earths 57–71	Hafnium 72	Tantalum 73	Tungsten 74	Rhenium 75	Osmium 76 Iridium 77 Platinum 78	
Gold 79	Mercury 80	Thallium 81	Lead 82	Bismuth 83	Polonium 84	Astatine 85		Radon 86
Francium 87	Radium 88	Actinide Series 89–100						

the number of the group in the Periodic Table of the Elements to which the atom belongs. Thus all atoms in group I may be represented by the sketch of Fig. 6.4*a*, all atoms in group II by Fig. 6.4*b*, and so on.

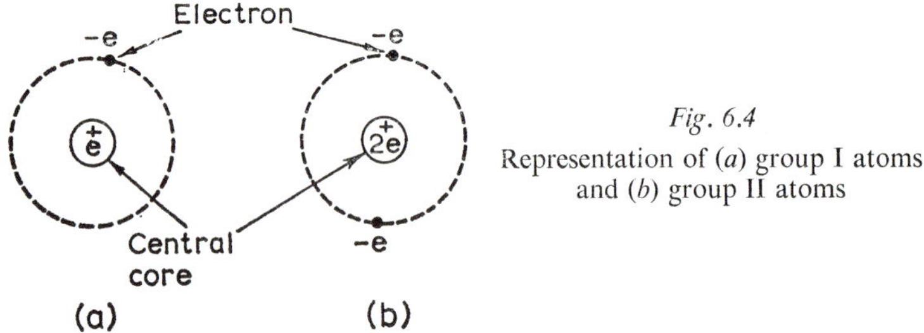

Fig. 6.4

Representation of (*a*) group I atoms and (*b*) group II atoms

The outer electrons are known as valence electrons and determine the chemical properties of the element.

Intrinsic Semiconductors

The two semiconductor materials that are employed in the manufacture of those semiconductor devices, such as diodes and transistors, of interest in this book are germanium and silicon. It can be seen from Table 6.1 that both these materials fall into group four of the Periodic Table of the Elements. An atom of either substance may be represented by a central core of positive charge $4e$ surrounded by four

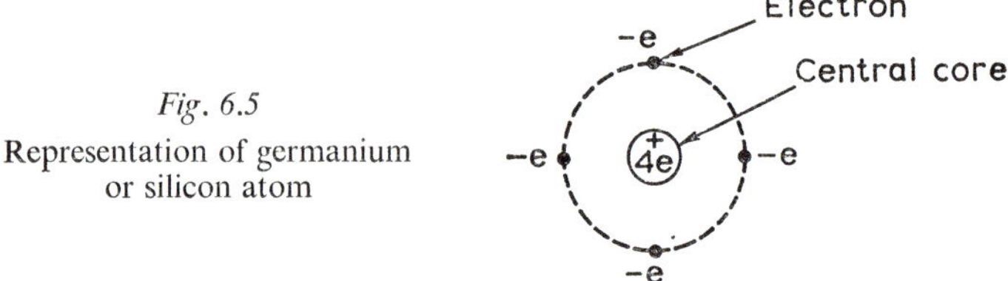

Fig. 6.5

Representation of germanium or silicon atom

orbiting electrons each of which has a negative charge of e (Fig. 6.5). In the remainder of this chapter the discussion of semiconductors will be with reference to germanium but will apply equally well to silicon; any differences between the two materials will be mentioned in the appropriate places. In its solid state, germanium forms crystals of the diamond type, that is, it forms a cubic lattice in which all the atoms (except those at the surface) are equidistant from their immediately neighbouring atoms. A study of crystal structures shows that the greatest number of atoms that can be neighbours to a particular atom at an equal distance away from that atom and yet be equidistant from one another is four. Hence each atom in a germanium crystal has four neighbouring atoms. In the crystal lattice each atom employs its four valence electrons to form covalent bonds with its four neighbouring atoms; each bond consisting of two electrons, one from each atom as shown in Fig. 6.6*a*. Each pair of electrons traverses an orbit around both its parent atom and a neighbouring atom. Each atom is effectively provided with an extra four electrons and these are sufficient to complete its final sub-shell. To simplify the drawings in the remainder of this chapter covalent bonding will be represented in the manner of Fig. 6.6*b*.

If the temperature of the crystal is raised above absolute zero the lattice is thermally excited and some of the valence electrons receive sufficient energy to allow them to break free from a covalent bond. When this occurs the liberated electrons wander randomly in the crystal and are free to accept further energy from an applied electric field and contribute to electrical conduction. With further increase in temperature more covalent bonds are broken, and the conductivity of the germanium is increased because of the increased number of free electrons. This means that a semiconductor material has a negative temperature coefficient of resistance.

When an electron escapes from a covalent bond it leaves behind it an "absence of an electron" which, since it consists of a missing negative charge *e*, is equivalent to a positive charge of magnitude *e*. Such a positive charge is known as a hole. A hole exerts an attractive force on electrons and can be filled by a nearby passing electron that has been previously liberated from another broken covalent bond. This process is known as recombination and it causes a continual loss of holes and free electrons. At any given temperature the rate of recombination of holes and electrons is always equal to the rate of production of new holes and electrons so that the total number of free electrons and holes is constant.

When a covalent bond is broken it is said that a hole-electron pair has been created; both holes and electrons are known as charge carriers. The lifetime of a charge carrier is the time that elapses between its creation and its recombination with a charge carrier of opposite sign.

MOVEMENT OF HOLES THROUGH THE LATTICE

Fig. 6.7 shows a part of a germanium crystal in which the breaking of covalent bonds by thermal agitation of the lattice is taking place. In Fig. 6.7*a* thermal agitation of the crystal lattice has caused a covalent bond to break and produce a hole-electron pair at point A. An instant later a second hole-electron pair is produced at point B

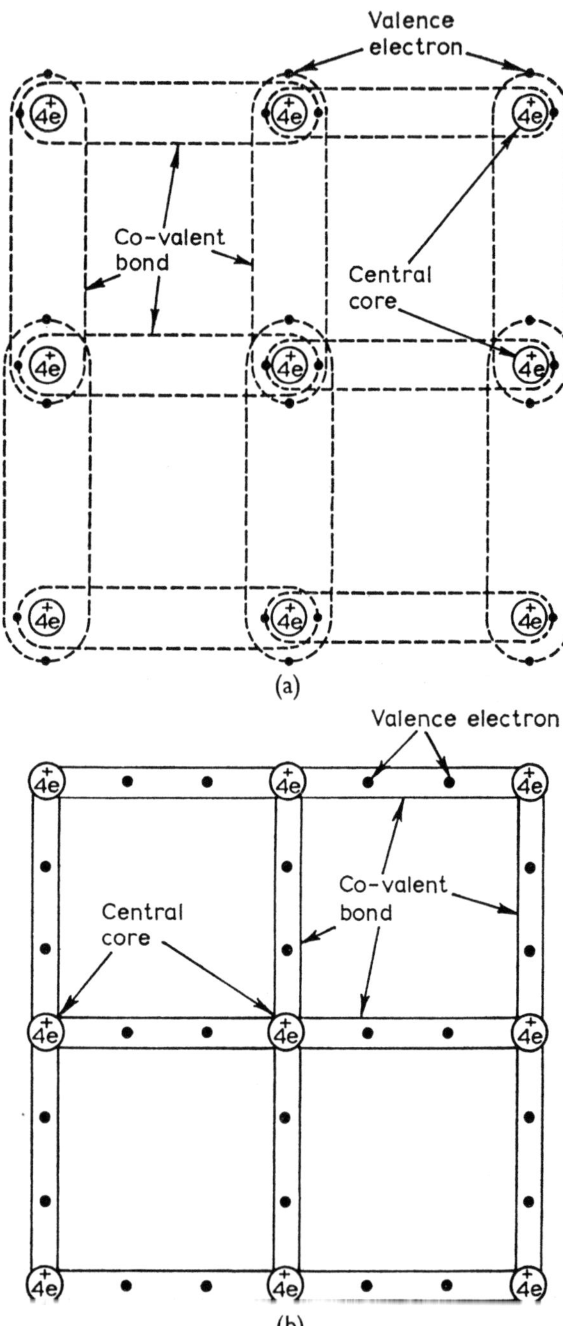

Fig. 6.6

Covalent bonding of atoms

and two electrons are free to wander in the lattice, Fig. 6.7*b*. In Fig. 6.7*c* electron 2 has wandered close enough to the first hole to be influenced by its electric field and recombination has occurred. The original hole has apparently moved from position A to position B, but at the same time another hole-electron pair has been created

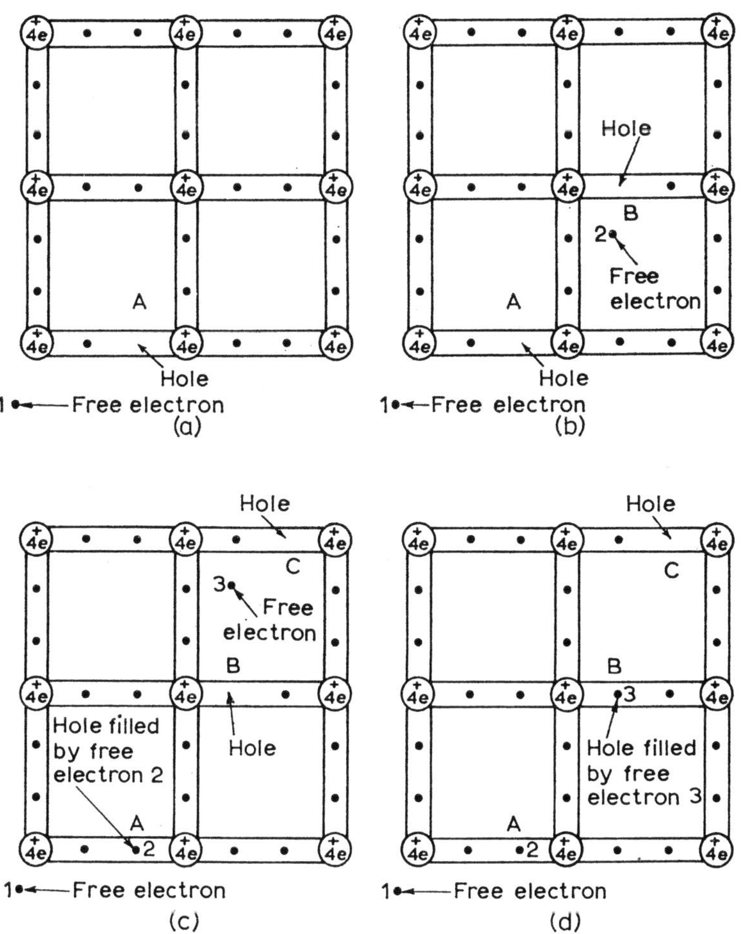

Fig. 6.7

The movement of holes through a crystal lattice

(*a*) Electron-hole pair created
(*b*) Second electron-hole pair created
(*c*) First hole disappeared and third electron-hole pair created
(*d*) Second hole disappeared

at point C. Finally, Fig. 6.7*d*, free electron 3 has travelled across the lattice and has recombined with the hole at point B and the effect is as though a hole has moved through the lattice from point A to point C.

The movement of both holes and electrons through the crystal is quite random but the holes appear to travel more slowly than do electrons. (This is because the movement of a hole in a particular direction actually consists of a series of discontinuous electron movements in the opposite direction.) If an electric field is set up in the crystal the electrons tend to drift in the direction of the field and the holes to drift in the opposite direction. Thus conduction in a pure semiconductor, known as intrinsic conduction, of current takes place (current flow is conventionally in the opposite direction to electron flow). Intrinsic conduction increases with increase in temperature at the approximate rate of 5% per degree centigrade for germanium and even higher for silicon.

Extrinsic (Impurity) Semiconductors

If an extremely small, carefully controlled, amount of an impurity element is introduced into a germanium crystal each of the impurity atoms will take the place of one of the germanium atoms in the lattice. Since the number of impurity atoms is very much smaller than the number of germanium atoms (approximately 1 in 10^8) it is reasonable to assume that the lattice is essentially undisturbed and that each impurity atom is surrounded by four germanium atoms. In practice, the impurities employed are always substances in either group III or group V of the Periodic Table of the Elements, and have either three or five valence electrons. Elements typically employed are arsenic, antimony and phosphorus in group V, and indium, aluminium and gallium in group III. The process of introducing impurity atoms into a germanium crystal is called "doping" and a treated crystal is said to be "doped".

N-TYPE SEMICONDUCTOR

Suppose a germanium crystal has been doped with a small quantity of arsenic, a substance having five valence electrons. Each arsenic atom will set up covalent bonds with its four neighbouring atoms but, since only four of its valence electrons are required for this purpose, a spare electron exists, Fig. 6.8. This surplus electron is not bound to its parent atom and is free to wander in the lattice.

A free electron is created in the germanium crystal lattice for each impurity atom introduced without the creation of corresponding holes. Hole-electron pairs are, however, still produced by thermal agitation of the lattice. The number of free electrons in the crystal is much greater than the number of holes, negative charges predominate and so the crystal is said to be n-type. Since each impurity

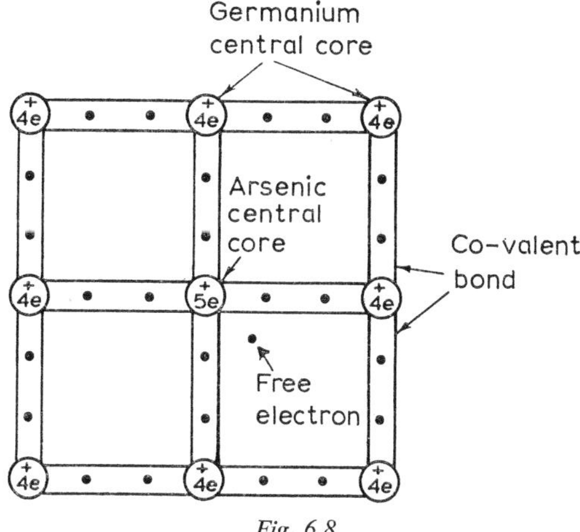

Fig. 6.8

Lattice of n-type germanium crystal

atom donates a free electron to the crystal the impurity atoms are known as donor atoms.

P-TYPE SEMICONDUCTOR

If, instead of arsenic, a group III element, such as indium, is introduced into a germanium crystal, each indium atom will attempt to form a covalent bond with each of its four neighbouring germanium atoms. Indium, however, only has three valence electrons and so only three of the bonds can be completed, Fig. 6.9. One hole is introduced into the lattice for each impurity atom and is able to move about in the crystal in the same way as a hole produced by thermal agitation. In this case holes are in the majority and the material is known as p-type, while the impurity atoms are called acceptor atoms.

A crystal of n-type or p-type germanium is electrically neutral because each impurity atom introduced into the lattice is itself neutral. In n-type material, electrons are the majority charge carriers and holes are the minority charge carriers. In p-type material the reverse applies.

CURRENT FLOW

If a potential difference is maintained across an extrinsic semiconductor (Fig. 6.10), a current will flow into the material at one end and out of the material at the other. The positively charged central cores cannot move from their positions in the crystal lattice and so the current flowing *into* the material can only consist of electrons

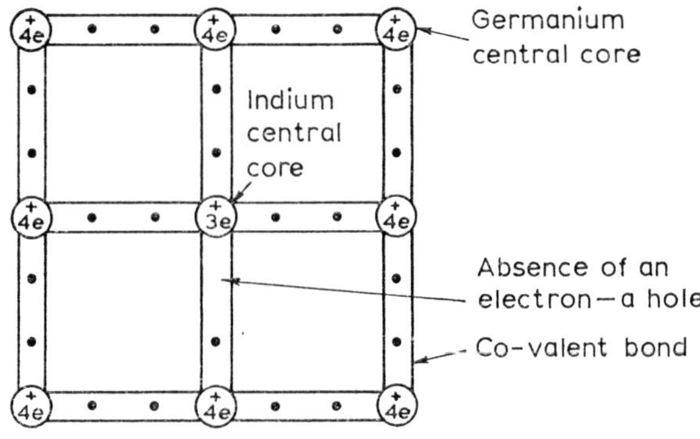

Fig. 6.9

Lattice of p-type germanium crystal

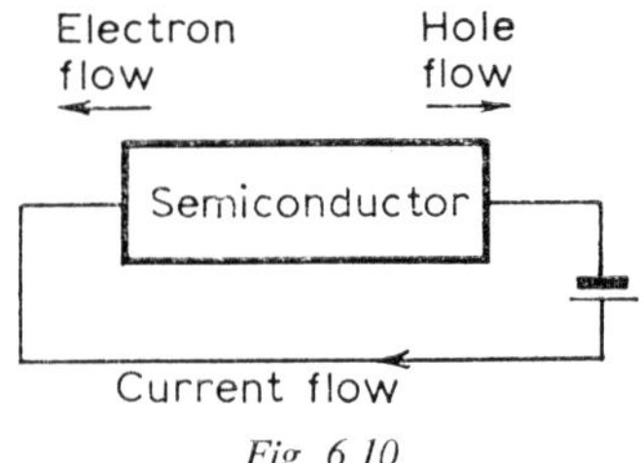

Fig. 6.10

Current flow in an extrinsic
semiconductor

flowing *out* and the current flowing *out* of the material is actually an inward flow of electrons.

The p-n Junction

If a germanium crystal is doped with donor atoms at one end and acceptor atoms at the other the crystal will have both p-type and n-type regions and there will be a junction between them. In Fig. 6.11 the plane AA′ is the p-n junction; only the free electrons and holes have been shown, to clarify the drawing. Both regions include charge carriers of either sign but in the n-type region electrons are in the majority and in the p-type region holes predominate. In both regions the probability of a minority charge carrier meeting and recombining with a majority charge carrier is high and the minority charge carrier lifetime is short.

The free electrons and holes have completely random motions and wander freely in the lattice. However, since there are more electrons to the left of the p-n junction than to the right and more holes to the right of the junction than to the left, *on average* more electrons cross the junction from left to right than from right to left, and more holes cross from right to left than from left to right. On average, therefore, the n-type region gains holes and loses electrons and the p-type region gains electrons and loses holes. This process is known as diffusion and may be defined as the tendency for charge carriers to move away from areas of high density. Since the n-type region loses negative charge carriers and gains positive charge carriers and

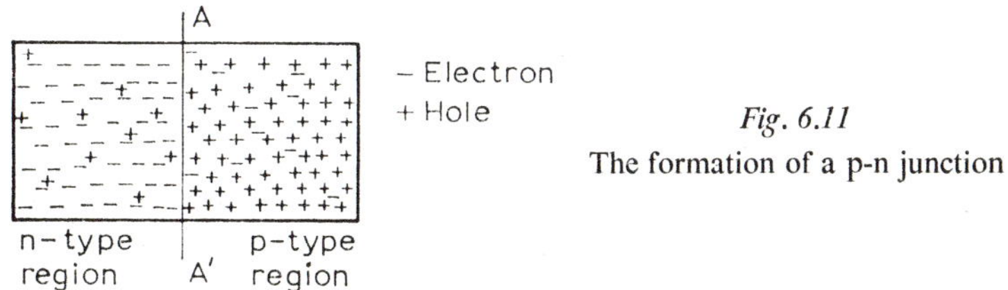

Fig. 6.11

The formation of a p-n junction

the p-type region loses positive charge carriers and gains negative charge carriers, the region just to the left of the junction becomes positively charged and the region just to the right of the junction becomes negatively charged. A hole passing into the n-type region, or an electron passing into the p-type region, becomes a minority charge carrier and will probably recombine with a carrier of opposite sign and disappear; however, one region has still lost a positive (or negative) charge and the other region has gained a positive (or negative) charge. The movement of holes and electrons across the junction constitutes a current and this is known as the diffusion current. If the crystal was neutral before diffusion took place it must be neutral afterwards. Further, because both regions were also originally neutral they must contain equal and opposite charges after diffusion. These charges have an attractive electric force between them and are not able to diffuse away from the vicinity of the junction. The two charges are concentrated immediately adjacent to the junction, and produce a potential barrier across the junction. The polarity of the potential barrier is such as to oppose the further diffusion of majority charge carriers across the junction, but to aid the movement of minority charge carriers; this gives rise to a minority charge carrier current in the opposite direction to the diffusion current.

The difference in potential from one side of the junction to the other is called the height of the potential barrier and is measured in volts. The height of the potential barrier attains such a value that the diffusion (majority charge carrier) and minority charge carrier currents are equal and so the net current across the junction is zero. Any charge carriers entering the region on either side of the

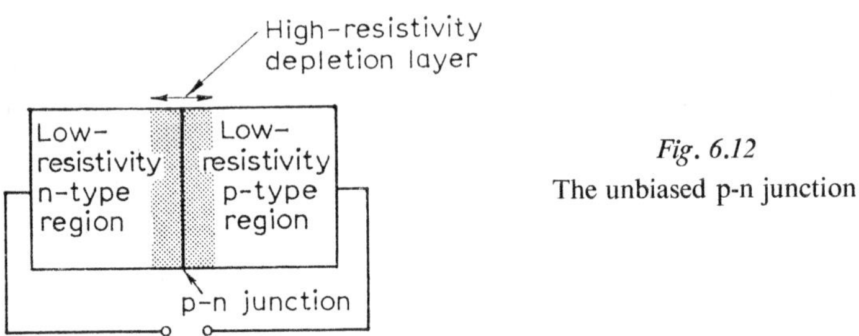

Fig. 6.12

The unbiased p-n junction

junction over which the barrier potential is effective are rapidly swept out of it, and hence this region is depleted of charge carriers. The depletion layer, as it is called, is a region of relatively high resistivity and is approximately 0·001 mm in width.

If an external source of e.m.f. is applied across the p-n junction, the equilibrium state of the junction is disturbed and the potential barrier is either increased or decreased according to the polarity of the external e.m.f. The germanium crystal consists of two regions of low resistivity separated by a region of high resistivity, the depletion layer, and the application of an e.m.f. across the crystal is effectively the same as placing it across the depletion layer, see Fig. 6.12.

THE FORWARD-BIASED P-N JUNCTION

If a battery is connected across the crystal in the direction shown in Fig. 6.13, holes are repelled from the positive end of the crystal and are caused to drift towards the junction, and electrons are repelled from the negative end of the crystal and also drift towards the junction. This drift of holes and electrons towards the junction reduces both the width of the depletion layer and the height of the potential barrier and the junction is said to be forward biased. The reduction in the height of the potential barrier allows majority charge carriers of lower energy to cross the junction and, since the minority charge carrier current remains constant, there is a net majority charge carrier current across the junction from the p-type region to the n type region. This current increases very rapidly with increase in the forward bias voltage as can be seen from the typical current/voltage characteristic shown in Fig. 6.14.

The holes drifting through the p-type region towards the p-n junction may be considered to have been injected by the positive terminal of the battery. Some of these holes may recombine with electrons diffusing across the junction in the other direction and so the hole current across the junction is slightly less than the injected hole current. After they have passed across the junction the holes recombine with the excess electrons in the n-type region. Similarly,

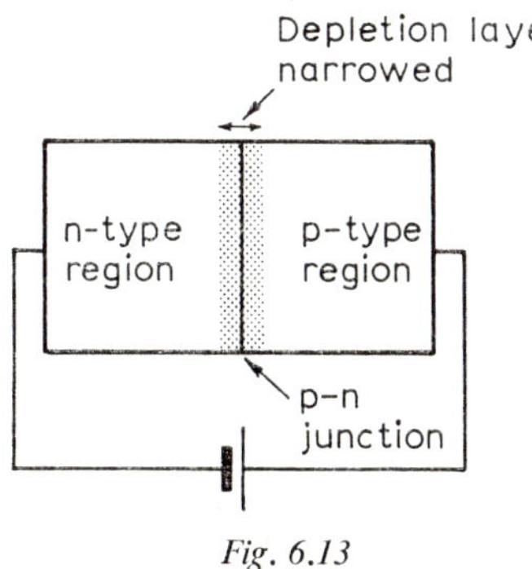

Fig. 6.13

The forward-biased p-n junction

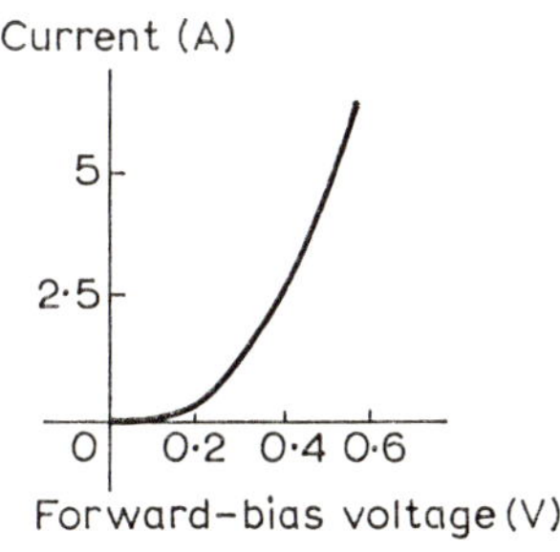

Fig. 6.14

The current/voltage characteristic of a forward-biased p-n junction

the negative battery terminal injects electrons into the n-type region and most of these electrons cross the junction. The total current is the sum of the electron and hole currents and is constant throughout the crystal. The current enters the p-type region as a hole current and leaves the n-type region as an electron current, i.e. in the forward direction current flow is by majority charge carriers.

THE REVERSE-BIASED P-N JUNCTION

Fig. 6.15 shows a p-n junction biased in such a direction as to attract majority charge carriers away from the junction and thus increase both the height of the potential barrier and the width of the depletion layer. Fewer majority charge carriers now have sufficient energy to be able to surmount the potential barrier and the majority charge carrier current decreases. The minority charge carrier current has remained constant and so a net current flows across the junction from n-type region to p-type region. This current increases with increase in the reverse bias voltage until the point is reached where almost no majority charge carriers possess sufficient energy to be able to cross the junction. The current flowing across the junction is then constant and equal to the minority charge carrier current and it is then known as the reverse saturation current.

If the reverse bias voltage is increased beyond a certain value a rapid increase in current occurs, this critical voltage is the breakdown voltage of the junction. Two effects are responsible for breakdown:

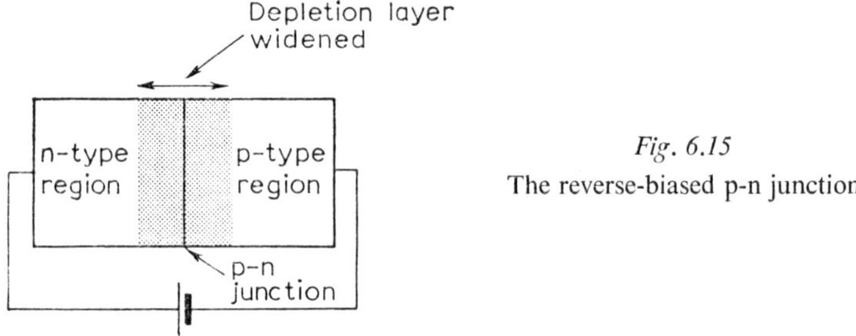

Fig. 6.15

The reverse-biased p-n junction

(*a*) the Zener effect in which the electric field across the junction is strong enough to break some of the covalent bonds and (*b*) the avalanche effect in which charge carriers are accelerated to such an

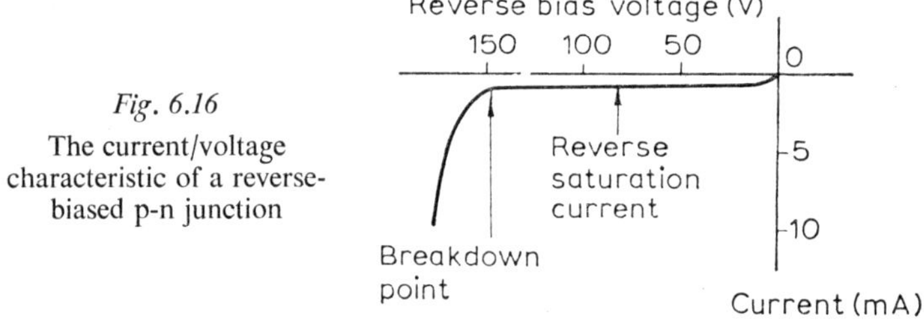

Fig. 6.16

The current/voltage characteristic of a reverse-biased p-n junction

extent that they are able to break covalent bonds by collision. A typical current/voltage characteristic for a reverse-biased p-n junction is shown in Fig. 6.16.

Construction and Characteristics of Semiconductor Diodes

Germanium and silicon for use in the manufacture of semiconductor diodes must be first purified until an impurity concentration of less than 1 part in 10^{10} is achieved. The wanted impurity atoms, donors and/or acceptors, are then added in the required amounts and the material is made into a single crystal.

A p-n junction may be formed in a number of different ways but three basic techniques are generally employed, either singly or in combination. An example of the first method is outlined in Fig. 6.17 and consists of alloying an indium pellet on to a n-type germanium wafer.

To make the n-type germanium wafer some intrinsic germanium and a small amount of impurity are melted in a crucible in a vacuum, and a seed crystal is lowered into the melt to a depth of a few millimetres. The temperature of the molten germanium is just above the melting point of the seed crystal and the few millimetres of seed immersed in the melt, melt also. The seed is rotated at a constant velocity and at the same time is slowly withdrawn from the melt,

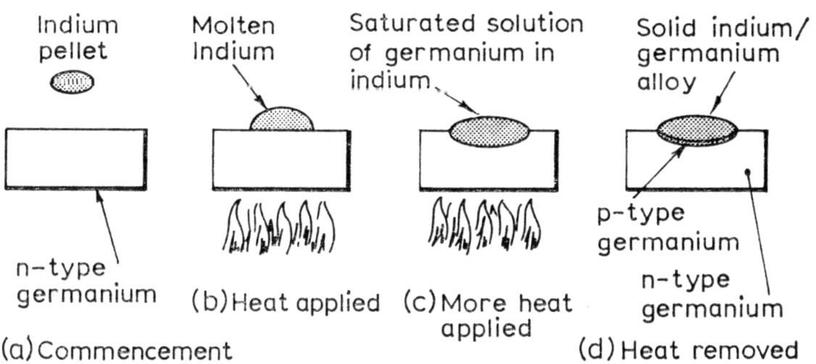

Fig. 6.17

The "alloying" method of forming a p-n junction

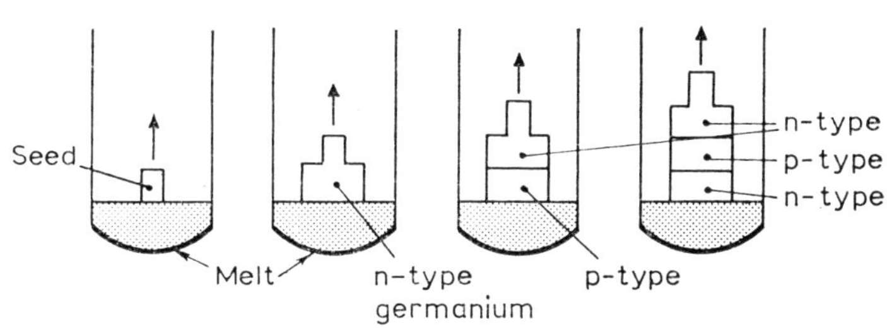

Fig. 6.18

The "rate growing" method of forming a p-n junction

thus forming an n-type crystal. By careful control of the process the required impurity concentration can be achieved. A pellet of indium is placed on the germanium wafer and is heated to a temperature above the melting point of indium but below the melting point of germanium. The indium melts and dissolves the germanium until a saturated solution of germanium in indium is obtained. The wafer is then slowly cooled and in the cooling a region of p-type germanium is produced in the wafer, and an alloy of germanium and indium (mainly indium) is deposited on the wafer.

Fig. 6.18 shows the rate growing method of forming a p-n junction. Some intrinsic germanium and some impurities of both types,

i.e. donor and acceptor (typically antimony and gallium), are placed
in a crucible in a vacuum, and then heated until a melt is produced.
A seed crystal is then lowered into the melt to a depth of a few
millimetres, and is then rotated and withdrawn steadily at one of two
different rates, either 1 cm/hour or 20 cm/hour. During its with-
drawal the seed pulls, or grows, a crystal of doped germanium with
it. In the periods of slow growing the germanium is primarily
doped with acceptor atoms and is p-type; during periods of fast
growth the donor atom concentration is greater than the concentration of acceptor atoms and a n-type region is produced. By inter-
spersing periods of slow growth with periods of fast growth an n-p-n,
etc., structure can be produced from which the required p-n junction
can be obtained.

The third method of producing a p-n junction to be considered
here is diffusion and it is outlined in Fig. 6.19. Some elements will

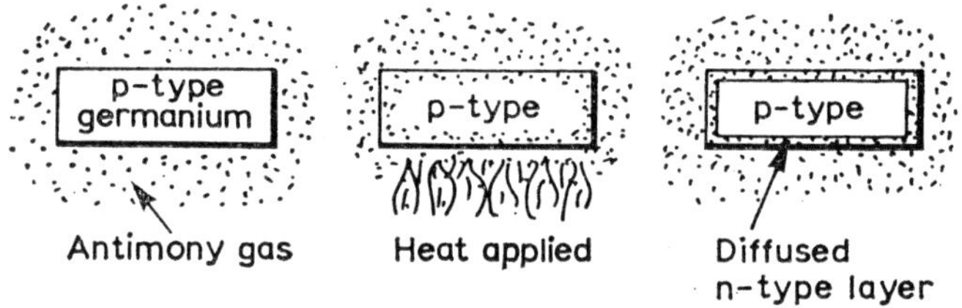

Fig. 6.19

The "diffusion" method of forming a p-n junction

diffuse distances of the order of tens of microns (1 micron $= 10^{-3}$
millimetre) into heated germanium, or silicon, in quite a short time,
the germanium remaining in the solid form throughout. A crystal
of p-type germanium is placed in a gas containing the required
impurity and is then heated. The atoms of the impurity gas diffuse
into the crystal and produce a layer of n-type germanium as shown.

A junction diode consists of a crystal having both p-type and n-type
regions. Junction diodes are made from either germanium or
silicon, the former having the advantage of a lower forward resistance
and the latter the advantages of a higher breakdown voltage and a
lower reverse saturation current. Connection to the junction is
made by wires fixed to each of the two regions. The complete device
is usually enclosed in a hermetically sealed container to prevent the
entry of moisture.

THE CAPACITANCE OF A P-N JUNCTION

When a p-n junction is reverse biased the depletion layer is a high-
resistance region with low-resistance regions either side and so it

acts as though it were a parallel-plate capacitor, the capacitance of which is a function of the magnitude of the applied bias voltage. A p-n junction can be made with the transition from the p-type region to the n-type region either abrupt or gradual. For an abrupt junction the depletion capacitance is proportional to the square root of the bias voltage, and for a gradual junction, to the cube root. Typical values for the depletion capacitance of small-signal germanium transistors are 30–100 pF for an abrupt junction and 10–50 pF for a gradual junction. Small-signal silicon planar transistors may have a capacitance in the range 1–15 pF.

For forward bias an even larger capacitance, several hundred picofarads in value, appears across the p-n junction and this capacitance is known as the diffusion capacitance.

The depletion and diffusion capacitances limit the usefulness of a junction diode to relatively low frequencies and for frequencies in the u.h.f. and higher-frequency bands the point-contact diode is employed.

THE POINT-CONTACT DIODE

Point-contact diodes are made from either germanium or silicon, the construction of the two types being shown in Fig. 6.20. It can be

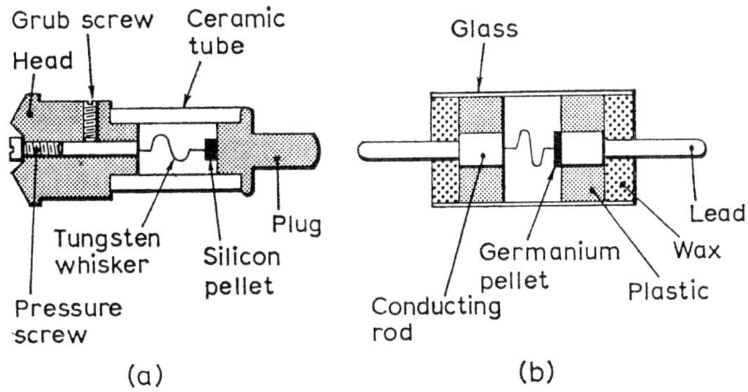

Fig. 6.20

The construction of (*a*) a silicon point-contact diode and (*b*) a germanium point-contact diode

seen that a point-contact diode consists essentially of a pellet of n-type semiconductor, germanium or silicon, that has the tip of a tungsten wire, or whisker, pressing on to its surface. Connection to the whisker is via the silver-plated plug and head ends in the case of the silicon diode, and via two copper leads for the germanium diode. In the silicon diode the pressure exerted by the whisker on the silicon pellet is adjustable by means of the pressure screw.

The silicon diode is successfully employed for many applications at microwave frequencies but since it is not robust, electrically or mechanically, it is little used at lower frequencies where the germanium diode can be employed.

During the manufacture of a point-contact diode a pulse of current, produced by the discharge of a capacitor, is passed through the diode and this results in the area of the pellet immediately adjacent to the tip of the whisker becoming a p-type region. A low capacitance p-n junction is then produced in the pellet.

SEMICONDUCTOR DIODE CURRENT/VOLTAGE CHARACTERISTICS

The current/voltage characteristic of a semiconductor diode is a graph of the current flowing in the device plotted against the voltage applied across it. At any point along the characteristic the ratio (voltage applied/current flowing) is a measure of the d.c. resistance of the diode for that voltage. If the characteristic is linear this ratio will be a constant quantity but should the characteristic be non-linear

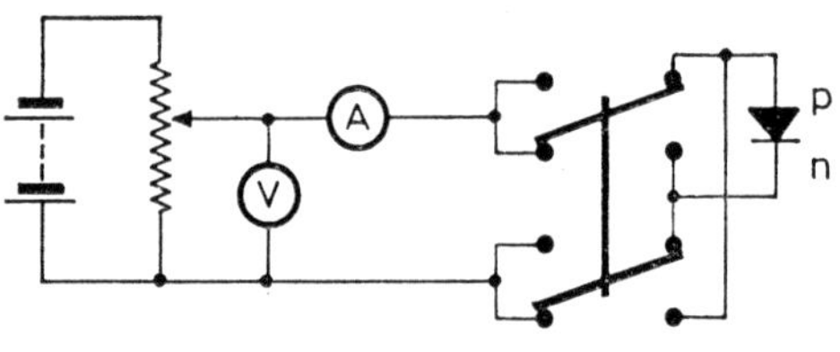

Fig. 6.21

Circuit for measuring the current/voltage characteristic of a semiconductor diode

the d.c. resistance will vary with the point of measurement. The a.c. resistance r_{ac} of a diode at a point is equal to the reciprocal of the slope of the characteristic at that point, that is

$$r_{ac} = \frac{\text{change in voltage}}{\text{resulting change in current}}$$

$$= \frac{\delta V^*}{\delta I}$$

The current/voltage characteristic of a semiconductor diode can be measured with the aid of the circuit arrangement of Fig. 6.21.

---

* The Greek letter δ will continually be employed throughout the remainder of this book to denote "a change of." Thus

$\dfrac{\delta V}{\delta I}$ means $\dfrac{\text{change of voltage}}{\text{change of current}}$ and $\dfrac{\delta I}{\delta T}$ means $\dfrac{\text{change of current}}{\text{change of time}}$, etc.

With the switch in the position shown the diode is reverse-biased; to forward bias the diode the switch is thrown to its other position to reverse the polarity of the applied voltage. For each position of the switch the applied voltage is increased from zero in a number of steps and the current flowing at each step is noted. The noted current values are then plotted to a base of voltage.

The magnitudes of the currents and voltages obtained in such a measurement can vary enormously according to the type of diode concerned; for example, Fig. 6.22 shows typical characteristics

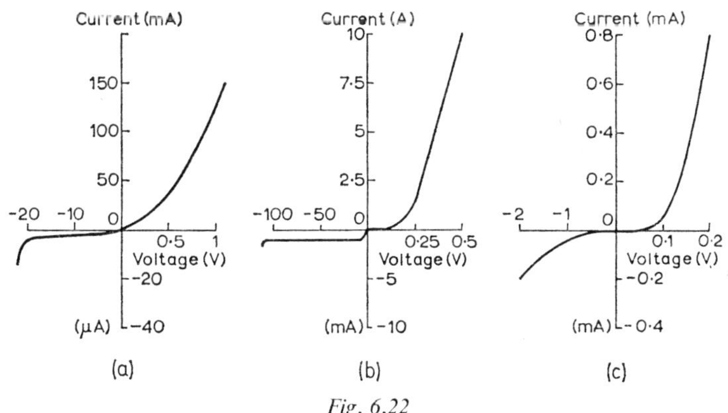

Fig. 6.22

Typical current/voltage characteristic for (*a*) a signal diode, (*b*) a power diode and (*c*) a point-contact diode

for (*a*) a signal diode, (*b*) a power diode and (*c*) a point-contact diode. It can be seen that the current through the diodes rises rapidly with increase in voltage when the diodes are forward-biased. For the junction diodes the current flowing in the reverse direction is approximately constant with increase in reverse bias voltage up to the breakdown point; for the point-contact diode, however, the reverse current increases steadily with increase in reverse bias voltage.

Example 6.1
Calculate the a.c. resistance of the semiconductor diode whose characteristic is shown in Fig. 6.23 at the point +0·1 V.

Solution.
The a.c. resistance r_{ac} is equal to $\delta V/\delta I$ and hence to find the forward resistance at the point $V = +0·1$ V it is necessary to select two equidistant points either side of this voltage and then, by projection to and from the characteristic, find the corresponding values of current.

Two points 0·02 V either side of +0·1 V have been selected, hence $\delta V = 0·04$ V. Projection upwards from these points to the curve and then from the curve to the current axis, as shown by the dotted lines, shows that the corresponding values of current are 15·5 mA and 5·5 mA, i.e.

$$\delta I = 10 \text{ mA}$$

$$\therefore \qquad r_{ac} = 0\!\cdot\!04/10 \times 10^{-3} = 4 \text{ ohms} \qquad Ans.$$

The slope of the reverse saturation current curve is very small and cannot be measured from the characteristic; this means that the reverse a.c. resistance of the diode is high, of the order of several thousands of ohms.

The diagrams show that the forward resistance of a semiconductor diode is much less than its backward resistance. This means that the

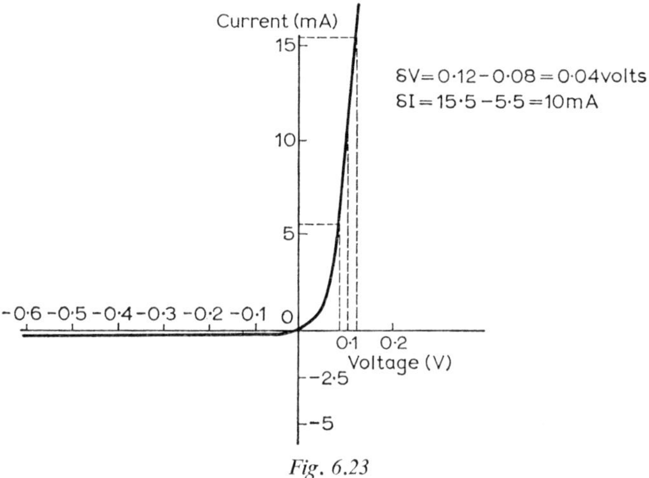

Fig. 6.23

device will pass current more readily in one direction than in the other, that is, it will act as a rectifier of alternating currents.

The Effect of Temperature

At any given temperature hole-electron pairs are continually created in the two regions of a semiconductor diode and the charge carriers produced contribute to the currents crossing the junction. For a forward-biased junction the effect on the current flowing is small since the forward current is relatively large, but the reverse saturation current of a reverse-biased germanium junction is found to be approximately doubled for every 8 degC rise in temperature and for a silicon junction for every 12 degC.

Exercises

1. Describe the construction and principle of operation of (*a*) a semiconductor diode, or (*b*) a copper-oxide metal rectifier.

   Include in your answer a sketch of a typical characteristic curve and a brief description with a circuit diagram of how such a curve could be obtained experimentally. (P1 1963)
2. Explain the rectifying action of a semiconductor diode. (part A 1963)
3. By reference to the formation of a potential barrier and a current/voltage characteristic, explain the rectifying property of a p-n junction. (A 1961)
4. Describe, with the aid of a sketch, the constructional features of either a silicon or a germanium crystal rectifier for use at very high frequencies. Quote a typical use for such a device. (A 1959)
5. By reference to the formation of a potential barrier and to a current/voltage characteristic explain the rectifying action of a p-n junction. (A 1965)

7 *The Transistor*

The transistor is a semiconductor device that can be employed to amplify an electrical signal, act as an electronic switch, and perform a number of other functions. Basically a transistor consists of a germanium or silicon crystal containing three separate regions. The three regions may consist of either two p-type regions separated by an n-type region, Fig. 7.1a, or two n-type regions separated by a p-type region, Fig. 7.1b. The first type of transistor is known as a

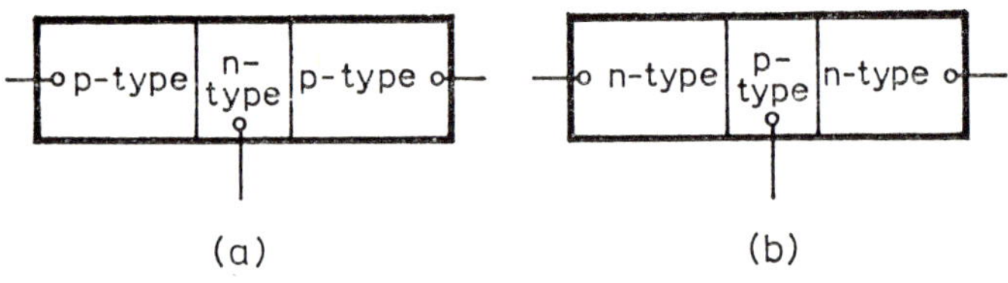

Fig. 7.1

(*a*) a p-n-p transistor (*b*) a n-p-n transistor

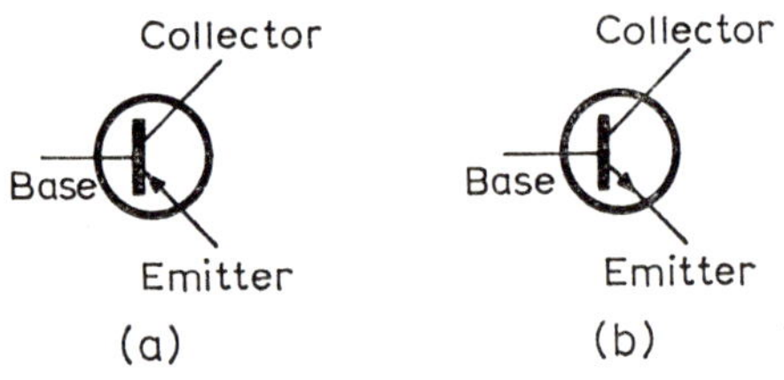

Fig. 7.2

Symbols for (*a*) a p-n-p transistor and (*b*) an n-p-n transistor

p-n-p transistor and the second type as an n-p-n transistor. Both types of transistor are employed, sometimes together in the same circuit, but since the p-n-p transistor is the more common the discussion in this chapter will be in terms of it. However, for the corresponding operation of an n-p-n transistor it is merely necessary to read electron for hole, hole for electron, negative for positive, and positive for negative.

The middle of the three regions in a transistor is known as the base and the two outer regions are known as the emitter and the collector. In most transistors the collector region is made physically larger than the emitter region because it will be expected to dissipate a greater power. The symbol for a p-n-p transistor is given in Fig. 7.2*a* and the symbol for an n-p-n transistor in Fig. 7.2*b*. Note that the emitter lead arrowhead is pointing in different directions in the two figures, pointing inwards for the p-n-p transistor and outwards for the n-p-n transistor. It will shortly become evident that the arrowhead indicates the direction in which holes travel in the emitter.

Transistors have a number of advantages over thermionic valves; perhaps the most important is that a transistor does not require a source of power to heat it before it can operate as does a thermionic valve. This makes transistorized equipment much faster than valve equipment to operate after being switched on. The power consumption is also much less and this is particularly important for large equipments such as computers. Further advantages of transistors are their smaller size and the much lower working voltages required.

The Action of a Transistor

A p-n-p transistor contains two p-n junctions and is normally operated so that one junction, the emitter/base junction, is forward-biased and the other, the collector/base junction, is reverse-biased. This is shown in Fig. 7.3 together with the directions of the various currents that flow in the transistor. The usual convention whereby

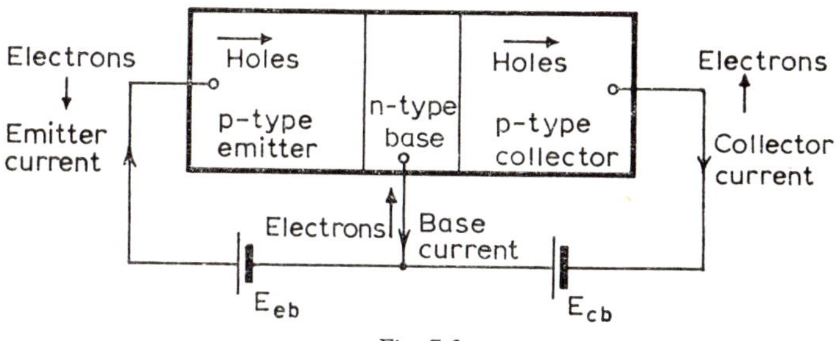

Fig. 7.3

Bias voltages for a p-n-p transistor and the currents flowing

the direction of current flow is opposite to the direction of electron movement has been employed. Consider that, initially, the emitter/base bias voltage E_{eb} is zero. Then the majority charge carrier current crossing the emitter/base junction is equal to the minority charge carrier current that is flowing in the opposite direction and the net junction current is zero. The collector/base junction is reverse-biased by the bias voltage E_{cb} and so a small minority charge carrier current flows in the collector lead. This current is the reverse saturation current discussed in the previous chapter but now it is known as the collector leakage current and is given the symbol I_{CBO}.

If the emitter/base bias voltage is increased in the positive direction a few tenths of a volt the emitter/base junction is forward-biased and a majority charge carrier current flows. This current consists of holes travelling from the emitter to the base and electrons passing from the base to the emitter. Only the hole current is useful to the action of the transistor, as will soon be evident, and it is therefore made much larger than the electron current by doping the base much more lightly than the emitter. The ratio of the hole current to the total emitter current is known as the emitter injection ratio or the emitter efficiency, symbol γ. Typically, γ is approximately equal to 0·99 and this means that only 1 % of the emitter current consists of electrons passing from the base to the emitter.

Immediately the holes cross the emitter/base junction, and are said to have been emitted or injected into the base, they become minority charge carriers and start to diffuse across the base towards the collector/base junction. Because the base is fairly narrow and is also lightly doped most of the emitted holes reach the collector/base junction and do not recombine with a free electron on the way. On reaching the junction the emitted holes augment the minority charge carrier current crossing the junction and cause an increase in the collector current. The ratio of the number of holes arriving at the collector to the number of emitted holes is known as the base transmission factor, symbol β. Typically $\beta = 0\cdot99$.

The collector current is less than the emitter current because (*a*) part of the emitter current consists of electrons that do not contribute to the collector current and (*b*) not all of the holes injected into the base are successful in reaching the collector. Factor (*a*) is represented by the emitter injection ratio and factor (*b*) by the base transmission factor; hence the ratio of collector current to emitter current is equal to $\beta\gamma$. Substituting the typical values quoted for γ and β shows that, typically, the collector current is about 0·98 times the emitter current.

The base current is small and has three components, (*a*) an electron current entering the base to replace the electrons lost by recombination with the diffusing holes, (*b*) the majority charge carrier electron current flowing from base to emitter, and (*c*) the collector leakage current I_{CBO}. The first two of these components are currents that flow out of the base and together are greater than I_{CBO} which flows into the base, and so the total base current flows out of the base. The total current flowing into the transistor must be equal to the total current flowing out of it and hence the emitter current I_e is equal to the sum of the collector and base currents, I_c and I_b respectively, that is

$$I_e = I_c + I_b \qquad (7.1)$$

Typically, I_c is equal to $0.98I_e$ so that I_b is equal to $0.02I_e$.

If the emitter current is varied by some means the number of holes arriving at the collector, and hence the collector current, will vary accordingly. The magnitude of the collector/base voltage V_{cb} has relatively little effect on the collector current as will be seen shortly. Control of the output (collector) current can thus be obtained by means of the input current to the emitter and this, in turn, can be controlled by variation of the bias voltage applied to the emitter/base junction. An increase in the bias voltage (which is in the forward direction) lowers the height of the potential barrier and allows an increased emitter current to flow; conversely, a decrease in the bias voltage reduces the emitter current.

The ratio of the output current of a transistor to its input current in the absence of an a.c. signal is known as the d.c. current gain of the transistor. In the previous discussion the output current has been the collector current I_c and the input current has been the emitter current I_e. Thus,

$$\text{d.c. current gain, } - h_{FB} = \frac{I_c}{I_e} \qquad (7.2)$$

The minus sign indicates that the input and output currents are flowing in opposite directions. By convention, a current flowing into a transistor is taken to be positive and a current flowing out is taken to be negative.

A transistor may be connected in a circuit in one of three ways and in each case one electrode is common to both input and output. The connection is then described in terms of the common electrode; for example, the common-base connection has the base common to both input and output, the input signal is fed between the emitter and the base and the output signal is developed between the collector and the base. In all connections, the base/emitter junction is always forward-biased and the collector/base junction is always reverse-biased.

THE COMMON-BASE CONNECTION

The basic arrangement of the common-base connection, or configuration, is shown in Fig. 7.4. The transistor has an alternating source of e.m.f. E_s volts r.m.s. and internal resistance R_s ohms connected to its input terminals. The alternating source is connected in series with the emitter/base bias voltage E_{eb} and varies the forward bias applied to the emitter/base junction.

During positive half-cycles of the source e.m.f. the forward bias applied to the junction is increased, the potential barrier is lowered, and an enhanced emitter current flows into the transistor. Conversely, during negative half-cycles the emitter current is decreased and in this way the collector current is caused to vary in accordance with the waveform of the alternating source. The collector/base bias battery E_{cb} has negligible internal resistance and so the collector/base voltage remains constant as the collector current varies. The collector circuit is said to be short-circuited so far as alternating currents are concerned.

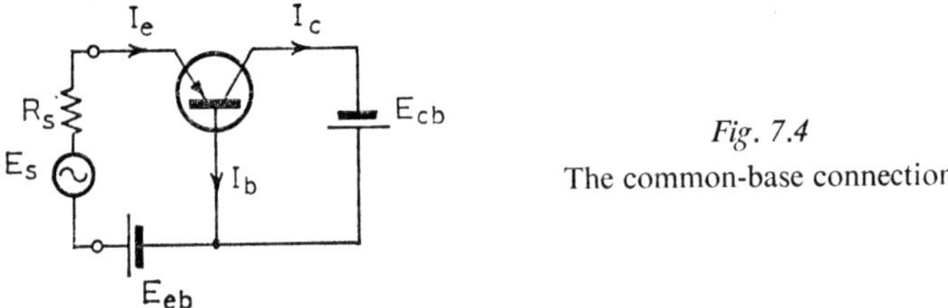

Fig. 7.4

The common-base connection

In a common-base amplifier circuit an important parameter is the short-circuit current gain of the transistor, symbol h_{fb} or, more commonly, α. The short-circuit current gain is defined as the ratio of a change in collector current to the change in emitter current producing it, with the collector/base voltage maintained constant, that is

$$\alpha = \frac{\delta I_c}{\delta I_e} \text{ when } V_{cb} \text{ is constant} \tag{7.3}$$

The *short-circuit* current gain is specified since analysis shows that the current gain is a function of the value of any resistance placed in the collector circuit. For the common-base circuit, however, the difference between the short-circuit current gain and the current gain for any particular collector load resistance is very small for all resistance values used in practical circuits and will be neglected in this book.

Example 7.1
In a certain transistor a change in emitter current of 1 mA produces a change in collector current of 0·98 mA. Determine the short-circuit current gain of the transistor.

Solution.

$$\text{Current gain } \alpha = \frac{\delta I_c}{\delta I_e} = \frac{0{\cdot}98}{1} = 0{\cdot}98 \qquad Ans.$$

This is a typical value for the short-circuit current gain of a transistor connected in the common-base configuration. It should be evident that α must be less than unity, because the emitter current is the sum of the base and collector currents. Clearly, then, a common-base transistor must have a current gain of less than unity but, if a resistor is connected in the collector circuit, as shown in Fig. 7.5, both voltage and power gains are possible.

The output voltage is developed across the collector load resistor and since the internal resistance of the collector supply is negligible the top end of the resistor is effectively at earth potential so far as

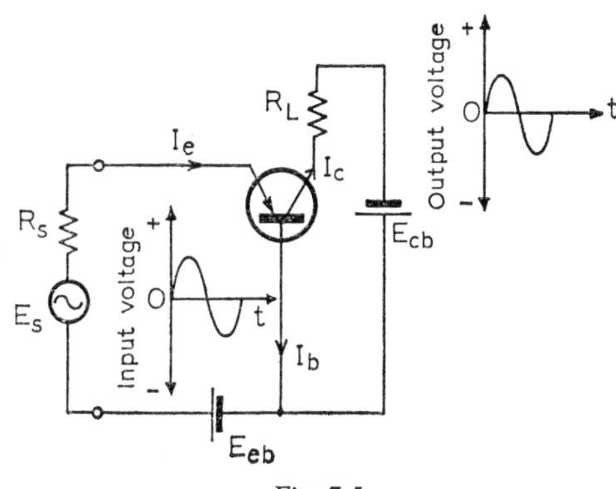

Fig. 7.5

The basic common-base amplifier

alternating currents are concerned. Thus the output signal voltage is taken from between the collector electrode and earth.

Expressions for the voltage gain and the power gain of a transistor connected with common base can be obtained.

Let the r.m.s. voltage and internal resistance of the voltage source applied to the input terminals of the transistor be E_s volts and R_s ohms respectively, and let the input resistance of the transistor be R_{IN} ohms.

Then, Fig. 7.6, the input current to the transistor, I_e, is

$$I_e = \frac{E_s}{R_s + R_{IN}}$$

and the voltage V_{IN} appearing across the transistor input terminals is

$$V_{IN} = I_e R_{IN}$$

or

$$I_e = \frac{V_{IN}}{R_{IN}}$$

The output or collector current I_c is

$$I_c = \frac{\alpha V_{IN}}{R_{IN}}$$

This current flows through the collector load resistance R_L and develops the output voltage V_{OUT} across it,

$$\therefore \qquad V_{OUT} = \frac{\alpha V_{IN}}{R_{IN}} R_L$$

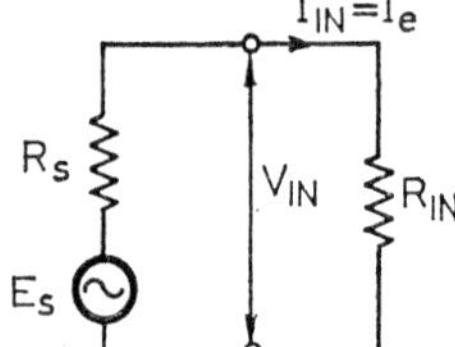

Fig. 7.6

Circuit for the calculation of the input current to a transistor

and the voltage gain A_v is

$$A_v = \frac{V_{OUT}}{V_{IN}} = \frac{\alpha R_L}{R_{IN}} \tag{7.4}$$

Since the short-circuit current gain α is approximately equal to unity and the collector load resistance R_L is always greater than the input resistance R_{IN} of the transistor, a voltage gain is readily achieved.

Now consider the power gain of a common-base transistor. A number of different definitions of power gain exist for transistors but this book will take power gain as meaning the ratio of the power delivered to the load to the power delivered to the transistor.

The input power P_{IN} to the transistor is (see Fig. 7.6)

$$P_{IN} = I_e^2 R_{IN}$$

and the output power P_{OUT} is

$$P_{OUT} = (\alpha I_e)^2 R_L$$

$$\therefore \qquad \text{Power gain } A_p = \frac{P_{OUT}}{P_{IN}} = \frac{\alpha^2 I_e^2 R_L}{I_e^2 R_{IN}}$$

$$= \frac{\alpha^2 R_L}{R_{IN}} \tag{7.5}$$

Again, since R_L is greater than R_{IN} a power gain is possible.

Example 7.2
A transistor is connected with common-base in a circuit and has a collector load resistance of 20,000 Ω. The short-circuit current gain of the transistor is 0·98

and its input resistance is 50 Ω. Calculate the voltage and power gains of the transistor. Quote the power gain in decibels.

Solution. From equation (7.4),

$$\text{Voltage gain} = \frac{\alpha R_L}{R_{IN}} = \frac{0.98 \times 20,000}{50} = 392 \qquad Ans.$$

From equation (7.5),

$$\text{Power gain} = \frac{\alpha^2 R_L}{R_{IN}} = 0.98 \times 392 = 384$$

and in decibels, power gain $= 10 \log_{10} 384 = 25.8$ dB $\qquad Ans.$

The source of the output power is the collector/base bias battery, the transistor effectively acting as a device for the conversion of d.c. power from the battery into the a.c. power supplied to the load.

In the common-base amplifier the input signal voltage and the output signal voltage are in phase, as shown by the waveforms of Fig. 7.5. Consider the input signal voltage to be passing through zero and increasing in the positive direction. The forward bias of the base/emitter junction is then increased and this results in an increase in the emitter current. The collector current is increased and the voltage drop across the collector load resistor R_L increases also and this makes the collector/base potential less negative. Thus a positive increment in the input signal voltage produces a positive increment in the output signal voltage.

THE COMMON-EMITTER CONNECTION
In practice, transistors are most often employed in the common-emitter configuration shown in Fig. 7.7.

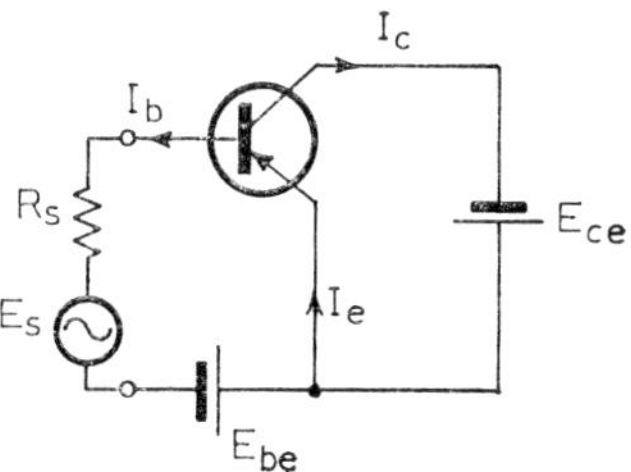

Fig. 7.7

The common-emitter connection

The emitter/base junction is forward-biased by the battery E_{be} and the collector/base junction is reverse-biased by a potential equal to $(E_{ce} - E_{be})$. However, since the voltage of the collector/emitter bias battery E_{ce} is much larger than the emitter/base bias voltage E_{be} the reverse bias voltage may be taken as merely equal to E_{ce} volts.

When a transistor is connected in this way the input current is the base current and not the emitter current as previously. During the negative half-cycles of the input signal voltage E_s, the forward bias of the emitter/base junction is increased, and so the emitter current I_e is increased by an amount δI_e. The collector current is also increased, by an amount $\delta I_c = \alpha \delta I_e$, and so is the base (input) current, by an amount

$$\delta I_b = \delta I_e - \delta I_c = \delta I_e(1 - \alpha)$$

Conversely, during positive half-cycles of the input signal voltage the three currents are reduced in magnitude.

The short-circuit current gain of a common-emitter connected transistor, symbol h_{fe}, β or α', is defined as the ratio of a change in collector current δI_c to the change in base current δI_b producing it, the collector/emitter voltage being maintained constant, that is

$$\alpha' = \frac{\delta I_c}{\delta I_b} \text{ when } V_{ce} \text{ is constant} \tag{7.6}$$

$$= \frac{\alpha \delta I_e}{\delta I_e - \alpha \delta I_e}$$

$$= \frac{\alpha \delta I_e}{\delta I_e(1 - \alpha)}$$

$$= \frac{\alpha}{1 - \alpha} \tag{7.7}$$

Typical values for the short-circuit current gain α of a common-base transistor are in the neighbourhood of unity and thus the common-emitter connection can give a considerable current gain.

Example 7.3
A transistor exhibits a change of 0·98 mA in its collector current for a change of 1 mA in its emitter current. Calculate (*a*) its common-base short-circuit current gain and (*b*) its common-emitter short-circuit current gain.

Solution.
(*a*) Common-base short-circuit current gain

$$\alpha = \frac{\delta I_c}{\delta I_e} = \frac{0·98}{1} = 0·98 \qquad \textit{Ans.}$$

(*b*) Common-emitter short-circuit current gain

$$\alpha' = \frac{\delta I_c}{\delta I_b} = \frac{\alpha}{1 - \alpha} = \frac{0·98}{1 - 0·98} = 49 \qquad \textit{Ans.}$$

When a load resistor R_L is connected in the collector circuit (Fig. 7.8) the current gain of the transistor is no longer equal to the short-circuit value but is somewhat less. The actual value of the current gain is dependent upon the value of the collector load resistor R_L, decreasing with increase in R_L.

The voltage and power gain of a transistor connected in the common-emitter configuration may be calculated using equations similar to equations (7.4) and (7.5), that is,

$$\text{Voltage gain } A_v = \frac{\text{current gain} \times R_L}{R_{IN}} \qquad (7.8)$$

and

$$\text{Power gain } A_p = \frac{(\text{current gain})^2 \times R_L}{R_{IN}} \qquad (7.9)$$

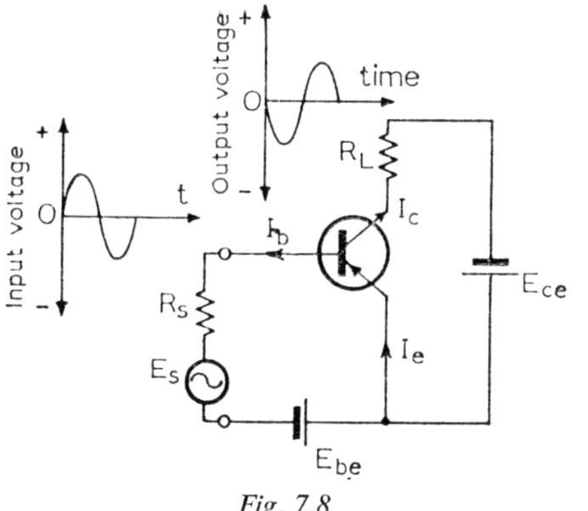

Fig. 7.8

The basic common-emitter amplifier

The current gain in these equations is the actual gain corresponding to the value of R_L employed and is larger than unity. Since the collector load resistance R_L is larger than the input resistance R_{IN} both voltage and power gain can be achieved.

THE COMMON-COLLECTOR CONNECTION

The third way in which a transistor may be connected is shown in Fig. 7.9. The collector is now common to both input and output

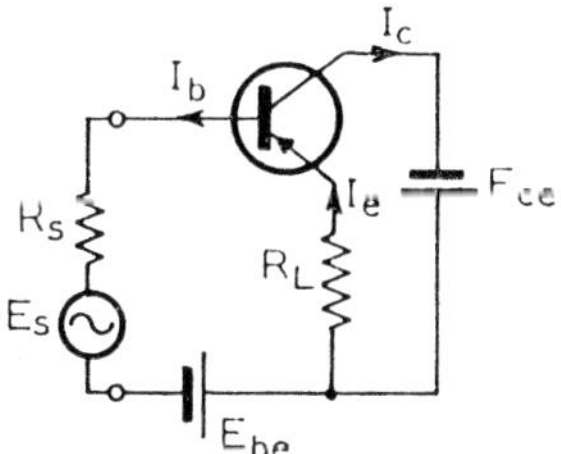

Fig. 7.9

The basic common-collector amplifier

circuits and the load is connected in the emitter circuit. With this configuration the base current is the input current and the emitter current is the output current. The short-circuit current gain is defined as

$$\text{Short-circuit current gain} = \frac{\delta I_e}{\delta I_b} \text{ when } V_{ce} \text{ is constant} \quad (7.10)$$

$$= \frac{\delta I_e}{\delta I_e - \delta I_c}$$

$$= \frac{\delta I_e}{\delta I_e(1 - \alpha)}$$

$$= \frac{1}{1 - \alpha} \quad (7.11)$$

Table 7.1

CHARACTERISTIC	COMMON-BASE	COMMON-EMITTER	COMMON-COLLECTOR
Short-circuit current gain	α	$\dfrac{\alpha}{1 - \alpha}$	$\dfrac{1}{1 - \alpha}$
Voltage gain	Good	Better than common-base	Unity or less
Power gain	15 to 30 dB	30 to 40 dB	10 to 15 dB
Input resistance	Low 30 to 100 Ω	Medium 200 to 5,000 Ω	High 5,000 to 500,000 Ω
Output resistance	High 10^5 to 10^6 Ω	High 10,000 to 100,000 Ω	Low 50 to 1,000 Ω

The short-circuit current gain of a transistor connected in the common-collector configuration is approximately equal to the short-circuit gain of the same transistor connected with common-emitter. The current gain when a load is connected in the emitter circuit is not equal to the short-circuit current gain but is reduced by an amount that is dependent on the value of the emitter load.

Expressions (7.8) and (7.9) may be employed to determine the voltage and power gains of a common-collector circuit. Now, however, the input resistance is considerably larger than the load resistance and a voltage gain of less than unity is obtained.

The main use of the common-collector circuit—or the "emitter-follower" as it is often called—is as a power-amplifying impedance transformer that can be connected between a high impedance source and a low impedance load.

A comparison between the main characteristics of the three transistor configurations is given in Table 7.1.

Transistor Static Characteristics

A number of current/voltage plots can be employed in the study of the operation of a transistor in a circuit. The resulting curves, which are known as the static characteristic curves, give information on the value of current flowing into, or out of, one electrode for either a given current flowing into, or out of, another electrode or a given voltage applied between two electrodes. Three sets of characteristics can be plotted for each configuration; these are (*a*) the input characteristic, (*b*) the transfer characteristic and (*c*) the output characteristic. In this book, however, the characteristics for the common-collector circuit will not be discussed.

COMMON-BASE STATIC CHARACTERISTICS
The method of determining the static characteristics of a transistor is to connect the transistor into a suitable circuit and then to vary the appropriate currents and/or voltages in a number of discrete steps, noting the corresponding values of other currents at each step. Fig. 7.10 shows a suitable circuit for the determination of the

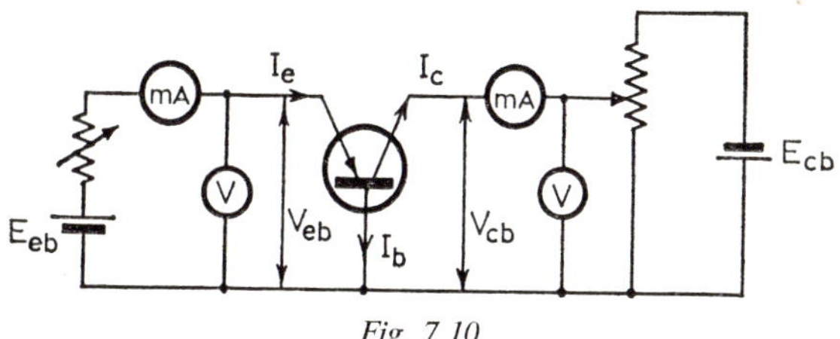

Fig. 7.10

Circuit for the determination of the static characteristics of a common-base connected transistor

characteristics of a p-n-p transistor in the common-base configuration.

The collector and base currents are shown as flowing *out* of the transistor and are therefore, by definition, negative; the emitter current is shown flowing *into* the transistor and must be taken as positive. If the characteristics of an n-p-n transistor had to be measured the polarities of the two batteries would have to be reversed.

(*a*) *The Common-base Input Characteristic*
The input characteristic of a transistor connected with common base shows how the emitter current varies with change in the emitter/base voltage, the collector/base voltage being maintained constant.

The method of determining the input characteristic is as follows. The collector/base voltage is adjusted to a convenient value and then the emitter/base voltage is increased in a number of discrete steps, the resulting value of emitter current being noted at each step. The values of emitter current obtained in this way are then plotted against the corresponding values of emitter/base voltage, Fig. 7.11

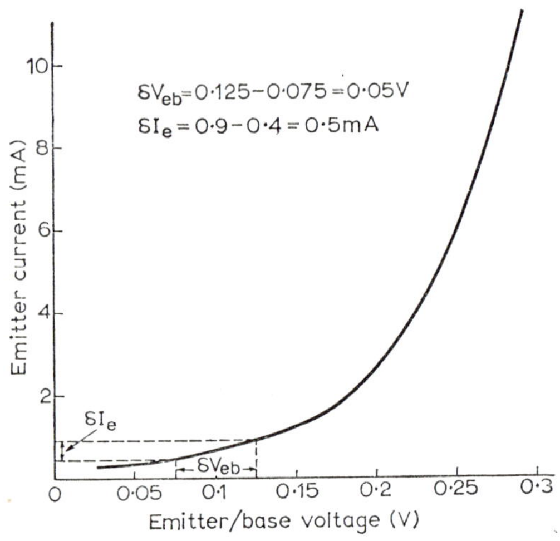

Fig. 7.11

Common-base input characteristic

showing a typical characteristic. The slope of the characteristic, that is $\delta I_e/\delta V_{eb}$, is the input conductance of the transistor with its output terminals short-circuited to alternating current, and the short-circuit input resistance R_{IN} is given by the reciprocal of the slope:

$$\therefore \qquad R_{IN} = \frac{\delta V_{eb}}{\delta I_e} \qquad (V_{cb} \text{ constant}) \qquad (7.12)$$

The short-circuit input resistance is a property of the transistor: when a load is connected to the input terminals, the input resistance depends on the load resistance.

Since the curve is not linear the value of R_{IN} will vary with the point of measurement. The input resistance R_{IN}, when the emitter/base voltage V_{eb} is 0·1 V, is from Fig. 7.11,

$$R_{IN} = \frac{0 \cdot 05}{0 \cdot 5 \times 10^{-3}} = 100 \ \Omega$$

It is evident that at the point $V_{eb} = 0 \cdot 25$ V the input resistance will be less than 100 ohms because at this point a change in V_{eb} of 0·05 volt gives a change in emitter current in excess of 0·5 mA. The change in the input resistance with change in the emitter/base voltage gives rise to distortion of signals handled by the transistor. The effect can be minimized by connecting a relatively high resistance in series with the input terminals of the transistor so that any change in input resistance is only a small percentage of the total resistance in the input circuit.

Very little change in the input characteristic is obtained when either the collector/base voltage or the temperature is varied.

(b) The Common-base Current-transfer Characteristic

The transfer characteristic shows how the collector current varies with changes in emitter current, the collector/base voltage being held constant. The collector voltage is set to a convenient value

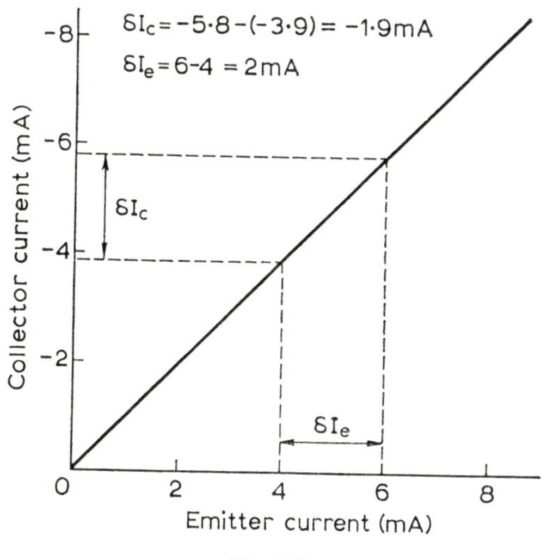

Fig. 7.12

Common-base current-transfer characteristic

and then the emitter current is increased in a number of steps; at each step the corresponding value of collector current is noted. A typical transfer characteristic is to be seen in Fig. 7.12 and is more or less independent of changes in either the collector/base voltage or the temperature.

The slope of the transfer characteristic gives the short-circuit current gain α of the transistor. At the point $I_e = 5$ mA,

$$\alpha = \frac{\delta I_c}{\delta I_e} \quad (V_{cb} \text{ constant})$$

$$= 1 \cdot 9/2 = 0 \cdot 95$$

Although the short-circuit current gain of a transistor can be determined from its transfer characteristic, the transfer characteristic is not often used because the same information can be obtained from the output characteristic.

(c) The Common-base Output Characteristic

The output characteristic indicates the way in which the collector current varies with change in collector/base voltage, the emitter current being maintained constant. The emitter current is set to a convenient, low, value and the collector/base voltage is increased from zero in a number of discrete steps and at each step the collector

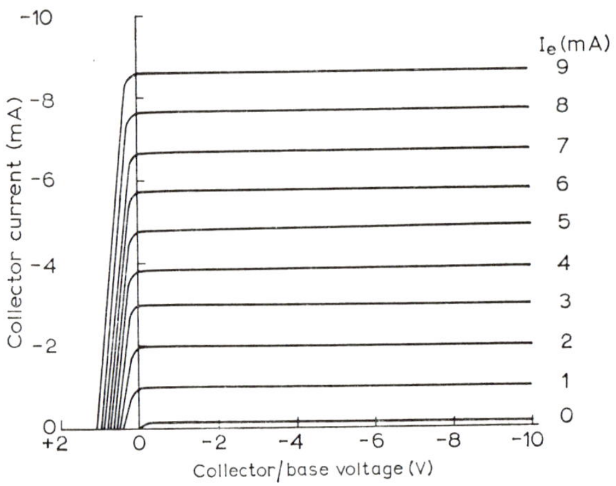

Fig. 7.13

Common-base output characteristics

current that flows is noted. The collector/base voltage is then restored to zero and the emitter current is increased to another convenient value and the procedure repeated. In this way a whole family of curves relating collector current to collector/base voltage can be obtained, a typical family being shown in Fig. 7.13.

The reciprocal of the slope of the output characteristic, $\delta V_{cb}/\delta I_c$, gives the output resistance of the transistor, when the input terminals are open-circuited to alternating current, at the point where the measurement is made. The open-circuit output resistance is a property of the transistor: when a signal is applied to the input terminals, the output resistance depends on the resistance of the signal source. Since the curves are linear over most of their length the output resistance is fairly constant, and since the curves are nearly parallel to the collector/base voltage axis the output resistance is very high, of the order of 1 megohm or more.

It can be seen that some collector current still flows when the collector/base voltage has been reduced to zero. This is because the potential barrier across the collector/base junction must be reduced to zero before the collector current ceases to flow (because the potential barrier aids the passage of minority charge carriers). Another, more important, feature of the characteristics is the collector current that flows for all negative values of collector/base voltage when the emitter current is zero. This current is the minority charge carrier current that passes across the collector/base junction (similar to the reverse saturation current in a junction diode) and is known as the collector leakage current, symbol I_{CBO}.

The short-circuit current gain α of a transistor can be estimated from the output characteristic since it is easy to determine the change in collector current resulting from a change in emitter current, for a constant value of collector/base voltage. Thus, referring to Fig. 7.13, when the emitter current changes from 5 mA to 7 mA the collector current changes from 4·9 mA to 6·8 mA and so α is equal to 1·9/2 or 0·95.

COMMON-EMITTER STATIC CHARACTERISTICS

To determine the static characteristics of a transistor connected in the common-emitter configuration the transistor must be connected, with common emitter, in a circuit similar to that given in Fig. 7.10; the only circuit alteration necessary consists of the removal of the milliammeter from the emitter circuit and the insertion of a microammeter in the base circuit.

(*a*) *Common-emitter Input Characteristic*

The input characteristic shows the way in which the base current varies with change in the base/emitter voltage, the collector/emitter voltage remaining constant. The method of determining the input characteristic is to maintain the collector/emitter voltage constant at a convenient value and increase the base/emitter voltage in a number of discrete steps, noting the base current at each step. The procedure is then repeated for a different, constant, value of collector/emitter voltage V_{ce}, since change in this voltage has an effect on the input characteristic. A typical pair of input characteristics are shown in Fig. 7.14. The input resistance, for a given base/emitter voltage V_{be} is given by the reciprocal of the slope of the curve at that point. For example, consider the short-circuit input resistance of the transistor at the point $V_{be} = 0{\cdot}1$ V for both $V_{ce} = 0$ and $V_{ce} = -4$ V.

For $V_{ce} = 0$ V

$$R_{IN} = \frac{\delta V_{be}}{\delta I_b} \qquad (V_{ce}\text{ constant}) \qquad (7.13)$$

$$= \frac{0{\cdot}05}{53 \times 10^{-6}} = 943\ \Omega$$

and for $V_{ce} = -4$ V

$$R_{IN} = \frac{0{\cdot}05}{9 \times 10^{-6}} = 5{,}556\ \Omega$$

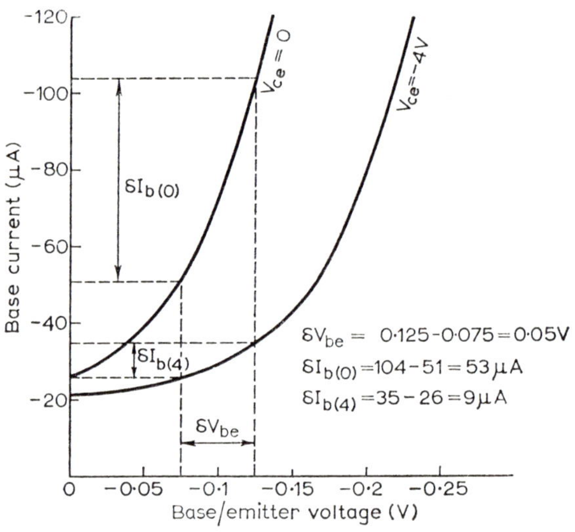

Fig. 7.14

Common-emitter input characteristic

(b) Common-emitter Current-transfer Characteristic

The transfer characteristic shows how the collector current changes with changes in the base current, the collector/emitter voltage being maintained at a constant value. For this measurement the collector/emitter voltage is kept constant and the base current is increased in a number of discrete steps and at each step the collector current is noted. Finally a plot is made of collector current against base current. Since the transfer characteristic is not independent of the value of the collector/emitter voltage the procedure is repeated for a number of different collector/emitter voltages to give a family of curves, Fig. 7.15.

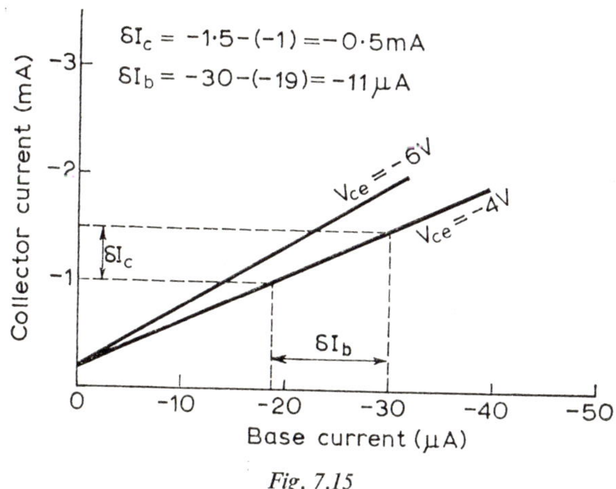

Fig. 7.15

Common-emitter current-transfer characteristic

The slope of the transfer characteristic gives the short-circuit current gain α' of the transistor when it is connected in the common-emitter configuration. Thus for the curve marked $V_{ce} = -4$ V,

$$\alpha' = \frac{\delta I_c}{\delta I_b} = \frac{0{\cdot}5 \times 10^{-3}}{11 \times 10^{-6}} = 45{\cdot}45$$

(c) Common-emitter Output Characteristic

The output characteristic illustrates the changes that occur in collector current with changes in collector/emitter voltage, for constant value of base current. The base current is set to a convenient value and is maintained constant and the collector/emitter voltage is increased from zero in a number of discrete steps, the collector current being noted at each step. The collector/emitter voltage is then restored to zero and the base current increased to another convenient value and the procedure repeated. In this way a family of curves, see Fig. 7.16, can be obtained.

The open-circuit output resistance of the transistor is equal to the reciprocal of the slope of the output characteristic.

The open-circuit output resistance at the point $V_{ce} = -6$ V and $I_b = -60\ \mu$A is

$$R_{OUT} = \frac{\delta V_{ce}}{\delta I_c} \qquad (I_b \text{ constant}) \qquad (7.14)$$

$$= \frac{2}{0.2 \times 10^{-3}} = 10,000\ \Omega$$

When a characteristic is non-linear its slope will vary according to the point of measurement and therefore the point of measurement

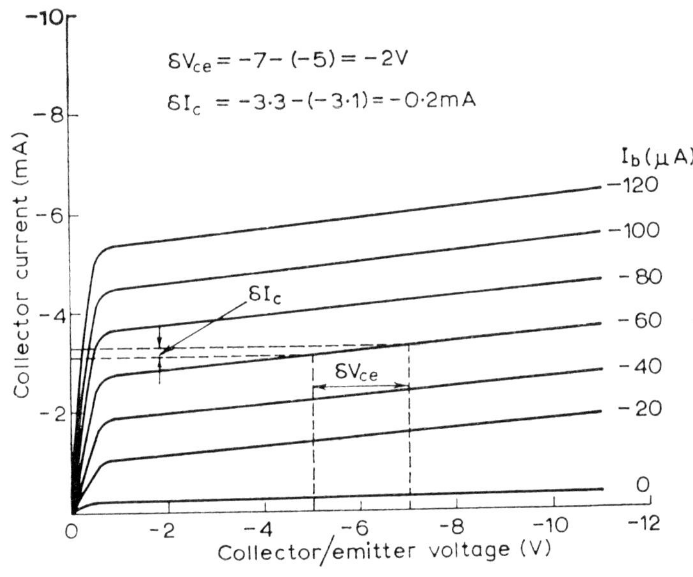

Fig. 7.16

Common-emitter output characteristics

should always be quoted. It is usual, unless specified otherwise, to measure the slope of the most linear portion of a characteristic. For the greatest accuracy the increments taken either side of the chosen point should be as small as possible although this has not been done in this chapter in order to clarify the diagrams.

It will be seen from Fig. 7.16 that a collector current flows even when the input or base current is zero. This current is the common-emitter leakage current, symbol I_{CEO}. I_{CEO} is related to the common-base leakage current I_{CBO}, according to the expression.

$$I_{CEO} = I_{CBO} (1 + \alpha') \qquad (7.15)$$

The collector leakage current I_{CBO} of a common-base transistor is extremely temperature sensitive and is approximately doubled for every 12 degC rise in temperature for silicon transistors and every 8 degC rise for germanium transistors. However, the leakage current of a silicon transistor at a given temperature is much less than the leakage current of a germanium transistor at the same temperature. As a result silicon transistors can be successfully operated at higher temperatures than are possible for germanium transistors, approximately 150°C as opposed to approximately 85°C.

THERMAL RUNAWAY

An increase in the temperature of the collector/base junction will cause the collector leakage current I_{CBO} to increase. The increase in collector current produces an increase in the power dissipated at the junction and this, in turn, further increases the temperature of the junction and so gives further increase in I_{CBO}. The process is cumulative and, particularly in the common-emitter connection $(I_{CEO} \gg I_{CBO})$, may lead to the eventual destruction of the transistor. In practice, thermal runaway is prevented in well-designed circuits, by the use of stabilization circuitry that compensates for any increase in I_{CBO} and also, for power transistors, by the use of a heat sink to provide rapid conduction of heat away from the junction.

The Construction of Transistors

A number of different methods of manufacturing transistors have been developed since the invention of the transistor in 1948 and a description of many of them is outside the scope of this book. The most commonly employed type of transistor is the alloy junction transistor and only the construction of this and two other important types will be briefly discussed.

The construction of a germanium alloy junction transistor is shown in Fig. 7.17; the method employed is an extension of the method previously described for the alloy junction diode. In Fig. 7.18 is shown the construction of a grown junction transistor; the method of construction has also been previously discussed. In recent years the silicon planar transistor has been increasingly employed because of its excellent performance and reliability. The construction of a transistor of this type is shown in Fig. 7.19 and involves the use of a diffusion technique. The steps involved in the manufacture of a silicon planar transistor are as follows: a wafer of n-type silicon is oxidized to a depth of approximately 1 micron (Fig. 7.20a) and then the oxide is partially etched off (Fig. 7.20b). Next the wafer is exposed to a vapour of the acceptor element boron and the impurity is allowed to diffuse into the wafer to a predetermined depth. At the same time the wafer surface is reoxidized (Fig. 7.20*c*). A part of the reoxidized surface is then etched away (Fig. 7.20*d*) and the wafer exposed to a vapour of the donor element phosphorus and is also reoxidized again (Fig. 7.20*e*). The wafer now contains a layer of

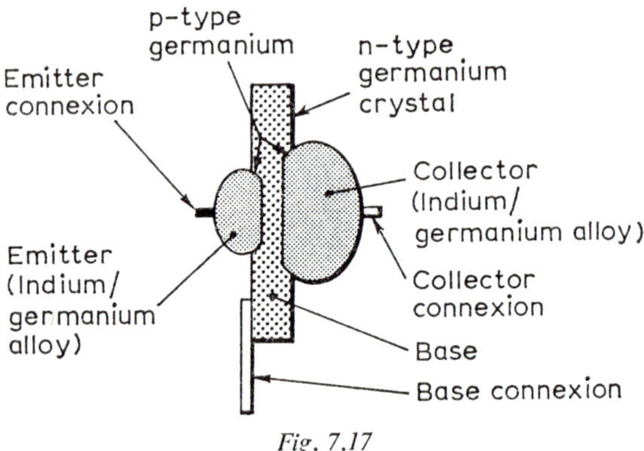

Fig. 7.17

Construction of an alloy junction transistor

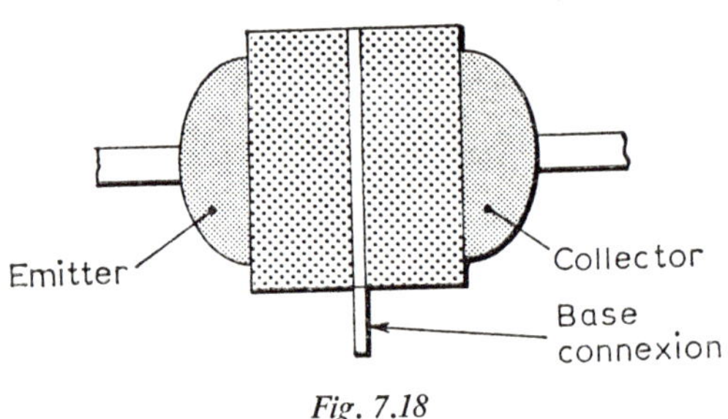

Fig. 7.18

Construction of a grown junction transistor

p-type material that will form the base of the transistor and a layer of n-type material that provides the emitter. The wafer is then etched to separate the base and emitter regions on the surface of the wafer (Fig. 7.20*f*) and finally (Fig. 7.20*g*) metal contacts are alloyed onto the etched areas.

The wafer is cut to the required size, mounted on a suitable collector contact and then leads are connected to the base and emitter contacts.

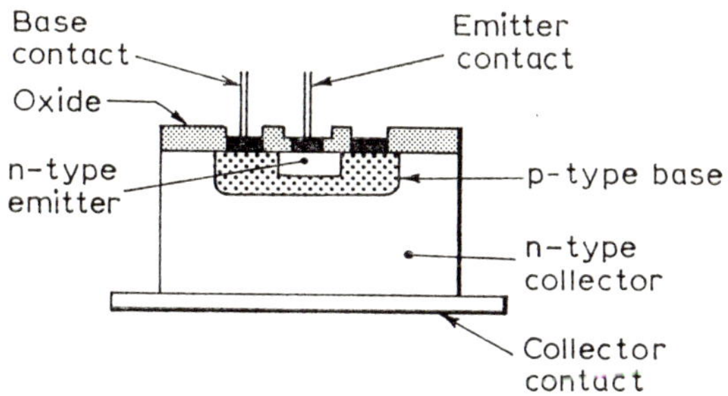

Fig. 7.19

Construction of a silicon planar transistor
(From *S.G.S. Fairchild Technical Report*)

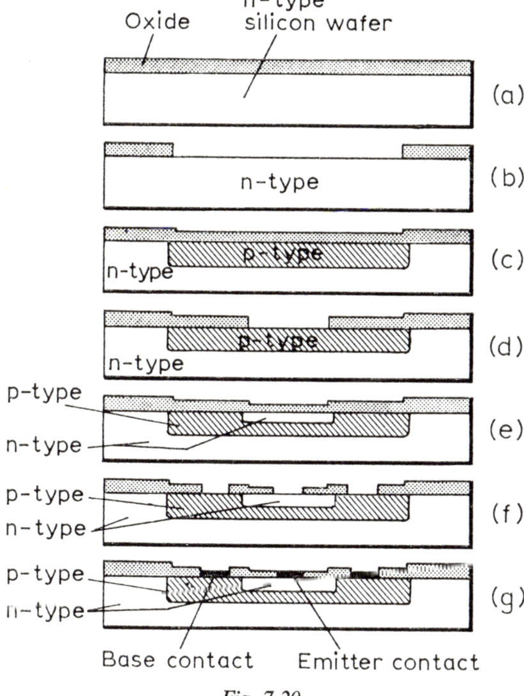

Fig. 7.20

The stages in the manufacture of a silicon planar transistor
(From *S.G.S. Fairchild Technical Report*)

The construction of a planar transistor requires a relatively thick collector wafer to give adequate mechanical support to the other layers and this increases the resistance of the collector. For some applications the collector resistance is too large and must be reduced by some means or another. If the resistance were to be reduced by the use of a low-resistivity material for the collector the breakdown voltage of the transistor would also be reduced and the collector/base capacitance would be increased, both undesirable effects. To overcome these effects an epitaxial layer is employed in the collector. This is a layer, approximately 0·1 mm thick, of high-resistivity material that is deposited on the main collector and that permits the collector to be made from a material of low resistivity.

Exercises

1. Describe briefly the operation of a transistor as an amplifying device. With the aid of a circuit diagram show how you would determine the basic static characteristics of a transistor.

     Sketch the current/voltage characteristics that would result from such measurements giving typical values on your axes.   (P1 1963)

2. Draw circuit diagrams showing how a junction transistor may be used in a single-stage audio-frequency amplifier (*a*) with common-base connection, (*b*) with common-emitter connection. Mark clearly the input, output, and biasing arrangements and the current flowing. State the input impedance, output impedance and current gain characteristics for each type of amplifier.

     Derive an expression showing how, for small input signals, the current gain of a transistor connected in a common-emitter circuit is related to the gain of the same transistor in a common-base circuit.

     A transistor connected in a common-emitter circuit shows changes in emitter and collector currents of 1·0 mA and 0·98 mA, respectively. What changes in base current produce these changes and what is the current gain of the transistor?   (A 1963)

3. Describe, in detail, with sketches, a junction transistor suitable for use in an audio amplifier.

     Sketch a typical family of characteristic curves for such a transistor when used in the common-emitter configuration. Scales must be clearly labelled.   (A 1962)

4. With the aid of diagrams showing suitable biasing arrangements, explain briefly the three ways in which a junction transistor may be connected in a small-signal amplifier. Compare the current gain, input impedance and output impedance of the three circuit arrangements.   (A 1961)

5. Describe, with the aid of sketches, the construction and features of a low-power alloy junction transistor. Discuss the materials used.   (A 1960)

6. The data given in Tables A and B refer to a transistor in the common-emitter configuration.

     Use the data in Table A to plot the collector-voltage/collector-current characteristics for $I_b = -40$, -80 and -120 μA, and from the characteristic for $I_b = -80$ μA deduce the output resistance of the transistor. Use the data in Table B to plot the collector-current/base-current characteristic for $V_{ce} = -4\cdot5$ V and from the graph deduce the current gain.   (A 1965)

Table A

Collector voltage V_{ce} (volts)		-2	-4	-6	-8	-10
Collector current (mA)	Base current $-120\ \mu$A	$-6\cdot4$	-7	$-7\cdot6$	$-8\cdot2$	$-8\cdot8$
	Base current $-80\ \mu$A	$-4\cdot4$	$-4\cdot8$	$-5\cdot2$	$-5\cdot6$	$-6\cdot0$
	Base current $-40\ \mu$A	$-2\cdot2$	$-2\cdot4$	$-2\cdot6$	$-2\cdot8$	$-3\cdot0$

Table B

	Collector voltage $V_{ce} = -4\cdot5$ volts				
Base current $I_b\ (\mu A)$	0	-5	-10	-15	-20
Collector current I_c (mA)	$-0\cdot15$	$-0\cdot45$	$-0\cdot75$	$-1\cdot1$	$-1\cdot4$

7. Draw circuit diagrams and briefly explain how a junction transistor may be used in a single-stage audio-frequency amplifier (*a*) with common-base connection, and (*b*) with common-emitter connection. Discuss the order of magnitude of the input and output impedances and the current gain characteristic for each type of amplifier. (A 1965)
8. Explain the principle of the transistor.

Why is it necessary to avoid excessive temperature rise in the transistors of an amplifier? (P1 1965)

8 *Thermionic Valves*

A thermionic valve consists of two or more electrodes contained within a glass envelope from which most of the air has been removed. A large number of different types of valve exist and are in common use but this chapter will be concerned only with the more simple types of valve, such as the diode, the triode, the tetrode, the pentode and the beam tetrode.

In all these types of valve one of the electrodes is the cathode and another is the anode and the basic principle of operation is that when the cathode is heated to a suitable temperature it emits large numbers of electrons and a proportion of these are collected by the anode, and constitute the anode current. In their passage from cathode to anode the emitted electrons may pass through one, two or three grids—depending on the type of valve—and potentials applied to these grids are able to control the flow of electrons and hence the anode current.

Thermionic Emission

At room temperatures the electrons in a metal are able to wander at random in the atomic structure of the metal, and some of the electrons near to the surface of the metal will pass out into the surrounding air. Now it is self-evident that at room temperatures ordinary metals do not lose their electrons in large numbers and so a force must exist that prevents electrons from permanently leaving the metal's surface. Immediately an electron leaves the metal the metal has lost a negative charge (the electronic charge) and this is equivalent to its gaining a positive charge. The positive charge exerts a force on the emitted electron that pulls the electron back towards the metal, and for an electron to be able to escape it must possess sufficient kinetic energy to enable it to overcome this force. At room temperatures very few electrons have sufficient energy and the number of electrons able to escape from the metal is very small. In order to considerably increase the number of electrons escaping from the metal it is necessary to give the electrons additional energy, and this can best be done by heating the metal. As the temperature of the metal is increased more electrons attain sufficient energy to enable them to overcome the retarding force and are able to escape from the metal.

For most metals the temperature necessary for adequate electron emission is too high if adequate life is to be obtained and/or the characteristics of the metal are not to be changed. In practice, the choice of cathode material is limited to tungsten, thoriated tungsten, and nickel coated with a mixture of barium oxide and strontium oxide to which some calcium oxide may also be added.

Once an electron has escaped from the cathode it is subjected to a retarding force set up by the negatively charged electrons that have already escaped. An electron is slowed down by this force and may well be returned to the cathode. An emitted electron can also be slowed down by coming into collision with a gas molecule but to minimize this effect the cathode is placed within an evacuated glass envelope.

Thus electrons are continually emitted from the surface of a heated cathode and are then subjected to forces which tend to return them to the cathode. Most of the emitted electrons travel only a short distance from the cathode before they are returned to it; this distance is proportional to the velocity with which an electron is emitted. The cathode is surrounded, therefore, by a cloud of electrons, some of which are moving away from the cathode and some moving towards it. Only those electrons having sufficient energy to overcome the restraining forces can escape from the vicinity of the cathode. The cloud of electrons around the cathode is known as the space charge. Obviously the space charge is negative.

If another electrode, the anode, is introduced into the evacuated space and is maintained at a positive potential with respect to the cathode it will exert an attractive force on the emitted electrons. The emitted electrons are then subjected to a retarding force produced by the electric fields set up by the positively charged cathode and the negatively charged space charge, and an attractive force produced by the electric field set up by the anode. Near to the anode the accelerating electric field is larger than the retarding electric field, while near to the cathode the retarding electric field predominates. At some point in between the cathode and the anode the two electric fields are equal and cancel out, and an electron reaching this point is neither accelerated nor retarded. Any emitted electrons that have sufficient energy to reach this point come under the influence of the anode field and are rapidly accelerated to the anode and there contribute to the anode current.

An increase in the positive potential of the anode moves the point of zero electric field nearer to the cathode and allows electrons of lower energy to reach the anode. The anode current thus increases with increase in anode voltage. If the anode voltage is increased to the point where the position of zero electric field occurs at the cathode, all the emitted electrons will reach the anode and the space charge is then non-existent. The anode current is then at its maximum value, since there are no more emitted electrons available for collection by the anode. A further increase in anode voltage does not produce a corresponding increase in anode current; an increased anode current can only be obtained by increasing the cathode temperature and thus increasing the number of emitted electrons.

CATHODE MATERIALS

Three different materials are employed for the cathodes of thermionic valves, the choice for a particular valve depending upon the range of anode potentials the valve will be expected to handle. For valves designed to handle only small powers, up to about 300 W dissipated at the anode, and to have voltages of up to some 2,000 V applied to their anodes, oxide-coated cathodes are employed. An oxide-coated cathode consists of a mixture of barium oxide and strontium oxide formed on the surface of a nickel base (sometimes materials such as tungsten and molybdenum are used for the base). Adequate electron emission occurs when a cathode of this type is heated to a temperature of the order of 800–1,000 °K. Medium-power valves, a maximum of some 1,000 W dissipated at the anode, require voltages of up to 5,000 V applied to their anodes and the use of such high potentials results in the bombardment of the cathode by ionized atoms. The severity of this bombardment is proportional to the magnitude of the anode voltage and would result in the rapid disintegration of an oxide-coated cathode. For medium-power valves thoriated-tungsten cathodes are employed. The emission efficiency of this type of cathode is less than the emission efficiency of the oxide-coated cathode, but adequate electron emission can be obtained if the cathode is heated to a temperature of approximately 2,000°K. Valves designed to handle very large powers, anode dissipation up to 100 kW or so, have up to 20,000 V applied to their anodes, and the severity of resulting ionic bombardment of the cathode makes the use of thoriated-tungsten cathodes impracticable. Consequently high-power valves employ cathodes made of pure tungsten. Pure tungsten is a much less efficient emitter than thoriated tungsten and must be operated at a temperature of some 3,000°K.

HEATING THE CATHODE

The cathode of a thermionic valve is always heated by the passage of an electric current. In a directly heated valve the heater current passes directly through the cathode, or filament, but in an indirectly-heated valve the heater current is passed through a resistance wire mounted inside a hollow cylindrical cathode, see Figs. 8.1*a* and *b*.

A directly heated cathode may be made of tungsten, thoriated-tungsten or tungsten coated with strontium oxide, while indirectly-heated cathodes are always made of nickel (or some other suitable base material) coated with a mixture of barium oxide and strontium oxide.

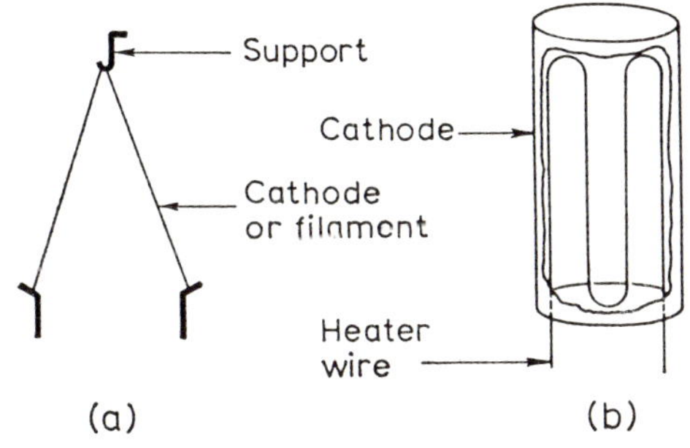

Fig. 8.1

The construction of (*a*) a directly heated cathode and (*b*) an indirectly heated cathode

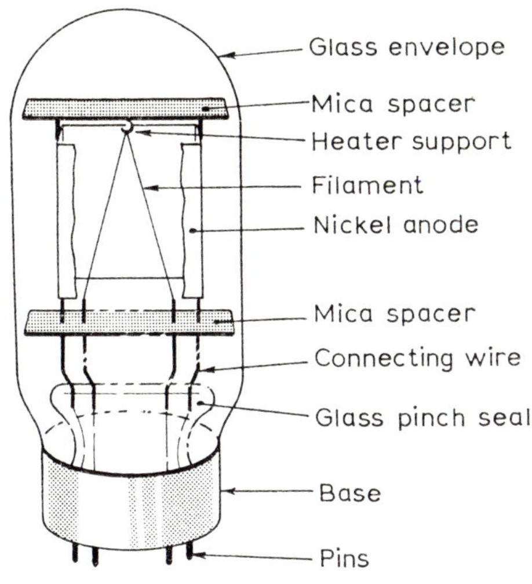

Fig. 8.2

A directly heated diode valve

The Diode Valve

A diode valve consists of two electrodes, the cathode and the anode, inserted inside an evacuated glass envelope. The construction of a typical directly heated valve is shown in Fig. 8.2. Often the anode is

cylindrical in shape and may be fitted with cooling fins to help
dissipate the heat produced at the anode. The construction of an
indirectly heated diode is very similar, the main difference being the
use of a cathode of the type shown in Fig. 8.3.

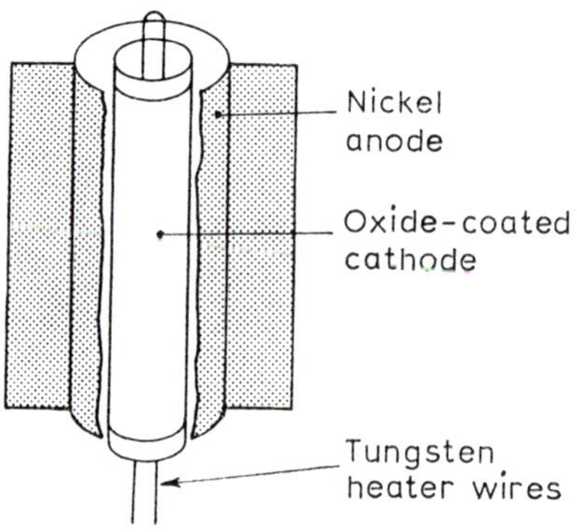

Fig. 8.3

The anode and cathode of an indirectly heated diode valve

THE STATIC CHARACTERISTIC

The static characteristic of a diode valve is a plot of anode current
to a base of anode voltage and may be determined with the aid of the
arrangement of Fig. 8.4. Meters A_1 and V are required to measure
values of anode current and anode voltage respectively and meter

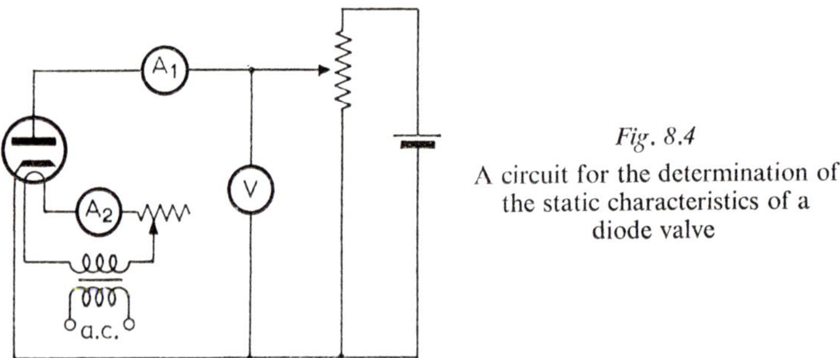

Fig. 8.4

A circuit for the determination of
the static characteristics of a
diode valve

A_2 reads values of heater current. The method of measurement is
briefly this: the heater current is set to a convenient value and then
the anode voltage is increased in a number of steps. At each step the
anode current flowing is noted and then the values of current obtained
are plotted against anode voltage to give a curve of the form shown
in Fig. 8.5. It can be seen that at first the anode current increases

rapidly with increase in anode voltage until a voltage is reached at which the current begins to increase less rapidly. This voltage is known as the knee, or saturation voltage, and when it has been reached the current is called the saturation current.

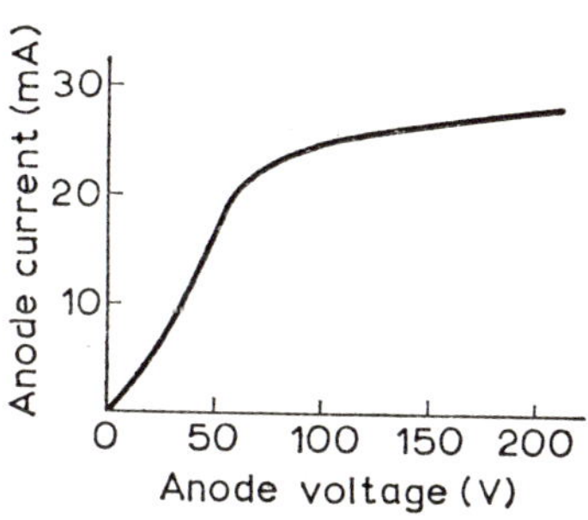

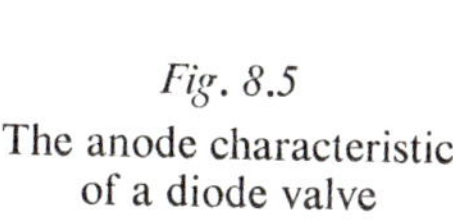

Fig. 8.5

The anode characteristic of a diode valve

Beyond the saturation point large changes in anode voltage produce only small changes in anode current because almost all the emitted electrons are already being collected by the anode. Although not shown in the figure a very small anode current, a few microamps, flows for zero and small negative anode voltages; this current indicates that a few electrons are emitted from the cathode with sufficient energy to reach the anode without the aid of the anode's accelerating field.

If the heater current is now altered and the measurement procedure repeated a similar characteristic is obtained, differing only in that the saturation point occurs for a different value of anode voltage.

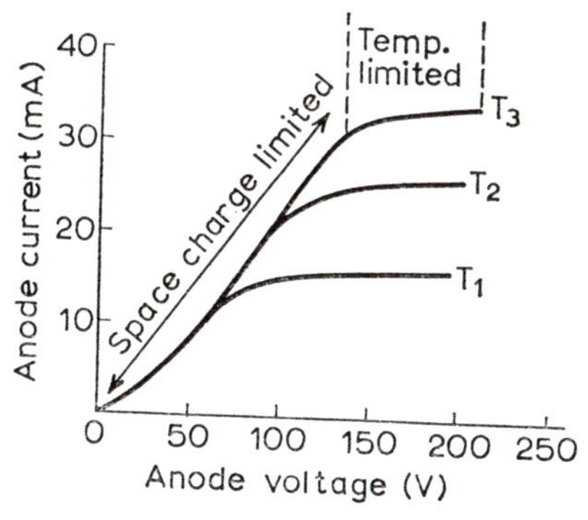

Fig. 8.6

Space-charge and temperature limitation of a diode valve

Fig. 8.6 shows the characteristics for cathode temperatures T_1, T_2 and T_3. When the anode current is less than the saturation value it is said to be space-charge limited since it can always be increased by increasing the anode voltage and so decreasing the space charge. Beyond the saturation point the anode current is said to be temperature limited since it can only be increased by increasing the temperature of the cathode.

The reciprocal of the slope of the diode static characteristic is the output resistance of the diode, generally known as the anode a.c. resistance, symbol r_a, but the term anode slope resistance is also used:

$$\text{Anode a.c. resistance } r_a = \frac{\delta V_a}{\delta I_a} \text{ ohms} \qquad (8.1)$$

where V_a = anode voltage and I_a = anode current.

Example 8.1
Measurement of a diode valve produced the following figures—

Anode voltage (volts)	0	0·5	1·0	1·5	2·0
Anode current (mA)	0	1·6	4·0	6·6	9·4

Plot the characteristic and determine the anode a.c. resistance of the diode over the straight portion of the curve.

Solution. The static characteristic is shown plotted in Fig. 8.7. Taking increments of 0·25 V either side of the point $V_a = 1·25$ V it is found, by projection from the characteristic, that the corresponding values of anode current

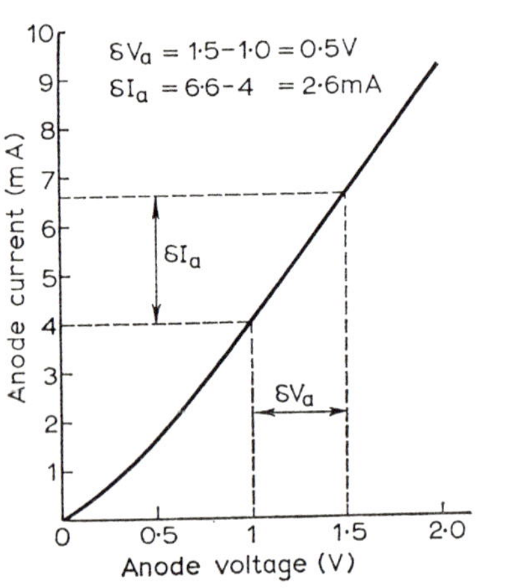

Fig. 8.7

I_a are 6·6 mA and 4 mA. Thus the change δV_a in anode voltage is 0·5 V and the change δI_a in anode current is 2·6 mA.

$$\therefore \quad r_a = \frac{\delta V_a}{\delta I_a} = \frac{0\cdot5}{2\cdot6 \times 10^{-3}} = 192\,\Omega \qquad Ans.$$

The Triode Valve

The triode valve has a third electrode, the control grid, inserted between the cathode and the anode near to the point of minimum electric field. The control grid consists of a fine mesh, and is normally maintained at a negative potential relative to the cathode by connecting a battery between grid and cathode, Fig. 8.8. The potential on the grid reduces the electric field of the anode and this reduces the number of electrons that are able to reach the anode. If the grid is made more negative with respect to the cathode, the electric field of the anode is further reduced and even fewer electrons possess sufficient energy to reach the anode. Conversely, if the grid is made less negative the electric field of the anode is increased and more

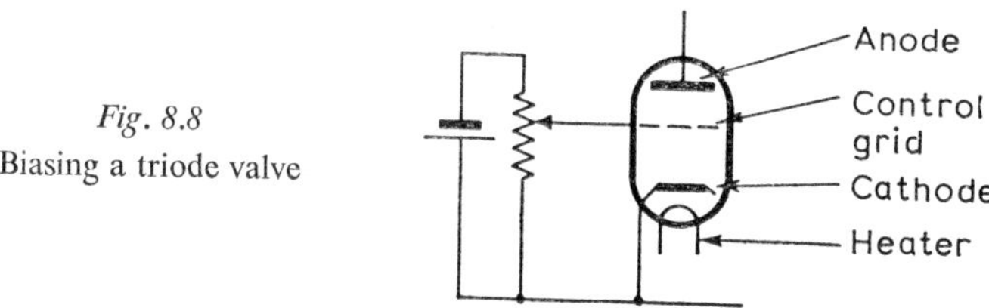

Fig. 8.8

Biasing a triode valve

electrons are able to reach the anode. The grid potential is able to control the number of electrons reaching the anode, and thus the anode current, in the same way as the anode potential. Since the grid is situated very much nearer to the cathode than is the anode the control per volt change in potential exerted by the grid is much greater than the control exerted by the anode. Continued increase in the negative potential on the control grid progressively reduces the anode current of a triode until the point is reached where the anode current ceases to flow. The grid/cathode voltage which just causes the anode current to cut off is known as the cut-off voltage of the valve.

In many circuits, the grid of a triode valve is not allowed to become positive relative to the cathode because this would allow grid current to flow, with detrimental effect on the operation of the circuit. To prevent the grid being made positive with respect to the cathode, the grid is generally biased negative a number of volts greater than the peak value of any signals to be applied to the grid. Grid bias will be discussed in the next chapter. The construction of a typical indirectly heated triode valve is given in Fig. 8.9. The whole assembly is placed inside an evacuated glass envelope and connections are made to pins in an insulated base. The structure of a directly heated triode differs only in the use of a cathode of the type shown in Fig. 8.1a. Sometimes the grid, or the anode, connection may be brought out to a pin on the top of the envelope, instead of in the valve base, and the anode may be provided with cooling fins.

THE PARAMETERS AND STATIC CHARACTERISTICS OF A TRIODE

The static characteristics of a triode valve may be experimentally determined in a manner similar to that employed for transistors. The resulting curves can be employed in the design of circuits or used to give information about the valve. Two families of static characteristics are generally drawn, (*a*) the anode characteristics which

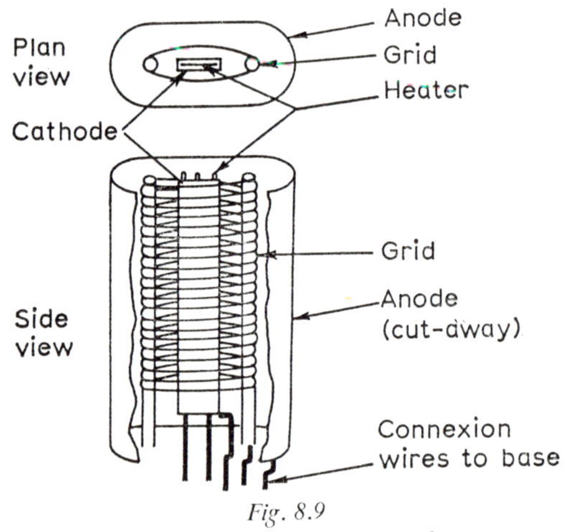

Fig. 8.9

An indirectly heated triode valve

illustrate the way in which the anode current varies with change in anode voltage for constant values of grid voltage, and (*b*) the mutual characteristics which show how the anode current varies with change in grid voltage for constant values of anode voltage. A third set of characteristics, the voltage transfer characteristics, can also be plotted, but since they are rarely employed in practice they will not be further mentioned.

A triode valve has three useful a.c. parameters and their values may be determined from either the anode characteristics or the mutual characteristics. These parameters are (*a*) the anode a.c. resistance r_a, (*b*) the mutual conductance g_m and (*c*) the amplification factor μ. They are defined thus:

The anode a.c. resistance r_a is the ratio of a change in anode voltage δV_a to the change in anode current δI_a produced by it, the grid voltage V_g remaining constant, i.e.

$$r_a = \frac{\delta V_a}{\delta I_a} \text{ ohms} \qquad (V_g \text{ constant}) \qquad (8.2)$$

The mutual conductance g_m is the ratio of a change in anode current δI_a to the change in grid voltage δV_g producing it, the anode voltage V_a remaining constant, i.e.

$$g_m = \frac{\delta I_a}{\delta V_g} \quad (V_a \text{ constant}) \tag{8.3}$$

With I_a measured in milliamperes and V_g in volts, g_m is expressed in millimhos, or milliamperes per volt (mA/V).

The amplification factor μ is the ratio of a change in anode voltage δV_a to the change in grid voltage δV_g producing it, the anode current I_a remaining constant, i.e.

$$\mu = \frac{\delta V_a}{\delta V_g} \quad (I_a \text{ constant}) \tag{8.4}$$

μ is the ratio of two voltage changes and is dimensionless. Multiplying equation (8.2) by equation (8.3) gives

$$r_a g_m = \frac{\delta V_a}{\delta I_a} \frac{\delta I_a}{\delta V_g}$$
$$= \frac{\delta V_a}{\delta V_g}$$

which, from equation (8.4), is equal to μ

$$\therefore \qquad r_a g_m = \mu \tag{8.5}$$

Since the three parameters, r_a, g_m and μ, of a triode valve are defined in terms of changes of anode current, anode voltage and grid voltage, it is to be expected that the values of these parameters for a particular valve can be found from the static characteristics of that valve.

THE ANODE CHARACTERISTICS

The anode characteristics are plots of anode current against anode voltage for constant values of grid voltage and can be determined experimentally with the circuit shown in Fig. 8.10.

To plot the anode characteristics the grid voltage must be maintained constant at a number of particular values and the anode voltage then varied in a number of steps; these requirements are readily satisfied by means of potential dividers P_1 and P_2. The measurement procedure is as follows. The grid voltage is set to a convenient value and the anode voltage is increased from zero in a number of discrete steps and at each step the anode current which flows is noted. The grid voltage is then set to a new value and the measurement repeated. The values of anode current which are obtained in this way are plotted to a base of anode voltage to give a family of anode characteristics, such as those shown in Fig. 8.11. The curves are linear over much of their range and the operation of the valve would normally be restricted to this linear part.

The reciprocal of the slope $\delta V_a/\delta I_a$ of any of the curves is a measurement of the anode a.c. resistance r_a of the valve at the point

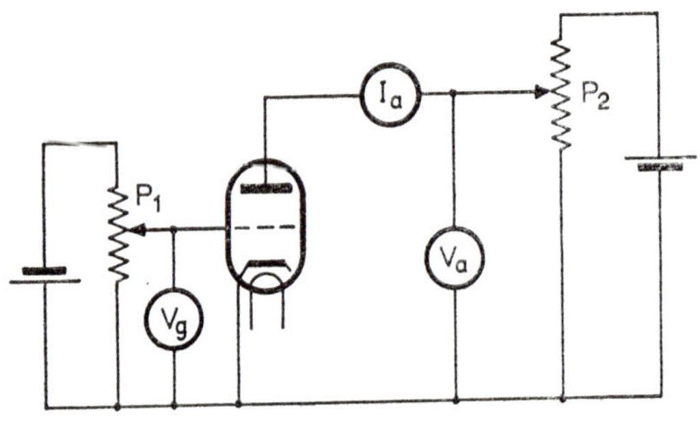

Fig. 8.10

A circuit for the determination of the static characteristics of a triode valve

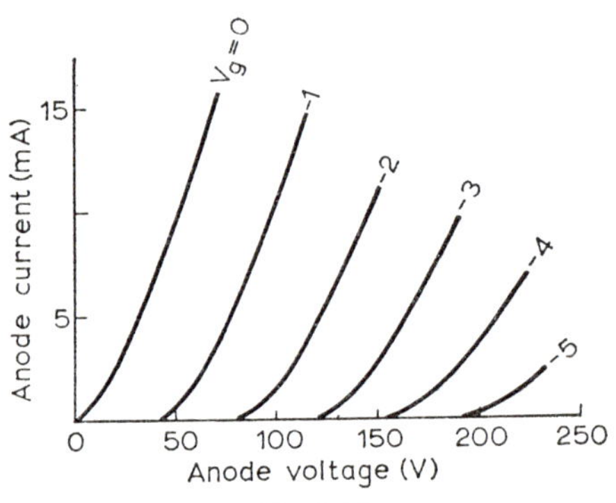

Fig. 8.11

Triode anode characteristics

(anode and grid voltage) considered. The method of calculating the value of r_a is identical with the method employed to calculate the output resistance of a transistor. The values of the mutual conductance g_m, and the amplication factor μ can also be determined from the anode characteristics, refer to Example 8.2.

THE MUTUAL CHARACTERISTICS

The mutual characteristics are plots of anode current against grid voltage for constant values of anode voltage. The mutual characteristics can be determined with the aid of the circuit shown in

Fig. 8.10, the measurement procedure being as follows. The anode voltage is set to a convenient value and the grid voltage is increased negatively, starting from zero, in a number of steps. At each step the anode current that flows is noted. The procedure is repeated for a number of different values of anode voltage to give a family of curves. A typical family of triode mutual characteristics is shown in Fig. 8.12. The slope $\delta I_a/\delta V_g$ of a mutual characteristic curve gives

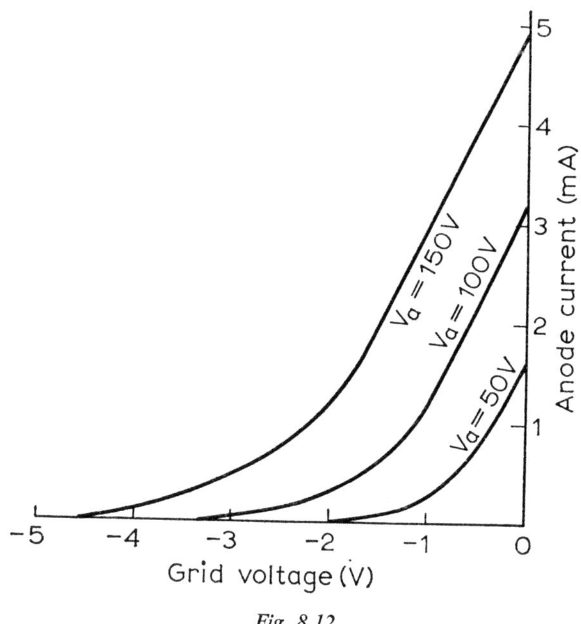

Fig. 8.12

Triode mutual characteristics

the mutual conductance g_m of the valve at the point of measurement. The anode a.c. resistance r_a and the amplification factor μ can also be determined from the mutual characteristic, see Example 8.2.

Example 8.2

A triode valve has the data shown overleaf.

Plot the anode characteristics and determine from them the values of the anode a.c. resistance, the mutual conductance and the amplification factor over the linear parts of the characteristics. Then plot the mutual characteristics and from them confirm the values for the anode a.c. resistance, mutual conductance and amplification factor previously obtained.

Solution. The anode characteristics are shown in Fig. 8.13.

ANODE VOLTAGE (volts)	ANODE CURRENT (mA)						
	Grid voltage (volts)	0	−1	−2	−3	−4	−5
50		3·5	2·0	—	—	—	
100		11·0	5·0	0·5	—	—	
150		20·0	12·5	5·5	1·5	—	
200		—	21·0	14·0	6·5	2·5	
250		—	—	—	15·0	8·5	3·0
300		—	—	—	—	—	9·5

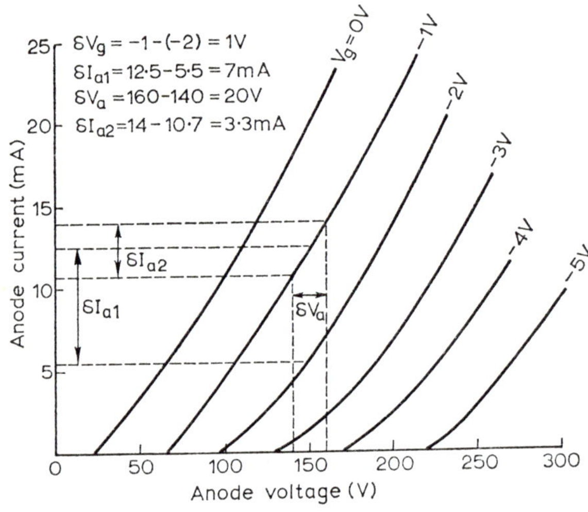

Fig. 8.13

The anode a.c. resistance r_a is given by $r_a = \delta V_a/\delta I_a$ (V_g constant) ohms. Hence, taking V_g as constant at -1 V and choosing increments of 10 V either side of the point $V_a = 150$ V, i.e. $\delta V_a = 20$ V, gives, by projection from the curve for $V_g = -1$ V, $\delta I_a = 3\cdot3$ mA.

$$\therefore \quad r_a = \frac{20}{3\cdot3 \times 10^{-3}} = 6{,}060 \ \Omega \qquad Ans.$$

The mutual conductance is given by $g_m = \delta I_a/\delta V_g$ mA/V (V_a constant). Hence, taking V_a constant at 150 V it can be seen from the characteristics that a change in grid voltage from -1 V to -2 V, $\delta V_g = 1$ V, produces a change in anode current of 7 mA.

$$\therefore \quad g_m = \frac{7}{1} = 7 \ \text{mA/V} \qquad Ans.$$

Also, $\mu = g_m r_a = 7 \times 10^{-3} \times 6{,}060 = 42\cdot42 \qquad Ans.$

The mutual characteristics of the triode in question are shown in Fig. 8.14.

The amplification factor is given by $\mu = \delta V_a / \delta V_g$ (I_a constant). Hence, taking I_a as constant at 12·5 mA it can be seen that if the grid voltage is increased from -1 V to $-2\cdot2$ V the anode voltage must be increased from 150 V to 200 V.

$$\therefore \qquad \mu = 50/1\cdot2 = 41\cdot66 \qquad Ans.$$

To find the mutual conductance from Fig. 8.14: If the anode voltage is taken as constant at 200 V and the grid voltage is increased from -1 V to $-2\cdot2$ V the anode current changes from 21 mA to 12·5 mA.

$$\therefore \qquad g_m = 8\cdot5/1\cdot2 = 7\cdot08 \text{ mA/V} \qquad Ans.$$

Also
$$r_a = \frac{\mu}{g_m} = \frac{41\cdot66}{7\cdot08 \times 10^{-3}} = 5{,}884\,\Omega \qquad Ans.$$

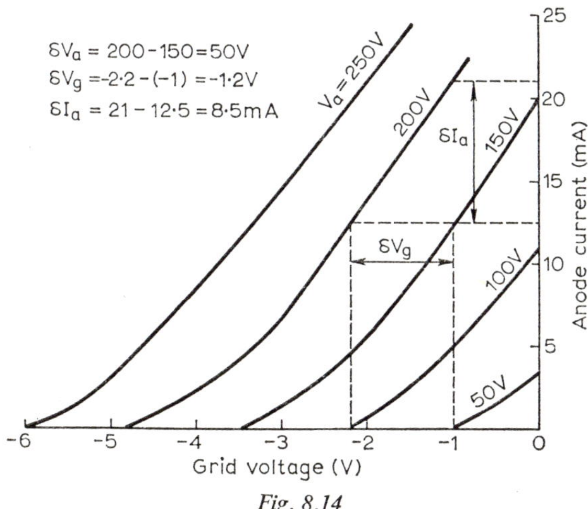

Fig. 8.14

The values obtained from the mutual characteristics for the three valve parameters differ slightly from the values obtained from the anode characteristics. This is only to be expected since the slightest error in the sketching of the curves will affect the values of the parameters derived from the curves. In general, an error of about 5% must be expected from a graphical determination of the valve parameters. If this should seem high it should be borne in mind that the resistors used in valve circuits may well have a tolerance of 10% or even 20%.

The value of the amplification factor μ could have been obtained directly from the anode characteristic and the value of the anode a.c. resistance r_a directly from the mutual characteristic instead of employing the relationship $\mu = r_a g_m$.

DYNAMIC MUTUAL CHARACTERISTICS

The static mutual characteristics of a triode valve are obtained by varying the grid voltage and maintaining the anode/cathode voltage at a constant value. In a practical circuit, however, the anode/cathode voltage is not a constant quantity because the anode circuit and/or the cathode circuit contains some resistance, and any change in the anode current produces a change in the voltage drop across this resistance and causes the anode/cathode voltage to vary.

The effect of a non-constant anode/cathode voltage can be taken into account either by drawing a load line on the static anode characteristics, or by drawing the dynamic mutual characteristic of the

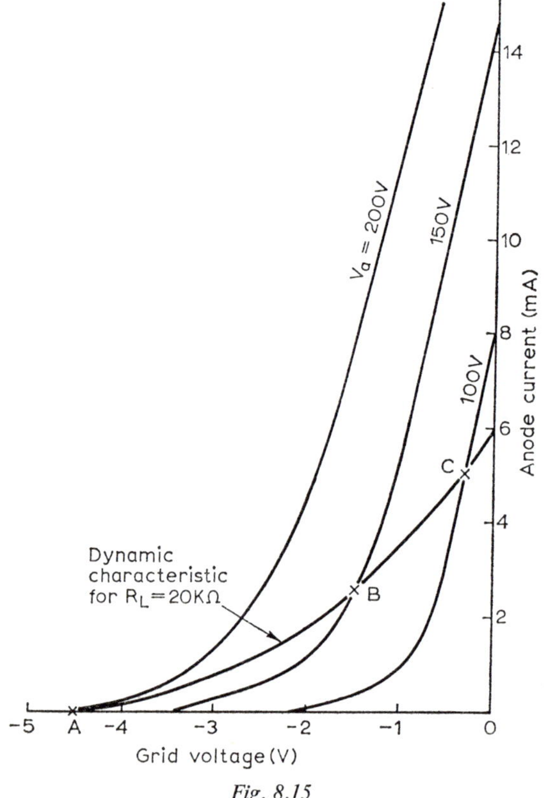

Fig. 8.15

Triode dynamic mutual characteristic

valve. The former method is beyond the scope of this book and a discussion must be left for the succeeding volume; the latter method will be discussed now. The dynamic mutual characteristic of a triode valve with some resistance in its anode circuit can be derived from the static mutual characteristics of the valve in the following manner. Suppose that the triode whose static mutual characteristics are shown in Fig. 8.15 is operated in a circuit with an anode resistor of 20,000 Ω and an h.t. supply of 200 V.

When the voltage applied to the grid of the triode is −4·5 V the anode current is cut off, point A, and the anode voltage is equal to the h.t. supply voltage. If the grid voltage is decreased the anode current will increase and the anode voltage will fall because of the increased voltage drop across the anode load resistor. Suppose the grid voltage to be decreased until the anode voltage has fallen to 150 V. The voltage drop across the 20,000 Ω anode load resistor must then be equal to (200 − 150) or 50 V and so the anode current flowing is equal to 50/20,000 or 2·5 mA. Thus point B is another point on the dynamic characteristic. If, now, the grid voltage is further decreased until the anode voltage falls to 100 volts the anode current flowing is (200 − 100)/20,000 or 5 mA giving a third

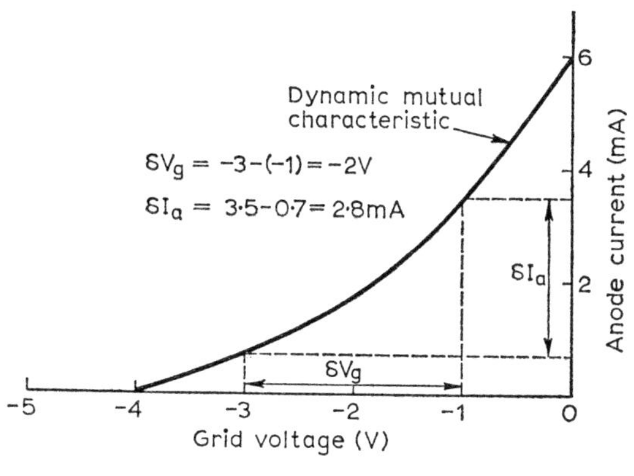

Fig. 8.16

Use of dynamic mutual characteristic to show changes in anode current

point, point C, on the characteristic. Drawing a smooth curve through the points A, B and C gives the dynamic mutual characteristic of the triode for an h.t. supply voltage of 200 V and an anode load resistance of 20,000 Ω. For any other values of h.t. supply voltage and/or anode load resistance a different dynamic characteristic would be obtained.

The dynamic mutual characteristic can be used to determine the changes in anode current that result from given changes in grid voltage. Thus, referring to Fig. 8.16, if the grid voltage varies 1 V either side of a mean value of −2 V the anode current varies between 3·5 and 0·7 mA. The voltage developed across the 20,000 Ω anode resistor varies between 70 and 14 V and so the voltage gain, defined as (change in anode voltage/change in grid voltage) is (70 − 14)/2 or 28, a value that is less than the amplification factor of the valve.

THE INTER-ELECTRODE CAPACITANCES OF A TRIODE VALVE

The grid/cathode structure of a triode valve consists of two conducting surfaces separated by an insulant and thus a capacitance exists between these two electrodes. Similarly, further capacitances exist between the anode and grid and the anode and cathode. A triode valve can be represented by the circuit of Fig. 8.17 in which C_{ga} represents the grid/anode capacitance, C_{gc} is the grid/cathode capacitance, and C_{ac} is the anode/cathode capacitance. These inter-electrode capacitances are very small, typical values being C_{ga}

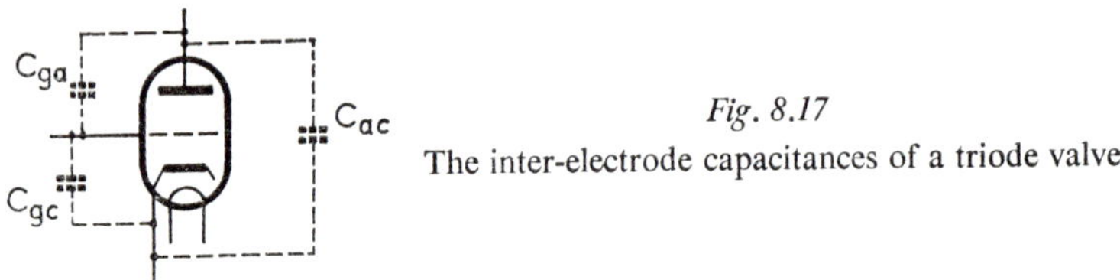

Fig. 8.17

The inter-electrode capacitances of a triode valve

$= 4$ pF, $C_{gc} = 2$ pF and $C_{ac} = 1$ pF, but at high frequencies their effect on the performance of the valve can be pronounced.

The anode/cathode C_{ac} and grid/cathode C_{gc} capacitances are not very important since, in practice, they can generally be incorporated into the associated circuitry. The grid/anode capacitance C_{ga}, however, is of utmost importance since it provides a path linking the anode and grid circuits through which a part of the output signal may be fed into the input circuit. This unwanted feedback is undesirable because it makes the circuit incorporating the triode unstable and may well produce self-oscillation.

The Tetrode Valve

The tetrode valve is a valve having four electrodes, a cathode, an anode, a control grid and a screen grid as shown symbolically in

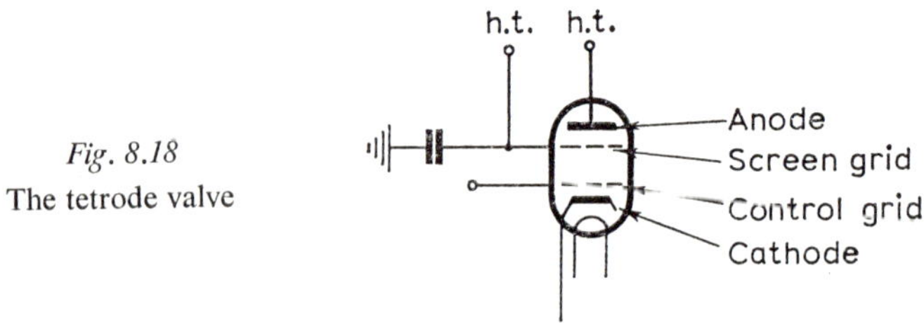

Fig. 8.18

The tetrode valve

Fig. 8.18. The extra grid is situated between the control grid and the anode and is connected, directly or via a resistor, to the h.t. supply and to earth via a capacitor. So far as alternating currents are concerned the screen is at earth potential and acts as an earthed screen between the anode and the control grid. The screen grid effectively divides the grid/anode capacitance C_{ga} into two capacitances, C_a and C_g, to earth, Fig. 8.19a. Ideally the control grid is now isolated from the anode and C_{ga} is zero, but, in practice, the

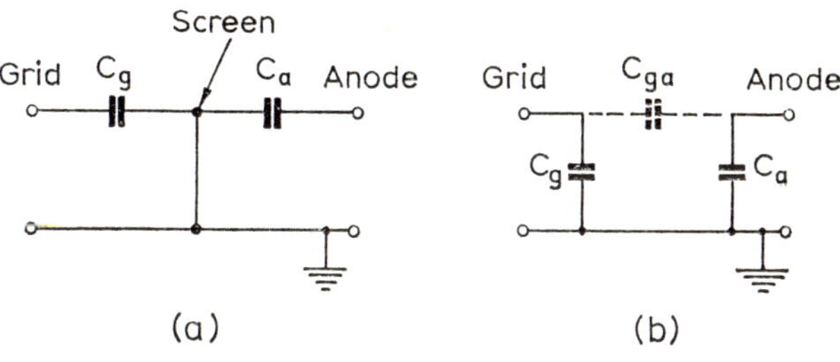

Fig. 8.19

The action of the tetrode screen grid

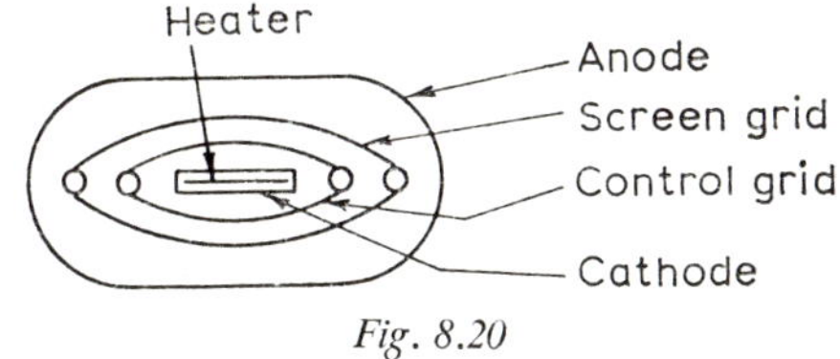

Fig. 8.20

The construction of a tetrode valve

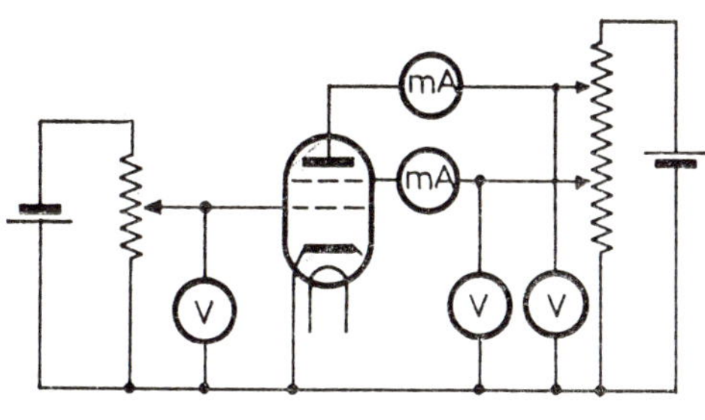

Fig. 8.21

A circuit for the determination of the static characteristics of a tetrode valve

screening is not perfect and some capacitance between anode and grid still exists, Fig. 8.19b. Typically, C_{ga} for a tetrode is of the order of 0·01 pF. A plan view of a typical tetrode valve is given in Fig. 8.20 and it is clear that the control grid is completely enclosed by the screen grid. The screen grid consists of fine wire mesh. The static characteristics of a tetrode valve can be obtained with the aid of the circuit shown in Fig. 8.21, using a procedure similar to that already described.

The anode characteristics consist of values of anode current plotted against anode voltage for fixed values of both screen grid voltage and control grid voltage. Typical anode characteristics are shown in Fig. 8.22, and four points should be noted: (*a*) the anode current rises rapidly from zero as the anode voltage is first increased, far more rapidly than in the case of a triode, (*b*) over a range of anode voltages from, approximately, 25 V to 75 V, the anode current decreases with increase in anode voltage, (*c*) the characteristics

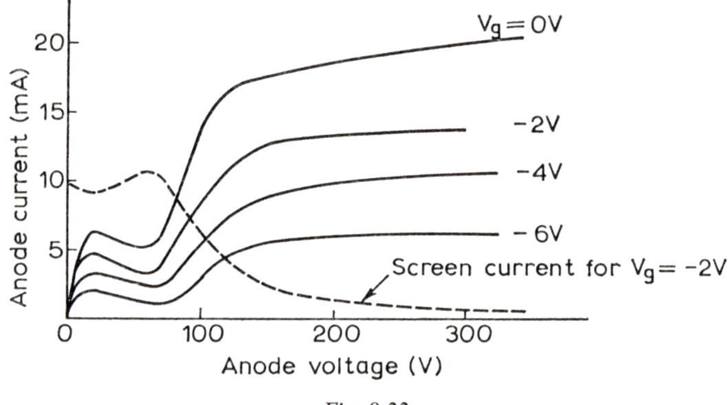

Fig. 8.22

Tetrode anode characteristics

shown relate to one particular value of screen grid voltage, there being a family of characteristics for each value of screen grid voltage chosen, and (*d*) above approximately 75 V the anode current increases with increase in anode voltage but the curve soon flattens out.

Over the flat part of each curve the slope $\delta V_a/\delta I_a$ is very small and hence the reciprocal of the slope, the anode a.c. resistance r_a of the valve is high, typically between 400,000 Ω and 1 MΩ. The influence of the control grid is more or less unaffected by the inclusion of the screen grid. Tetrodes, therefore, have mutual conductances similar to those of triodes, but their amplification factors ($\mu = g_m r_a$) are much larger.

The negative slope, or "kink," of part of the anode characteristics is a disadvantage of the tetrode since it drastically restricts the anode voltage swing that the valve is capable of handling without excessive distortion. The kink arises because of the onset of secondary emission of electrons from the anode.

SECONDARY EMISSION

Secondary emission of electrons from the anode in a tetrode valve takes place because the anode is subjected to impact by the electrons emitted from the cathode. The electrons emitted from the cathode are accelerated by the positive screen potential and may have considerable kinetic energy on their arrival at the anode and this energy is dissipated in the anode. Some of the electrons in the anode may absorb sufficient of this kinetic energy to permit them to escape from the anode. If the screen grid is at a higher positive potential than the anode it will attract these secondary electrons and the screen current will increase. The number of secondary electrons emitted from the anode increases with increase in the kinetic energy of the electrons arriving from the cathode and this energy, in turn, is proportional to the anode voltage. At a particular anode voltage the number of secondary electrons emitted from the anode becomes greater than the number of primary electrons emitted from the cathode, and beyond this point the anode current starts to decrease with increase in anode voltage. This reduction in anode current with increase in anode voltage continues until the anode voltage has been increased to the point where it is sufficiently positive, relative to the screen grid, to attract the secondary electrons back to the anode. Thereafter the anode current increases, and the screen current decreases, with increase in anode voltage.

The Pentode Valve

The pentode valve, which was originally developed to overcome the effect of secondary emission in the tetrode valve, has five electrodes, the additional electrode being another grid, the suppressor grid, situated between the anode and the screen grid, Fig. 8.23*a*. The suppressor grid is generally connected directly to the cathode, Fig. 8.23*b* and is thus many volts negative with respect to both anode and screen grid.

The electrons emitted from the cathode are accelerated by the screen grid potential to a sufficiently large velocity to enable them to overcome the retarding field of the suppressor grid and reach the anode. The secondary electrons emitted from the anode are emitted with relatively small kinetic energy and cannot overcome the retarding field of the suppressor grid. The secondary electrons are returned to the anode and the anode current is unaffected, there being no negative kink.

The suppressor grid also acts as another earthed screen between the anode and the control grid and the anode/grid capacitance C_{ga} is reduced to a very small value, typically 0·004 pF.

The static characteristics of a pentode valve can be obtained with the arrangement of Fig. 8.21. The mutual characteristics consist of anode current plotted against control grid voltage for constant values of screen grid voltage, the anode voltage being maintained constant throughout. Fig. 8.24 shows a typical family of pentode static mutual characteristics for a constant anode voltage of 200 V. Clearly the shape of the characteristics is similar to the shape of the triode static mutual characteristics. The characteristics for other values of anode voltage are almost identical with those shown in Fig. 8.24 provided that the anode voltage is not so small that the anode

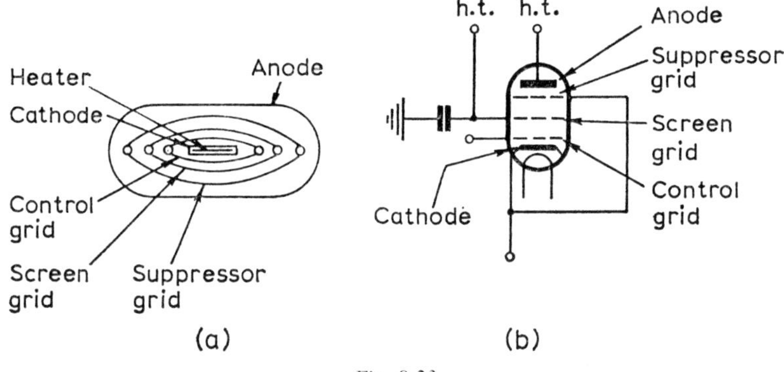

Fig. 8.23

(*a*) The construction of a pentode valve, (*b*) pentode symbol

cannot collect the electrons passing through the suppressor grid. The dynamic mutual characteristic for a particular anode load resistor and h.t. supply voltage can be drawn in the same way as described for a triode valve.

A typical family of pentode anode characteristics is shown in Fig. 8.25. The anode current increases rapidly with increase in anode voltage at first, and then becomes almost independent of further increase in anode voltage. Over the flat portion of the characteristic a large change in anode voltage is required to give only a small change in anode current, and this is an indication that the pentode valve has a large anode a.c. resistance, typically of the order of 1 megohm. Since the mutual conductance of a pentode is approximately the same as the mutual conductance of a triode very large values of amplification factor are common. Typically, μ may be from 1,000 to 8,000.

The pentode valve can be employed in a large number of different types of circuit, most of which are beyond the scope of this book, and it is sometimes necessary to know how the anode current varies

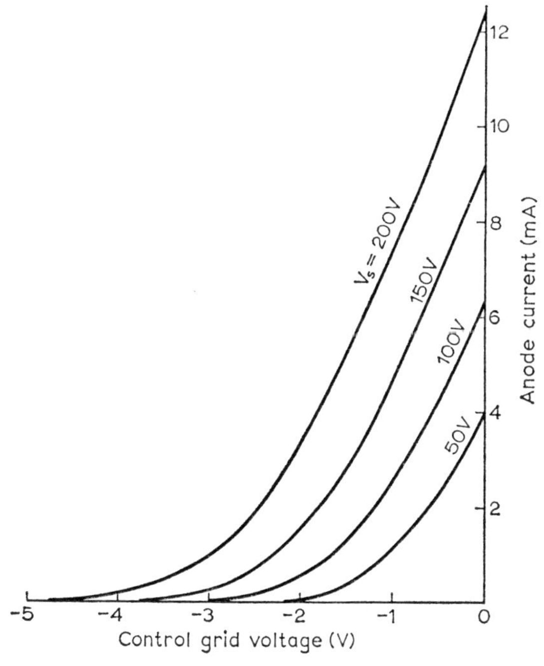

Fig. 8.24

Pentode mutual characteristics

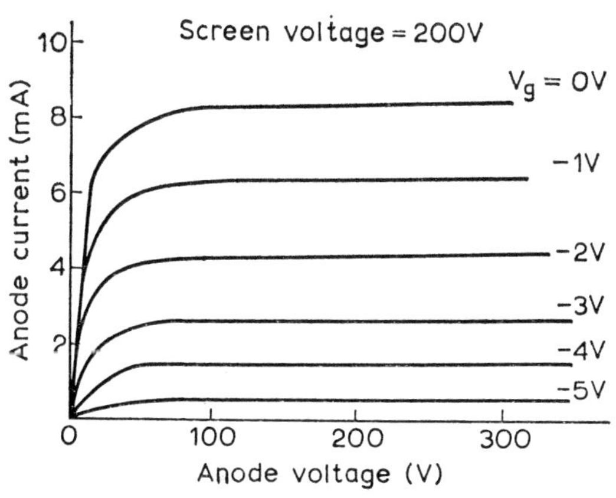

Fig. 8.25

Pentode anode characteristics

with change in the suppressor grid voltage. The suppressor grid characteristics of a pentode can be determined with the aid of the circuit of Fig. 8.21, suitably modified to allow the suppressor grid

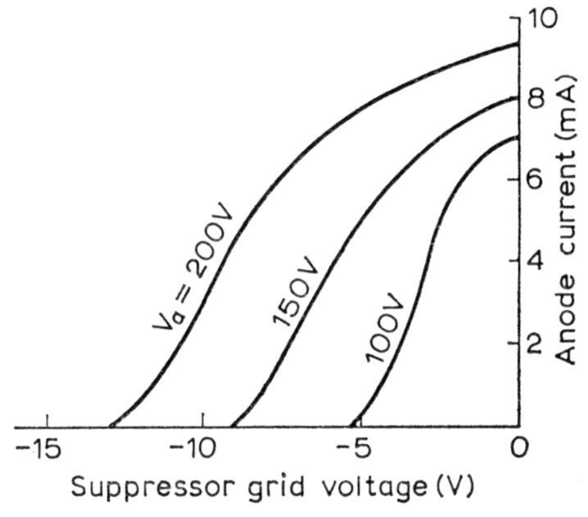

Fig. 8.26

Pentode suppressor grid characteristics

voltage to be varied. A typical family of suppressor grid characteristics is shown in Fig. 8.26 from which it can be seen that the curves are similar in shape to the mutual characteristics.

THE VARIABLE-μ PENTODE

In many circuits it is desirable to be able to control the voltage gain of a pentode valve by variation of the steady bias voltage applied to its control grid. This requirement can be satisfied by the use of a

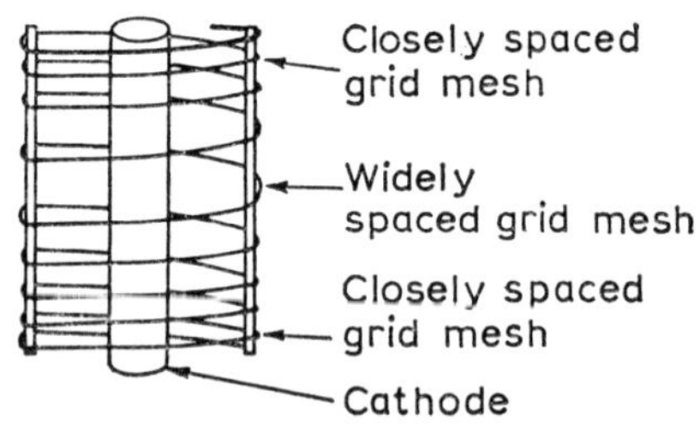

Fig. 8.27

The control grid of a variable-μ pentode valve

"variable-μ" pentode valve. A variable-μ pentode is one in which the pitch of the control grid varies along its length and hence its amplification factor varies at different parts of the valve.

Fig. 8.27 illustrates the construction of the control grid of a variable-μ pentode. The control grid consists of a wire mesh wound with wide spacing in the middle of the grid and close spacing at the ends. When a low negative voltage is applied to the grid the amplification factor is high and is the average of the effects of the widely and the closely spaced parts of the grid structure. For increased grid voltages the closely spaced parts of the grid mesh cut off the anode current and the amplification factor is reduced, since it is now the result of the effect of the widely spaced part of the grid only. Fig. 8.28

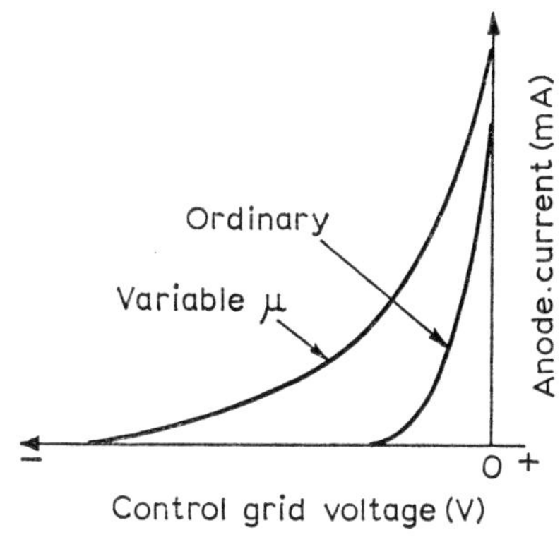

Fig. 8.28

Mutual characteristics of ordinary and variable μ pentode valves

compares the mutual characteristics of an ordinary pentode and a variable-μ pentode. Typically, a variable-μ pentode may have a mutual conductance of 4 mA/V when the grid voltage is -2 V and a mutual conductance of 0·05 mA/V when the grid voltage is -15 V.

The Beam Tetrode

Because of its large anode a.c. resistance a pentode valve has a large voltage gain and is eminently suited for use with small signals. When called upon to handle large signals and to deliver a power output the pentode has two disadvantages. Firstly, considerable power may be dissipated at the screen grid and this represents a loss of output power and, secondly, the anode characteristics are rounded at low anode voltages and this restricts the linear voltage swing that can be obtained. These disadvantages are overcome in the beam tetrode valve; this type of valve is so constructed that its screen grid intercepts only few electrons and so its screen dissipation is small, and (see Fig. 8.29) the "knee" of its anode characteristics occurs at a lower anode voltage than the knee of a pentode characteristic, permitting larger anode voltage swings to be handled without distortion.

The construction of a beam tetrode is shown in Fig. 8.30. The screen grid mesh has the same pitch as the mesh of the control grid and the two grids are accurately aligned. This arrangement,

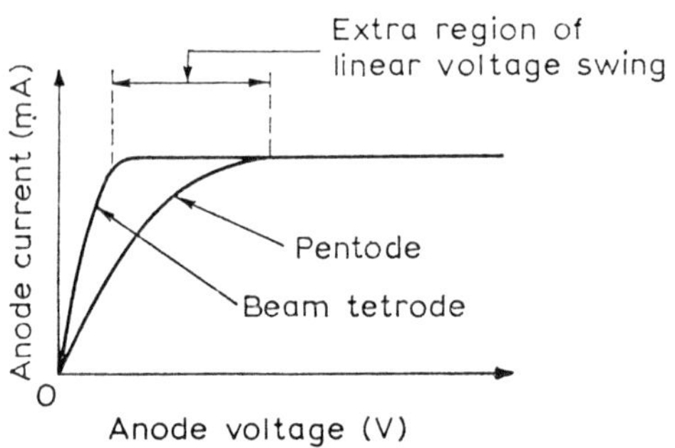

Fig. 8.29
Anode characteristics of pentode and beam tetrode valves

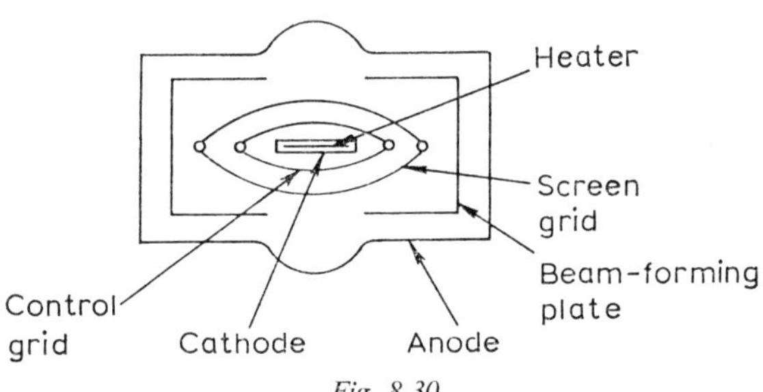

Fig. 8.30

Construction of a beam tetrode valve

together with the beam-forming plates, ensures that the electrons passing through the grids do so in a number of beams and that few electrons are intercepted by the screen grid. This is shown in Fig. 8.31 for a single electron beam passing through the two aligned holes, one in the control grid mesh, the other in the screen grid mesh. The current density in an electron beam is fairly high and secondary electrons emitted from the anode have to travel against the large retarding field of an electron beam to reach the screen. As a result the number of secondary electrons that succeed in reaching the screen is small and they have negligible effect on the anode characteristics of the valve.

Since adequate current density is required in the electron beams if secondary emission is to be prevented, the beam tetrode is only suitable for use as a power amplifier. A typical family of anode characteristics is shown in Fig. 8.32. It can be seen that for large

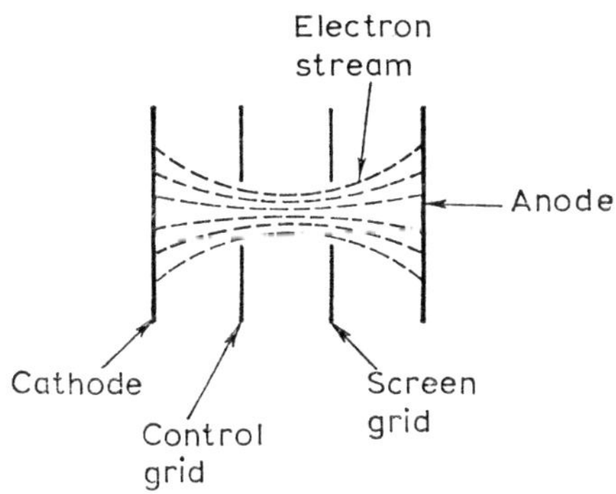

Fig. 8.31

The passage of an electron beam in a tetrode

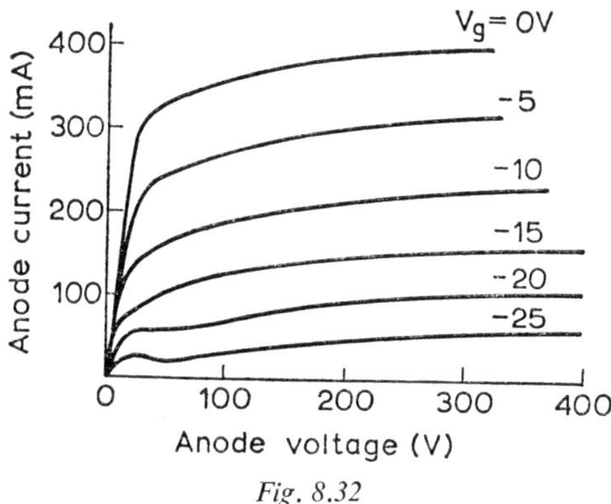

Fig. 8.32

Beam tetrode anode characteristics

negative grid voltages where the anode current, and hence the current density, in the electron beams is small, some degree of negative slope is evident at low anode voltages.

Applications of Triodes, Pentodes and Beam Tetrodes

Because of their lower amplication factor and their large anode/grid capacitance, triodes are generally only employed either in the first stage of low-noise amplifiers or in high-power amplifiers: the former because triodes generate less noise internally than either pentodes or beam tetrodes; the latter because in high-power amplifiers efficiency is all important and the triode does not have a screen grid to dissipate power. The beam tetrode is generally employed as a small- and medium-power amplifier and the pentode as a voltage amplifier and a small-power amplifier. The change from pentode to beam tetrode as a small power amplifier generally takes place in the region of 10 watts or so of output power.

Exercises

1. Sketch sets of characteristic curves for the following—

   (a) I_a against V_g for various values of V_a for a triode valve.
   (b) I_a against V_a for various values of V_g for a pentode valve.
   (c) I_c against V_{ce} for various values of I_b for a transistor in the common-emitter connection.
   (d) I_c against I_e for a given value of V_{cb} for a transistor in the common-base connection.

   State what information may be derived from the slope of the straight portion of each of these curves. (A 1964)

2. By reference to a sketch, explain the purpose of each of the three grids in a pentode valve.

   Briefly discuss the reasons for using pentode valves rather than triode valves in the high-frequency stages of medium-wave radio receivers. (A 1964)

3. What is meant by (a) the mutual conductance, (b) the anode a.c. resistance and (c) the amplification factor of a thermionic valve?

   The following data were obtained for a triode valve—

GRID VOLTAGE V_g (VOLTS)		0	$-0{\cdot}5$	$-1{\cdot}0$	$-1{\cdot}5$	$-2{\cdot}0$	$-2{\cdot}5$
Anode	$V_a = 225$ V	25	22·5	20	17·5	15	12·5
current	$V_a = 200$ V	20	17·5	15	12·5	10	7·5
I_a (mA)	$V_a = 175$ V	15	12·5	10	7·5	5	2·5

   Plot the I_a/V_g characteristic and determine the amplification factor, mutual conductance and anode a.c. resistance of the valve. (A 1963)

4. Describe, with the aid of a sketch, the constructional features of an indirectly heated pentode valve. What factors determine the electron emission from the cathode?

   How is a variable-μ characteristic obtained in such a valve? (A 1962)

5. Explain the operation of the thermionic valve diode as a rectifier of alternating currents. Sketch, and discuss the shape of, a typical I_a/V_a characteristic for such a valve over its full working range.

   What factors in the valve limit the current that can be passed through it? (P1 1961)

6. Measurements made on a triode valve are tabulated on the next page.

   Plot the mutual characteristics for $V_a = 200$, 250 and 300 V and the anode characteristics for $V_g = 0$, -1 and -2 V. Use these curves to determine the mutual conductance, the anode slope resistance, and the amplification factor of the valve over the straight parts of its characteristics.

ANODE VOLTAGE V_a (VOLTS)	ANODE CURRENT I_a (mA)						
	$V_g = 0$	$V_g = -1$	$V_g = -2$	$V_g = -3$	$V_g = -4$	$V_g = -5$	$V_g = -6$
100	8·8	3·6	0·8	—	—	—	—
150	15·2	7·6	2·4	0·4	—	—	—
200	22·8	13·2	5·6	2·0	—	—	—
250	32·0	20·0	10·0	5·0	2·0	—	—
300	—	26·8	16·7	9·0	4·8	0·4	—

(A 1961)

7. Sketch sets of curves showing the following valve characteristics.

    (a) I_a against V_a for various values of V_g for a triode valve.
    (b) I_a against V_g for various values of V_a for a triode valve,
    (c) I_a against V_a for various values of V_g for a pentode valve,
    (d) I_a against V_g for various values of V_s for a pentode valve.

where V_a = anode voltage; I_a = anode current; V_g = grid voltage; V_s = screen voltage.

    Explain, with reference to curves (b), what is meant by the term mutual conductance. (A 1960)

8. Discuss the shortcomings of triode valves and explain why pentode valves are commonly preferred to triode in the high-frequency sections of medium-wave radio receivers. (A 1960)

9. Describe, with sketches, the construction of either a beam tetrode valve suitable for use in the output stage of a domestic radio receiver, or a junction transistor suitable for use in a low-power audio amplifier.

    Say what materials are used for the various parts and sketch a typical family of characteristic curves for the device you have described. (A 1959)

10. The variation of anode current with grid and anode voltages for a triode valve are as follows—

GRID VOLTAGE (VOLTS)		0	−1	−2	−3	−4
Anode current (mA)	250 V anode voltage	24	20	16	12	8
	200 V anode voltage	16	12	8	4	0

    Plot the anode-current/grid-voltage curves, and determine the amplification factor, mutual conductance and anode slope resistance of the valve.

(1 1958)

11. Describe the construction of an indirectly heated triode valve. What factors determine the electron emission from the cathode? (1 1958)

9 *Resistance-loaded Amplifiers*

Thermionic valves (other than diodes) and transistors are able to amplify a.c. signals because their output current can be controlled by a signal applied to their input terminals. A change in the input signal causes a change in the output current, and if a voltage, or power, output is required the output current must be passed through a resistive load in the output circuit.

The thermionic valve has a very high input impedance and is a voltage-operated device and this means that a valve amplifier can provide a voltage gain only. Voltage gain is defined as the ratio of a change in output voltage to the change in input voltage producing it. The transistor, on the other hand, has a relatively low input impedance and requires an input current for its operation; a transistor amplifier, therefore, is capable of giving a voltage, current, or power gain. Current gain is defined as the ratio of a change in output current to the change in input current producing it, and power gain is defined as

$$\text{Power gain} = \frac{(\text{change in output current})^2 \times \text{load resistance}}{(\text{change in input current})^2 \times \text{input resistance}}$$

Choice of Operating Point

The dynamic mutual characteristic of a valve with a resistive anode load shows how its anode current varies with change in the voltage applied to its grid for the particular values of anode load resistance and h.t. supply voltage employed. Similarly, the dynamic transfer characteristic of a transistor shows how the collector current of the transistor varies with change in the base, or emitter, current for the particular values of collector load resistance and collector supply voltage employed. The slope of a dynamic characteristic is always less than the slope of the static characteristic from which it is derived.

A dynamic characteristic can be employed to determine, graphically, the waveform of the output current for a particular input signal waveform. Ideally, the two waveforms should be identical, but this requires the dynamic characteristic to be absolutely linear. In practice, some non-linearity always exists and for minimum signal distortion care must be taken to restrict operation to the most linear part of the characteristic. For this a suitable operating, or quiescent, point must be selected and the amplitude of the input signal must be limited. The chosen operating point is fixed by the application of a steady grid bias voltage for a valve amplifier, or a steady input current for a transistor amplifier. For optimum performance the operating point is usually placed in the centre of the linear portion of the dynamic characteristic. Then an alternating signal centred on

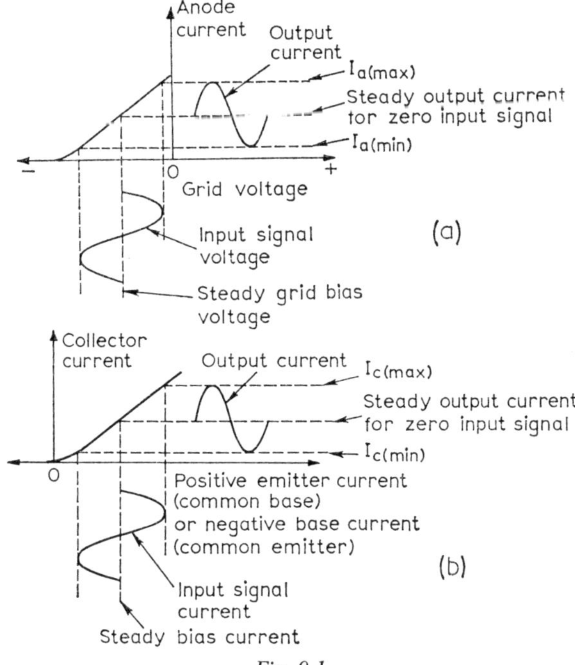

Fig. 9.1

Variations of output current with input signal for (*a*) a valve and (*b*) a transistor

this operating point produces equal swings of output current above and below the quiescent value, as shown in Figs. 9.1*a* and *b*.

The output current may conveniently be regarded as a direct current having an alternating current superimposed upon it. The direct current is equal to the current that flows when the input signal is zero, i.e. the quiescent current, and the alternating current has a peak-to-peak value of $I_{max} - I_{min}$.

Note that Fig. 9.1*b* shows the input current as having a sinusoidal waveform; however, the relationship between the input voltage and the input current of a transistor is not a linear one unless the steady bias current is much larger than the signal current. This is because the input resistance of a transistor is not a constant quantity but varies with change in input voltage (refer to Chapter 7). To minimize this distortion of the signal waveform the input signal should be fed from a source whose resistance is much larger than the input resistance of the transistor. Then any changes in input resistance, with change in input voltage, are but a small fraction of the total resistance in the input circuit and the input current has almost the same waveform as the input voltage.

VALVE BIAS

It is necessary to apply a negative bias voltage to the grid of a valve to ensure operation on the linear part of the dynamic characteristic of the valve, and also to ensure that the grid is not taken positive during positive half-cycles of the input signal. The second requirement is necessary to prevent the flow of grid current since this would produce further signal distortion.

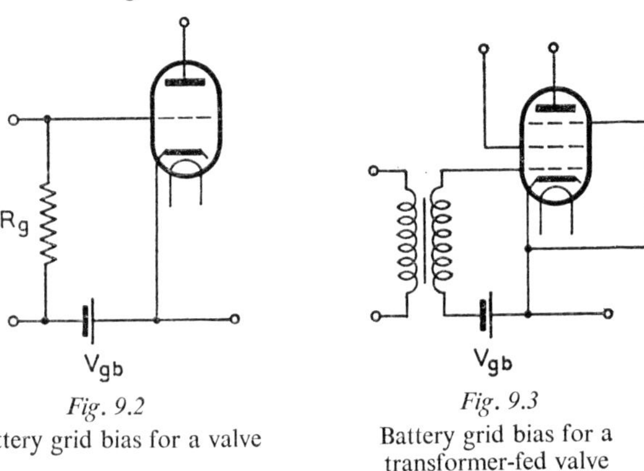

Fig. 9.2

Battery grid bias for a valve

Fig. 9.3

Battery grid bias for a
transformer-fed valve

The obvious method of providing a grid bias voltage is to connect a battery in the grid/cathode circuit as shown in Fig. 9.2. Here V_{gb} is the e.m.f. of the bias battery and R_g is a resistor that provides a d.c. path for the bias voltage to be applied to the grid. The resistance of R_g is made large, approximately 0·5 to 1 megohm, so that it does not shunt the signal path. No direct current flows in R_g and it does not affect the magnitude of the grid bias voltage. R_g is not necessary if the input signal is fed via a transformer, since the secondary winding of the transformer provides the necessary d.c. path, Fig. 9.3.

An alternative and very popular method of obtaining the grid bias voltage is the connection of a resistor in the cathode circuit, Fig. 9.4. The cathode current I_k of the valve* flows in the cathode

---

* For a triode the cathode current is equal to the anode current, for a tetrode or pentode the cathode current is equal to the sum of the anode and screen grid currents.

resistor R_k and a voltage, $I_k R_k$ volts, is developed, having the polarity shown. Assuming zero grid current, the grid of the valve is negative with respect to the cathode by the voltage developed across R_k. The value of R_k is chosen to give the desired grid bias voltage.

In the circuit of Fig. 9.4 the a.c. component of the anode current also passes through the cathode resistor R_k and causes the voltage developed across it to vary at the signal frequency. The a.c. voltage

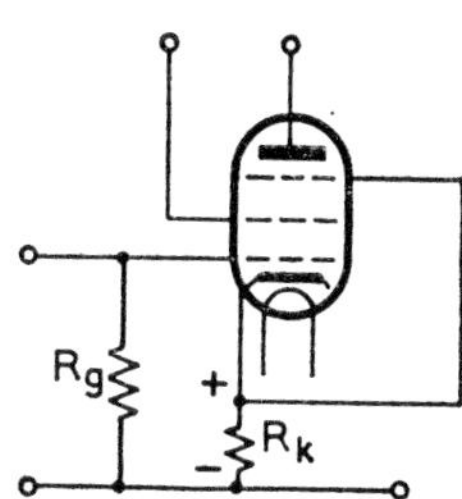

Fig. 9.4

Cathode bias

developed across R_k will subtract from the input signal and reduce the gain of the amplifier; in many circuits, but not all, this effect is undesirable and it is then necessary to prevent the a.c. component of the anode current from passing through R_k. This is achieved by connecting a capacitor C_k in parallel with R_k, and for the capacitor to adequately decouple R_k it must have a reactance at the lowest frequency of interest that is not greater than one-tenth of the value of R_k, i.e. $1/\omega C_k < R_k/10$. The capacitance value required of the cathode-decoupling capacitor C_k can be quite large and to provide it economically and in the minimum of space an electrolytic capacitor is often used.

TRANSISTOR BIAS

The input terminals of a transistor can be biased to give a required operating point by the use of a suitable battery connected as shown in Fig. 7.4 or Fig. 7.7. It is more convenient, however, to derive the input bias current from the collector supply voltage. A simple method of providing an emitter bias current for a common-base amplifier is shown in Fig. 9.5.

Resistors R_2 and R_3 form a potential divider across the collector supply voltage E_{cc}, and their junction is connected to the base of the transistor. The base of T_1 is held negative with respect to the emitter by a voltage

$$\frac{R_2}{R_2 + R_3} \times E_{cc} - \text{voltage drop across } R_1$$

and a steady current flows in the base/emitter circuit to bias the transistor. R_2 is decoupled by C_1 to prevent the a.c. component of the base current developing a voltage across R_2. R_1 is necessary to provide a d.c. path between the base and emitter electrodes and

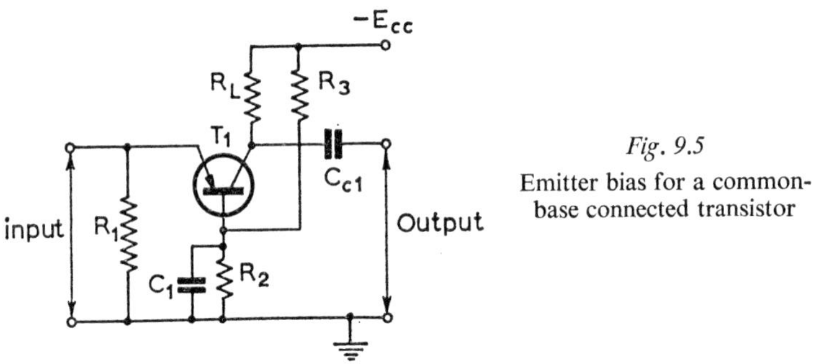

Fig. 9.5

Emitter bias for a common-base connected transistor

must have a value large enough to minimize its shunting effect on the signal path.

The simplest method of establishing the operating point of a transistor connected in the common-emitter configuration is shown in Fig. 9.6. The potential difference across a forward-biased p-n

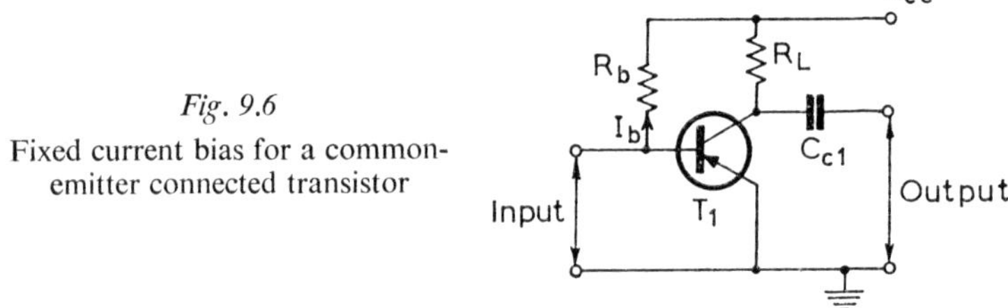

Fig. 9.6

Fixed current bias for a common-emitter connected transistor

junction is small and hence the base bias current I_b is given by

$$I_b \simeq \frac{E_{cc}}{R_b} \qquad (9.1)$$

The required bias current can be obtained by suitable choice of resistor R_b but, since a transistor bias arrangement is also required to stabilize the circuit against thermal runaway, this simple circuit is not very effective and is rarely used.

A better arrangement is shown in Fig. 9.7 and consists of connecting a bias resistor R_b between the collector and base terminals of the transistor. The base bias current I_b is given by the expression

$$I_b \simeq \frac{E_{cc} - R_L I_c}{R_b} \qquad (9.2)$$

where I_c is the steady collector current. The bias resistor R_b provides a path for the a.c. component of the collector current to feed into the base circuit, and if it is necessary to prevent this the a.c. component can be decoupled to earth as in Fig. 9.8. The bias resistor is replaced by two resistors giving the same total resistance and the decoupling capacitor C_1 is joined to their common point. It is necessary to

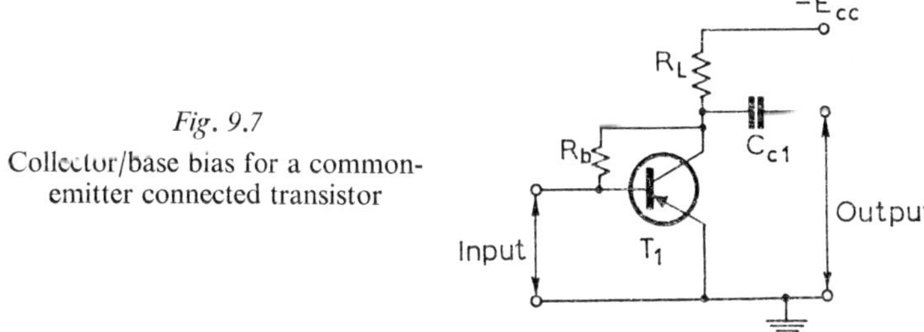

Fig. 9.7

Collector/base bias for a common-
emitter connected transistor

split the resistance in this way because if C_1 were connected to the collector end of R_b the output of the circuit would be short-circuited. The circuit provides some degree of d.c. stabilization against thermal runaway, the operation being, briefly, as follows. An increase in collector current caused by an increase in the temperature of the

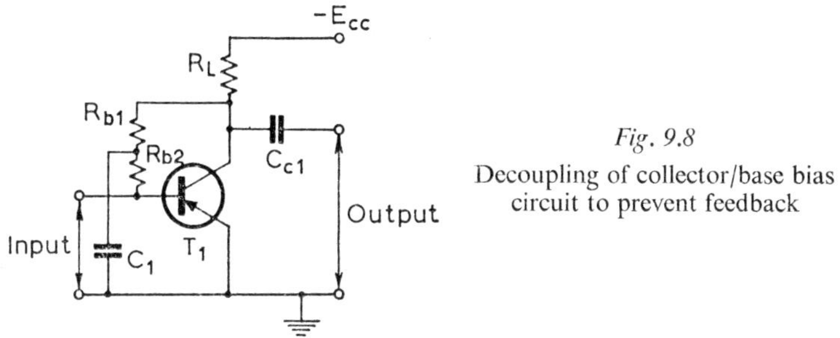

Fig. 9.8

Decoupling of collector/base bias
circuit to prevent feedback

collector/base junction causes the collector/emitter voltage to fall and, since this voltage is effectively applied across the base resistor R_b the base current falls also. The fall in base current results in a fall in the collector current which, to some extent, compensates for the original increase.

For an improvement in the d.c. stabilization the bias arrangement of Fig. 9.9 may be employed. The base of transistor T_1 is held at a negative potential V_b by the potential divider $(R_1 + R_2)$ connected across the collector supply, and the emitter is held at a negative potential V_e by the voltage developed across the emitter resistor R_3. The emitter/base bias potential is the difference between V_b and

V_e and the component values chosen are such that the junction is forward-biased by a fraction of a volt. A bias current therefore flows into the emitter. Emitter resistor R_3 is adequately decoupled

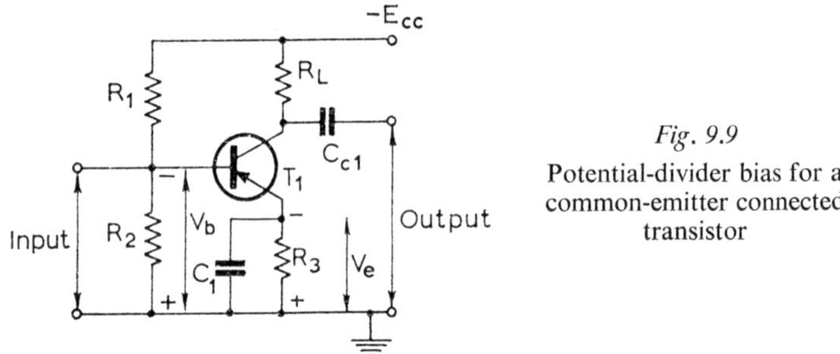

Fig. 9.9

Potential-divider bias for a common-emitter connected transistor

by capacitor C_1 to prevent an a.c. voltage at the signal frequency appearing across it and varying the emitter/base bias voltage. D.C. stabilization of the collector current is achieved in the following manner: an increase in the collector current, caused by an increase in the temperature of the collector/base junction, is accompanied by an almost equal increase in emitter current. This results in an increase in the voltage V_e developed across the emitter resistor and this, in turn, reduces the forward bias of the emitter/base junction. The base current is reduced causing a decrease in the collector current that compensates for the original increase.

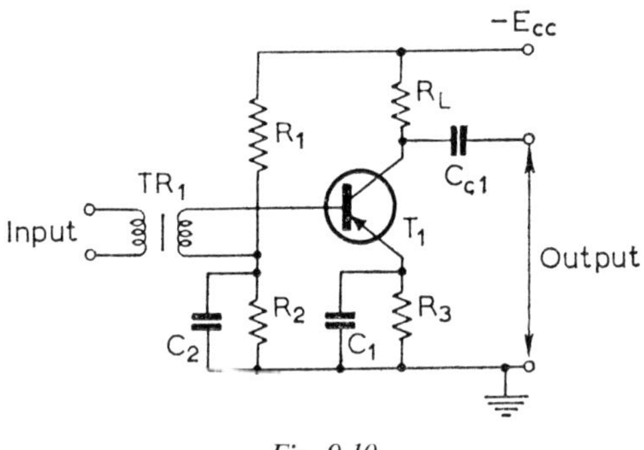

Fig. 9.10

Potential-divider bias for a transformer-fed common-emitter connected transistor

When a transformer-coupled input circuit is employed, Fig. 9.10, it is usual to decouple resistor R_2 to minimize the resistance (at signal frequencies) of the secondary circuit of the transformer.

Equivalent Circuits

A great deal of calculation on circuits incorporating thermionic valves or transistors can be carried out graphically with the aid of the static characteristics of the device, but very often the use of ordinary electrical theory, such as Ohm's law and Kirchhoff's laws, is more convenient. To apply electrical theory to a valve or transistor circuit the device must be replaced (on paper) by its equivalent circuit.

An equivalent circuit is a circuit that behaves electrically in exactly the same way as the device it is representing. Thus if the device were to be placed inside a sealed box, Fig. 9.11a, and its equivalent circuit were to be placed inside another, identical, sealed box, Fig. 9.11b, it would not be possible to distinguish between the two boxes by the application of electrical tests to their input and output terminals.

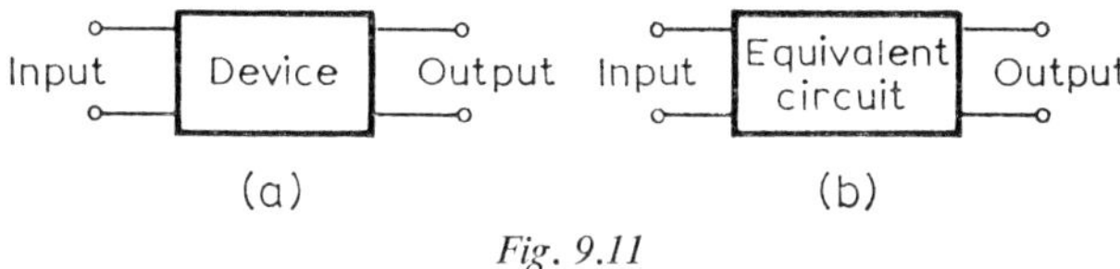

Fig. 9.11

Use of an equivalent circuit

An equivalent circuit is valid only over a limited frequency range, and provided the device is, or can reasonably be assumed to be, linear.

THERMIONIC VALVE EQUIVALENT CIRCUITS

Fig. 9.12 shows a simple triode circuit; the grid bias voltage is $-V_{gb}$ volts and may be provided by a battery as shown or by a

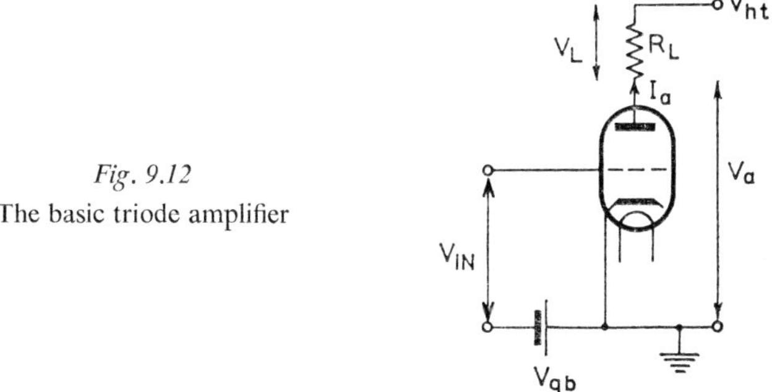

Fig. 9.12

The basic triode amplifier

cathode bias resistor. With no input signal applied to the circuit a steady anode current $I_{a(dc)}$ flows and the anode voltage V_a of the valve is

$$V_a = V_{ht} - I_{a(dc)}R_L$$

When a signal of r.m.s. voltage V_{IN} is applied to the input terminals of the circuit the grid voltage varies between the limits of $-V_{gb} - \sqrt{2}V_{IN}$ and $-V_{gb} + \sqrt{2}V_{IN}$. The anode current varies about its mean value $I_{a(dc)}$ attaining a maximum value $I_{a(max)}$ when the grid voltage is least negative, and a minimum value $I_{a(min)}$ when the grid voltage is most negative. The anode current may be considered to comprise a d.c. component, $I_{a(dc)}$, with an a.c. component

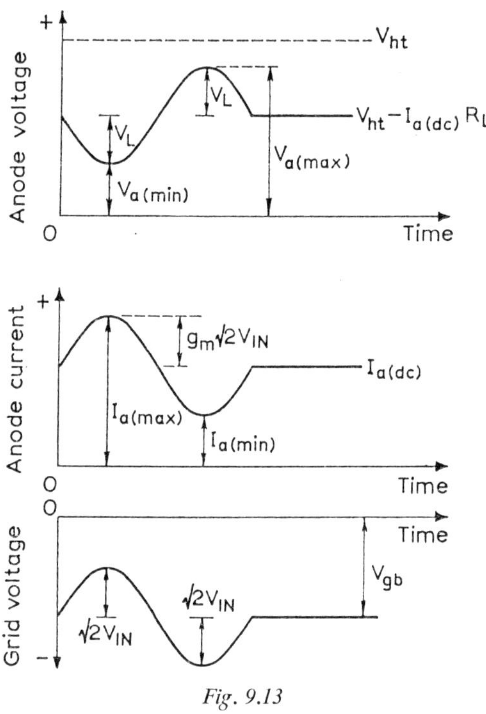

Fig. 9.13

Voltage and current waveforms in a triode amplifier

of peak value $I_a = g_m\sqrt{2}V_{IN}$ superimposed upon it. The peak value of the a.c. voltage V_L developed across anode load resistor R_L is then $V_L = g_m\sqrt{2}V_{IN}R_L$ volts. The anode voltage is equal to the h.t. supply voltage minus the voltage developed across R_L; hence when the input signal goes negative, the anode current falls and the anode voltage rises. The a.c. component of the anode voltage is therefore in antiphase with the input signal voltage. The relationships between the grid and anode voltages and the anode current are shown in Fig. 9.13.

The change in anode voltage caused by a change in anode current causes a further change, in the opposite direction to the original change, in the anode current. Now,

$$g_m = \frac{\delta I_a}{\delta V_g} \ (V_a \text{ constant})$$

$$\therefore \quad \delta I_a = g_m \, \delta V_g \ (V_a \text{ constant}) \tag{9.3}$$

and

$$r_a = \frac{\delta V_a}{\delta I_a} \ (V_g \text{ constant})$$

$$\therefore \quad \delta I_a = \frac{\delta V_a}{r_a} \ (V_g \text{ constant}) \tag{9.4}$$

The total change, ΔI_a, in anode current due to simultaneous changes in grid voltage and anode voltage is

$$\Delta I_a = g_m \delta V_g + \frac{\delta V_a}{r_a} \tag{9.5}$$

δV_g is the change in the grid voltage and is equal to $\sqrt{2}V_{IN}$ in this case. Henceforth the change in grid voltage will be written V_g, where V_g represents the r.m.s. value of the input signal voltage.

$$\therefore \quad I_a = g_m V_g + \frac{V_a}{r_a} \tag{9.5a}$$

where I_a is the r.m.s. value of the a.c. component of the anode current and V_a is the r.m.s. value of the a.c. component of the anode voltage.

V_a, however, is equal to the r.m.s. value of the a.c. voltage developed across the anode load resistor R_L but is in antiphase with it, i.e.

$$V_a = -I_a R_L \tag{9.6}$$

Substituting equation (9.6) into equation (9.5a),

$$I_a = g_m V_g - \frac{I_a R_L}{r_a}$$

$$I_a \left(1 + \frac{R_L}{r_a}\right) = g_m V_g$$

$$I_a = \frac{g_m V_g}{1 + (R_L/r_a)}$$

$$= \frac{r_a g_m V_g}{r_a + R_L}$$

$$= \frac{\mu V_g}{r_a + R_L} \tag{9.7}$$

The same current would flow in the anode load resistor R_L if the valve were to be replaced by an equivalent voltage generator of internal e.m.f. $-\mu V_g$ volts and internal resistance r_a ohms. Fig. 9.14, therefore, is the equivalent circuit of a resistance-loaded triode valve; the point A corresponds to the anode of the valve and the point B to its cathode. It is known, see Fig. 9.13, that the a.c. component of the anode voltage is in antiphase with the input signal

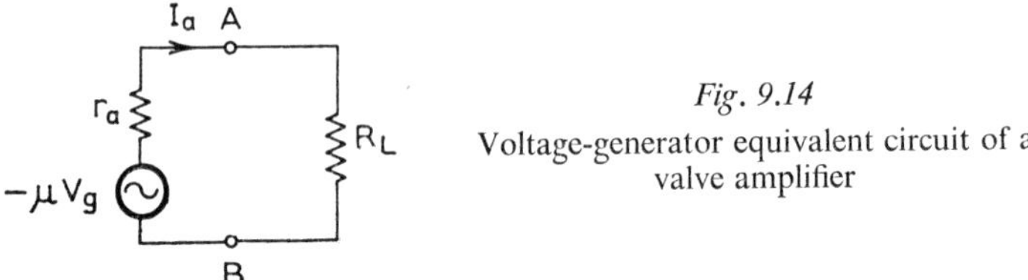

Fig. 9.14

Voltage-generator equivalent circuit of a valve amplifier

voltage and to obtain this result the equivalent voltage generator must have the polarity shown. This equivalent circuit is often known as the constant voltage equivalent circuit.

The output voltage V_{OUT} is the a.c. voltage developed across the anode load resistance R_L, i.e.

$$\therefore \qquad V_{OUT} = I_a R_L$$

$$= \frac{-\mu V_g R_L}{r_a + R_L} \tag{9.8}$$

and the voltage gain

$$A_v = \frac{V_{OUT}}{V_g} = \frac{-\mu R_L}{r_a + R_L} \tag{9.9}$$

It is usual to neglect the minus sign in equations (9.8) and (9.9) and remember that a phase shift of $180°$ occurs in a resistance-loaded amplifier.

It is evident from equation (9.9) that the voltage gain A_v of a resistance-loaded amplifier is always less than the voltage amplification factor μ of the valve employed, the reduction being in the ratio $R_L/(r_a + R_L)$. The larger the value of R_L employed the nearer will A_v approach μ, but increase in the value of R_L is limited by the h.t. supply voltage available, since an adequate anode voltage is essential.

Equation (9.8) may be written in the form

$$V_{OUT} = \frac{-g_m r_a V_g R_L}{r_a + R_L}$$

$$= -g_m V_g \times \frac{r_a R_L}{r_a + R_L} \tag{9.10}$$

Equation (9.10) can be recognized as representing the output voltage of the amplifier as the product of a current, $-g_m V_g$, and the total resistance of two resistances, r_a and R_L, connected in parallel. Fig. 9.15 is an alternative equivalent circuit for a triode valve and is known as the constant current equivalent circuit.

The equivalent circuits obtained for a triode valve give *only* the a.c. current and voltages produced in the anode circuit when an alternating signal is applied to the grid; d.c. values, such as the h.t supply voltage and the steady anode current, are not included. The equivalent circuits are therefore sometimes called the a.c. equivalent circuits. The equivalent circuits apply equally well to tetrode and

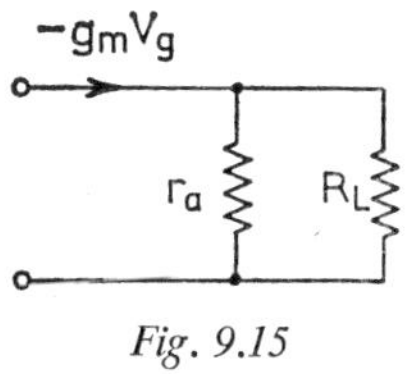
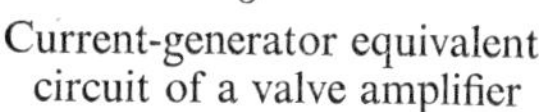
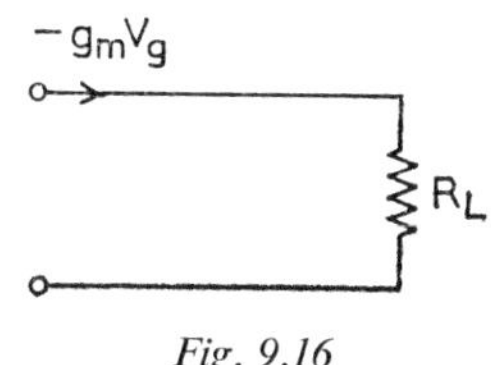

Fig. 9.15

Current-generator equivalent
circuit of a valve amplifier

Fig. 9.16

Approximate current-generator
equivalent circuit of a pentode amplifier

pentode valves also and either circuit may be employed for the solution of a particular valve problem.

The calculations are simplified, however, if the constant voltage circuit is employed for valves having a low to medium anode a.c. resistance, e.g. triodes, and the constant current circuit is employed for valves having a high anode a.c. resistance, e.g. pentodes.

The reason for using the constant current equivalent circuit for the solution of a pentode circuit is simply that r_a and R_L then appear in parallel and since r_a is very much larger than R_L it may be neglected. The simplified constant current equivalent circuit of a pentode amplifier is shown in Fig. 9.16. From Fig. 9.16 the output voltage V_{OUT} of a pentode amplifier is

$$V_{OUT} = g_m R_L V_g$$

and the voltage gain is

$$A_v = \frac{V_{OUT}}{V_g} = g_m R_L \qquad (9.11)$$

Example 9.1
A triode valve has an anode a.c. resistance of 6,000 Ω and a mutual conductance of 2 mA/V. The valve is connected in a circuit with an h.t. supply voltage of 200 V and an anode load resistance of 7,000 Ω. Calculate the voltage gain of the circuit. What would be the voltage gain if the triode were replaced by a pentode having the same mutual conductance and an anode a.c. resistance of 1 megohm?

Solution. From equation (9.9),

$$A_v = \frac{\mu R_L}{r_a + R_L}$$

$$= \frac{6 \times 10^3 \times 2 \times 10^{-3} \times 7 \times 10^3}{6 \times 10^3 + 7 \times 10^3} = \frac{84}{13} = 6\cdot46 \qquad Ans.$$

When the triode is replaced by the pentode, μ is increased to $2 \times 10^{-3} \times 1 \times 10^6$ or 2×10^3 and r_a to 1×10^6:

$$A_v = \frac{2 \times 10^3 \times 7 \times 10^3}{1 \times 10^6 + 7 \times 10^3} = 13\cdot9 \qquad Ans.$$

Or, using equation (9.11),

$$A_v = g_m R_L$$
$$= 2 \times 10^{-3} \times 7 \times 10^3 = 14 \qquad Ans.$$

Clearly little error is introduced by the use of equation (9.11).

TRANSISTOR EQUIVALENT CIRCUITS
Several possible equivalent circuits for a transistor exist and various advantages are claimed for each of them, but only two, the T equivalent circuit and the h parameter equivalent circuit, are nowadays

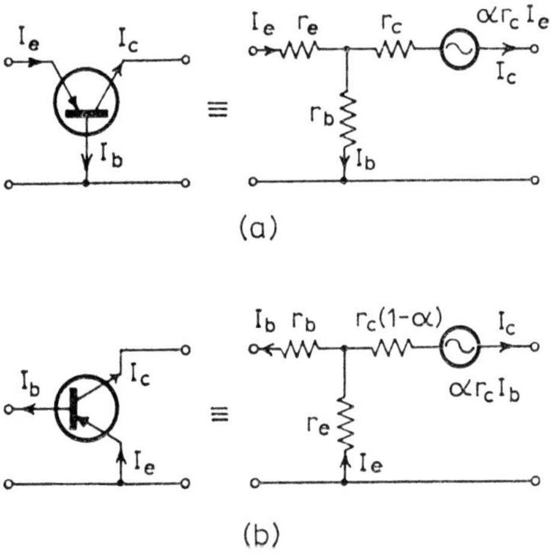

(a)

(b)

Fig. 9.17

T equivalent circuit of (a) a common-base connected transistor, and (b) a common-emitter connected transistor

used to much extent. It is beyond the scope of this book to derive any of the transistor equivalent circuits, and discussion will be limited to a brief description of the T equivalent circuit and the h parameter equivalent circuit.

The T equivalent circuits for a transistor connected in (*a*) the common-base configuration and (*b*) the common-emitter configuration are shown in Fig. 9.17*a* and *b* respectively (r.m.s. values of current are shown). The T equivalent circuits are based upon the physical effects taking place inside the transistor. In these circuits r_e is the a.c. resistance of the forward-biased emitter/base junction, r_c is the a.c. resistance of the reverse-biased collector/base junction, and r_b is the base resistance. The value of r_b depends upon the base width, the resistivity of the base material and the method of making the base connection. The values of r_e and r_c are independent of the circuit configuration employed, depending solely on the junction biasing. It is clear that the input circuit of a transistor is coupled

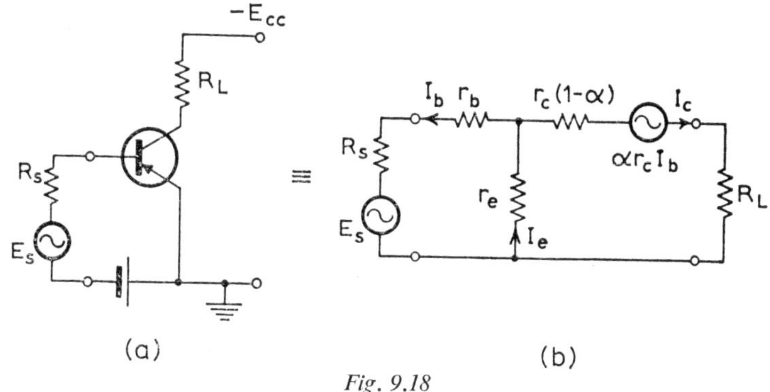

Fig. 9.18

A simple common-emitter transistor amplifier and its T equivalent circuit

to the output circuit by r_b for the common-base connection, or by r_e for the common-emitter connection, and this means that the input conditions are not independent of the output conditions and vice versa.

For a transistor fed by a source of e.m.f. E_s and internal resistance R_s, and terminated in a collector load R_L, the T equivalent circuit must be extended to include these items. This has been done in Fig. 9.18.

If the equivalent circuit given in Fig. 9.18 is analysed, using Kirchhoff's laws, expressions can be obtained for the current gain I_c/I_b the input resistance R_{IN} and the output resistance R_{OUT}. These expressions show that both the current gain and the input resistance are dependent upon the value of the collector load resistance R_L and that the output resistance R_{OUT} is dependent upon the value of the source resistance R_s (see Fig. 9.19). If the current gain, input resistance, and load resistance are known the voltage and power gains can be determined in the manner described in Chapter 7.

The T equivalent circuits shown in Fig. 9.17 give a useful representation of a transistor at audio frequencies but become increasingly inadequate as the frequency is increased. To overcome this disadvantage the circuits must be modified to take account of the changes

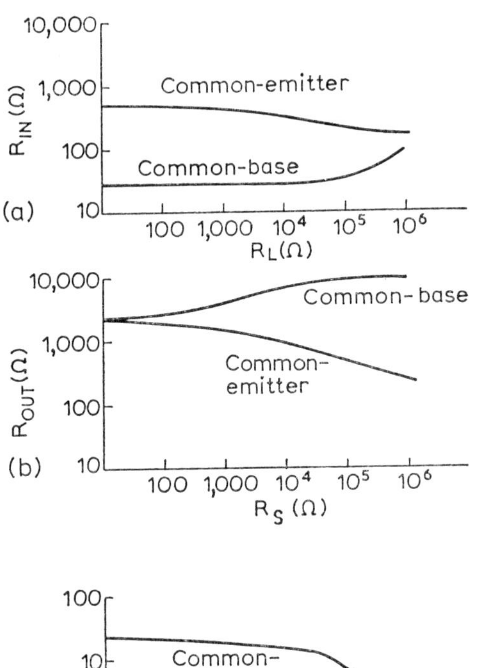

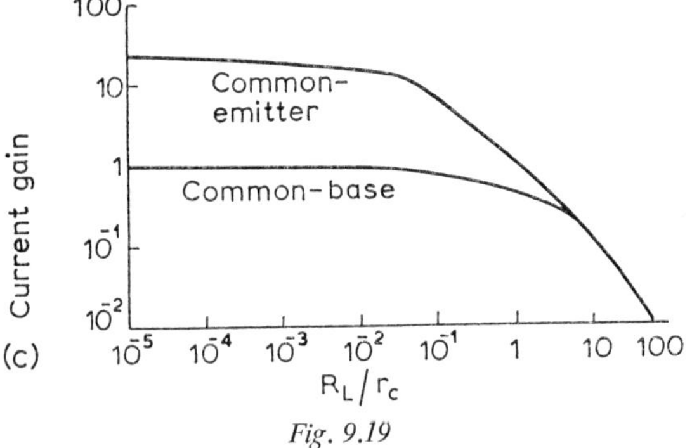

Fig. 9.19

Curves showing how (a) the input resistance of a transistor varies with collector load resistance, (b) the output resistance of a transistor varies with source resistance, and (c) the current gain of a transistor varies with collector load resistance

in transistor performance that occur at higher frequencies but these modifications will not be discussed here.

It is difficult to measure, directly, the component values of the T equivalent circuit, and an alternative approach is to treat the transistor as a four-terminal network and to determine its h parameters. The h parameters of a transistor are easily measured and are defined in Table 9.1.

Table 9.1

h PARAMETER		MEANING
COMMON-BASE	COMMON-EMITTER	
$h_{ib} = \dfrac{V_{eb}}{I_e}$	$h_{ie} = \dfrac{V_{be}}{I_b}$	Input impedance with output terminals short-circuited (ohms)
$h_{rb} = \dfrac{V_{eb}}{V_{cb}}$	$h_{re} = \dfrac{V_{be}}{V_{ce}}$	Reverse voltage ratio with input terminals open-circuited (dimensionless)
$h_{fb} = \dfrac{I_c}{I_e}$	$h_{fe} = \dfrac{I_c}{I_b}$	Current gain with output terminals short-circuited (dimensionless)
$h_{ob} = \dfrac{I_c}{V_{cb}}$	$h_{oe} = \dfrac{I_c}{V_{ce}}$	Output admittance with input terminals open-circuited (mhos)

In Table 9.1 all currents and voltages are r.m.s. values and the symbol meanings are as follows: V_{eb} = emitter/base voltage; V_{be} = base/emitter voltage; V_{cb} = collector/base voltage; V_{ce} = collector/emitter voltage; I_e = emitter current; I_c = collector current; and I_b = base current.

The h parameter equivalent circuits of (*a*) a common-base amplifier, and (*b*) a common-emitter amplifier are shown in Figs. 9.20*a* and *b* respectively.

The h parameters have values that differ for each transistor configuration and they must be measured, or specified, for the particular configuration of interest. Relationships exist between the h parameters and the components of the T equivalent circuit and these may be employed to transform from one equivalent circuit to the other. The choice between using the T equivalent circuit or the h parameters for a particular analysis is purely arbitrary and depends upon the personal preferences of the user.

EQUIVALENT CIRCUITS FOR SINGLE-STAGE AMPLIFIERS
The equivalent circuits of simple resistance-loaded amplifiers, valve and transistor, have already been given; here it is the intention to consider the effect of the bias component(s) on the a.c. signal.

Fig. 9.21*a* shows a single-stage transistor amplifier, the base bias current of which is defined by a resistor R_b connected between the collector and base terminals of the transistor. Neglecting capacitor C_{c1}, the T equivalent circuit shows clearly that the bias resistor R_b

provides a path for a fraction of the output signal to be fed back into the input circuit.

As an example of the use of the h parameter equivalent circuit

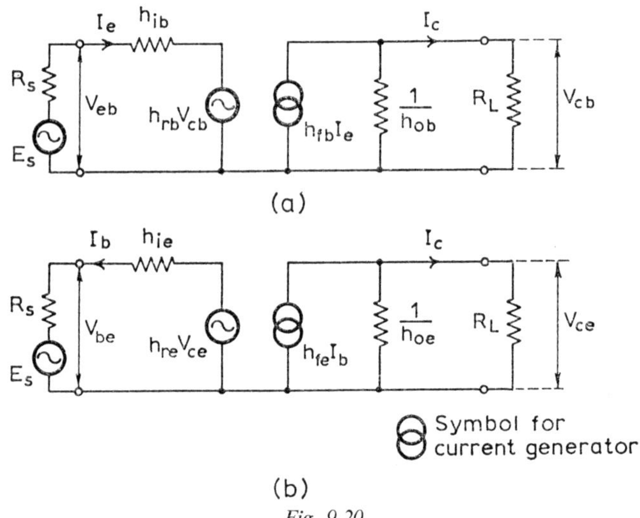

Fig. 9.20

h parameter equivalent circuit of (a) a common-base amplifier and (b) a common-emitter amplifier

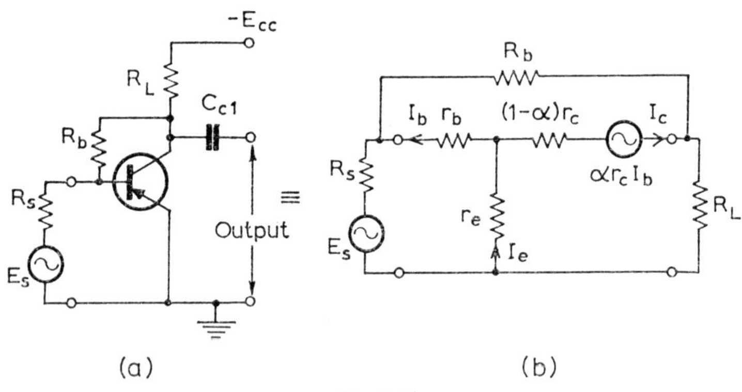

Fig. 9.21

A common-emitter amplifier and its T equivalent circuit, showing the bias component

consider Fig. 9.22. Fig. 9.22a shows a common-emitter amplifier employing the most popular bias and stabilization circuit, and Fig. 9.22b shows its h parameter equivalent circuit (assuming capacitors

C_{c1} and C_1 have negligible reactance). Clearly the potential divider resistors, R_1 and R_2, are each in shunt across the signal path.

Fig. 9.23a shows the circuit of a single-stage pentode amplifier. The screen voltage is fed via screen resistor R_s to provide the correct

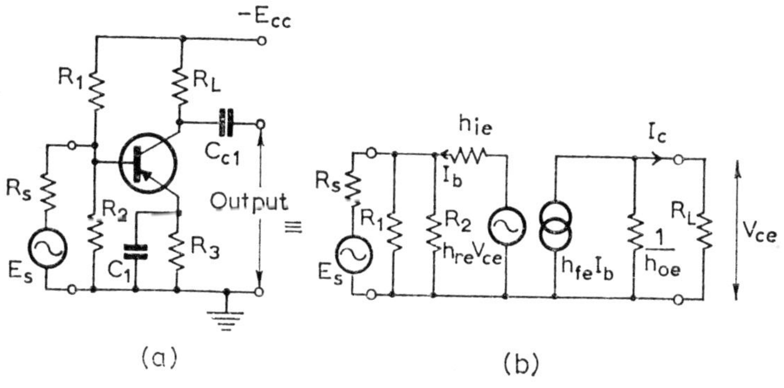

(a)   (b)

Fig. 9.22

A common-emitter amplifier and its *h* parameter equivalent circuit, showing the bias components

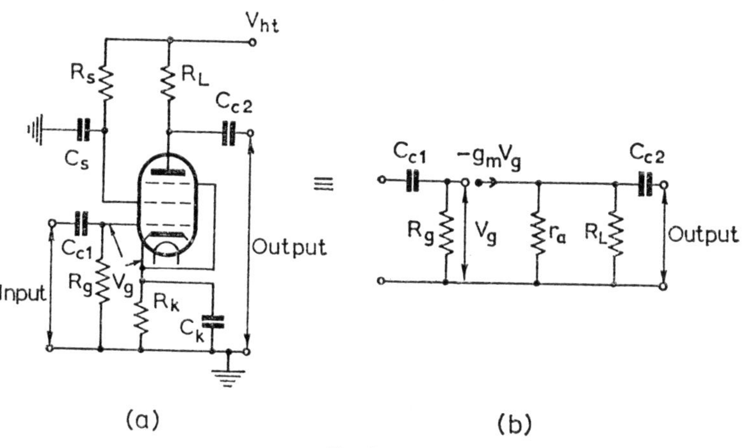

(a)   (b)

Fig. 9.23

A pentode amplifier and its equivalent circuit

value and is decoupled to earth by capacitor C_s. Resistor R_s and capacitor C_s do not appear in the equivalent circuit, Fig. 9.23b, because they play no part in the a.c. operation of the amplifier. R_g and C_{c1}, however, provide a potential divider across the input circuit and must be included.

Multi-stage Amplifiers

Very often the gain required from an amplifier is greater than the gain obtainable from a single stage and then two or more stages may be coupled in cascade to produce the required gain.

The overall gain A of a multi-stage amplifier is the product of the individual stage gains. For example, for a two-stage amplifier $A = A_1 A_2$ and for an n-stage amplifier $A = A_1 A_2 \ldots A_n$. The signal is applied to the input terminals of the first stage, amplified, and then applied to the input terminals of the next stage and so on until the required gain has been obtained. The output signal of one stage is the input signal of the next and so some form of coupling network must be provided between stages. The coupling network must couple stages with the minimum loss and must also block d.c. voltages to prevent the bias arrangements being upset.

The three types of coupling network to be discussed in this chapter are shown in Fig. 9.24.

Transistors connected in the common-base configuration have a current gain of less than unity and so the cascading of common-base stages using resistance-capacitance (R-C) coupling is not employed. Common-base amplifiers always employ transformer coupling. A transistor connected with common-emitter has a current gain of some 30–50 times and either R-C or transformer coupling may be employed; R-C coupling is most often employed since it avoids the use of transformers. Transformers have the disadvantages that they are relatively bulky and expensive and have a limited frequency response, particularly the miniature types used in transistor circuits.

RESISTANCE-CAPACITANCE (R-C) COUPLING

The circuit of a R-C coupled triode amplifier is shown in Fig. 9.25a and its equivalent circuit is shown in Fig. 9.25b. Capacitor C_{c2} and resistor R_{g2} couple the output voltage of the first stage to the input of the second stage. C_{c2} is required to have a negligible reactance over most of the frequency band to be amplified and must block the h.t. voltage from the grid circuit of V_2. The dotted shunt capacitance C_{s1} represents the anode/cathode capacitance C_{ac} of V_1 plus the stray wiring capacitances to the left of coupling capacitor C_{c2}; the shunt capacitance C_{s2} represents the input capacitance of V_2 plus the stray wiring capacitances to the right of C_{c2}. Note that the network coupling V_1 to the stage preceding it, or to the input terminals of the amplifier, has not been shown in the equivalent circuit since it is usual to regard it as a part of the preceding stage. If required, the input coupling network can be added to the equivalent circuit in the manner shown in Fig. 9.23b.

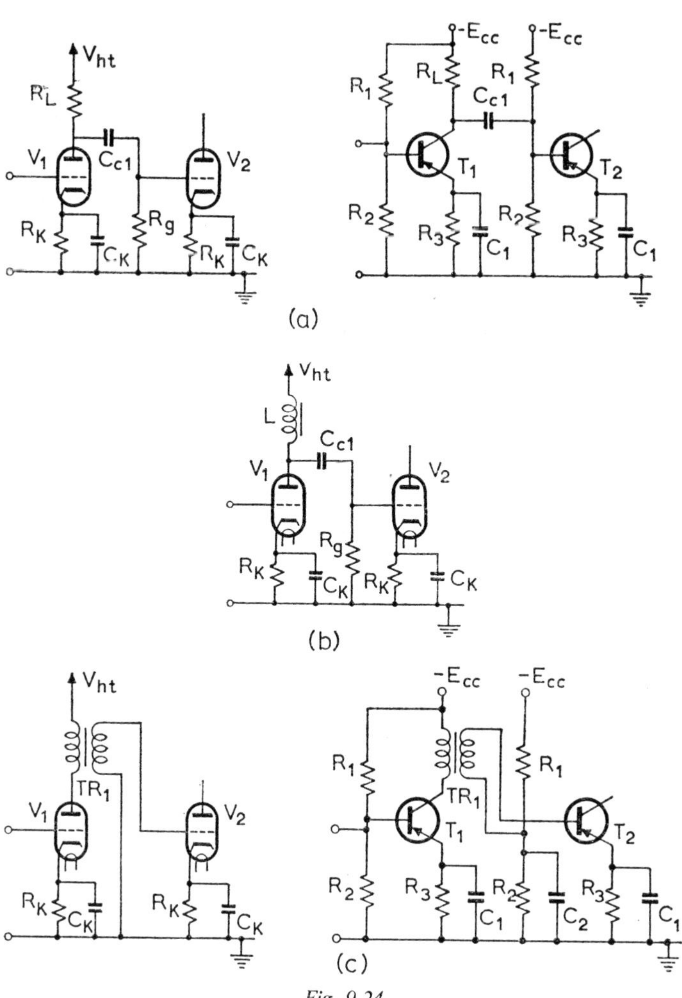

Fig. 9.24

Coupling methods for multi-stage amplifiers

(*a*) Resistance-capacitance coupling
(*b*) Choke-capacitance coupling
(*c*) Transformer coupling

Resistance-loaded Amplifiers

The coupling capacitor and the shunt capacitances have an adverse effect on the gain/frequency characteristic of the amplifier, and an analysis of the amplifier performance can be carried out using the equivalent circuit of Fig. 9.25*b*. This circuit is complex and it is customary to consider the performance of the amplifier in three distinct stages: (*a*) middle frequencies, where the reactance of C_{c2} is so small and the reactances of C_{s1} and C_{s2} are so large* that all three capacitances can be neglected, (*b*) low frequencies, where the

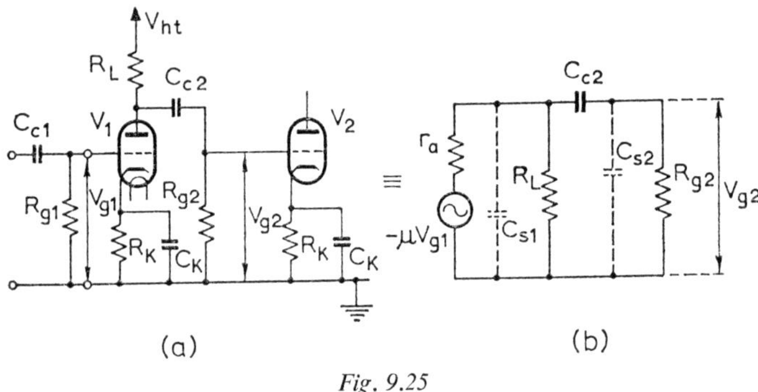

(a)   (b)

Fig. 9.25

An *R-C* coupled triode amplifier and its equivalent circuit

reactance of C_{c2} is comparable with the resistance of the grid-leak resistor R_{g2}, and so C_{c2} must be considered, and (*c*) high frequencies, where the reactance of C_{c2} is negligible but the capacitances C_{s1} and C_{s2} have a low reactance and a pronounced shunting effect on the signal path. The simplified equivalent circuits of a *R-C* coupled triode amplifier at middle, low and high frequencies are shown in Fig. 9.26*a*, *b* and *c* respectively.

At middle frequencies, Fig. 9.26*a*, grid-leak resistor R_{g2} is effectively in parallel with the anode load resistor R_L and the voltage gain A_v of the stage is

$$A_v = \frac{V_{g2}}{V_{g1}}$$

$$= \left(\frac{\mu R_L R_{g2}}{R_L + R_{g2}}\right) \bigg/ \left(r_a + \frac{R_L R_{g2}}{R_L + R_{g2}}\right)$$

$$= \frac{\mu R_L R_{g2}}{r_a R_L + r_a R_{g2} + R_L R_{g2}} \tag{9.12}$$

* This implies that the capacitance of C_{c2} is very much larger than the shunt capacitances, C_{s1} and C_{s2}. Typically, C_{c2} would be 0·1 μF and C_{s1} and C_{s2} a few picofarads.

If the gain of the stage is to be as large as possible R_{g2} should shunt R_L to a negligible extent, and typically R_{g2} is of the order of 0·5 to 1 megohm.

As the frequency is reduced the reactance of the coupling capacitor C_{c2} increases and can no longer be ignored and the capacitor forms, with R_{g2}, a potential divider across the anode load resistor, Fig. 9.26b. Decrease in frequency increases the reactance of C_{c2} and decreases the voltage V_{g2} applied to the grid of V_2 until at very low frequencies the voltage transferred to the grid of V_2 is negligible. The voltage gain of an *R-C* coupled triode amplifier thus falls with decrease in frequency at low frequencies.

At high frequencies, Fig. 9.26c, the reactance of the total shunt

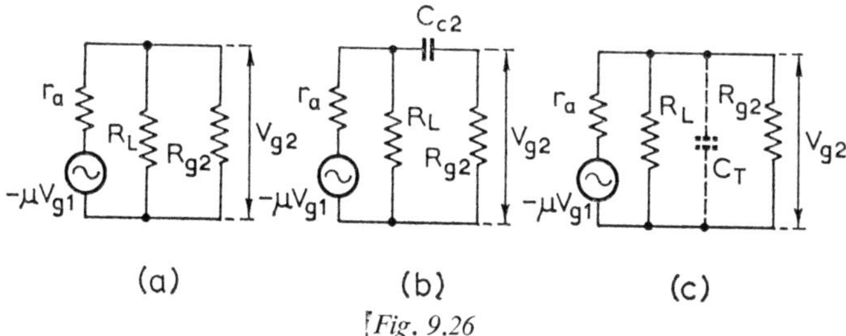

(a)   (b)   (c)

[*Fig. 9.26*

Simplified equivalent circuits of an *R-C* coupled triode amplifier

(a) Middle frequencies
(b) Low frequencies
(c) High frequencies

capacitance, $C_T = C_{s1} + C_{s2}$, is low enough for the capacitance to have a pronounced shunting effect on the anode load resistor and the effective anode load impedance Z_L for V_1 is reduced. Increase in frequency increases the shunting effect of C_T and reduces the value of Z_L and, since the voltage gain is proportional to Z_L, this causes the gain to fall. This means that the voltage gain of an *R-C* coupled triode amplifier falls with increase in frequency at high frequencies.

The equivalent circuit of a pentode *R-C* coupled amplifier is also best considered in three separate stages. Fig. 9.27a shows an *R-C* coupled pentode amplifier and Figs. 9.27b, c and d show its equivalent circuits. The input coupling network, C_{c1} and R_{a1}, has been taken as part of the preceding stage and omitted from the equivalent circuits.

From Fig. 9.27b, the voltage gain, A_v, at middle frequencies is

$$A_v = \frac{V_{g2}}{V_{g1}} = \frac{g_m}{\dfrac{1}{r_a} + \dfrac{1}{R_L} + \dfrac{1}{R_{g2}}} \tag{9.13}$$

Usually, both R_{g2} and r_a are much greater than R_L, and when this is the case

$$A_v \simeq g_m R_L \tag{9.14}$$

Coupling capacitor C_{c2} and the total shunt capacitance C_T affect the gain at low and high frequencies in the same way as for a triode amplifier, but since the inter-electrode capacitances of a pentode are smaller than those of a triode, the reduction in gain due to C_T

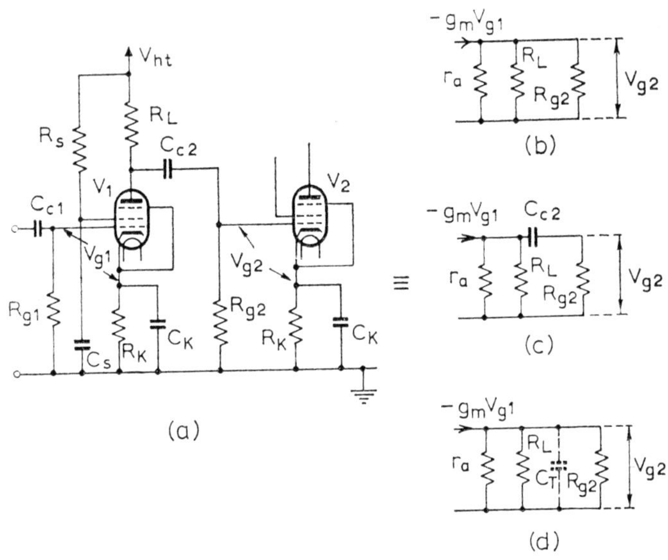

Fig. 9.27

An R-C coupled pentode amplifier and its simplified equivalent circuits

(*b*) At middle frequencies
(*c*) At low frequencies
(*d*) At high frequencies

does not occur until higher frequencies are reached. Thermionic valve *R-C* coupled amplifiers almost always employ pentodes because of their greater voltage gain. Fig. 9.28 shows a typical gain/frequency characteristic for a pentode *R-C* coupled amplifier. The frequency axis has been plotted to a logarithmic scale so that equal distances on the axis represent equal frequency ratios. This practice gives a better idea of the usefulness of the amplifier for the amplification of sound waveforms since the human ear is sensitive to frequency ratios rather than absolute frequency changes.

If the cathode and screen decoupling is not perfect as, of course, is the case in practice, the cathode and screen circuits will also affect the low frequency end of the gain/frequency characteristic, making the gain fall off even more rapidly.

The bandwidth of an amplifier is usually quoted as the frequency band between the 3 dB points; these are the two frequencies, one high and one low, at which the voltage gain has fallen by 3 dB (or to 0·707 times) from the gain at middle frequencies.

An *R-C* coupled common-emitter transistor amplifier is shown in Fig. 9.29*a*, and its *h* parameter equivalent circuit is shown in Fig. 9.29*b*. Resistors R_1, R_2, R_3, R_4, R_5 and R_6, comprise the bias and d.c. stabilization circuit components and electrolytic capacitors C_1 and

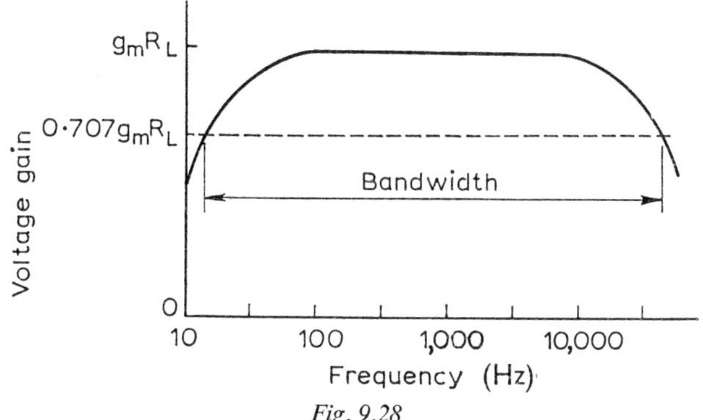

Fig. 9.28

Gain/frequency characteristic of an *R-C* coupled pentode amplifier

C_2 decouple R_3 and R_6 respectively. For alternating currents the collector supply is effectively short-circuited and so R_4 and R_5 appear in the equivalent circuit in parallel with one another and shunted across the signal path. Similarly, bias resistors R_1 and R_2 also shunt the signal path but, together with capacitor C_{c1}, are regarded as part of the previous stage and are not shown. R_{IN} is the input resistance of transistor T_2, and electrolytic capacitor C_{c2} is employed to couple the collector of T_1 to the base of T_2. Capacitor C_{s1} represents the output capacitance of T_1 plus stray capacitances to the left of C_{c2}, and C_{s2} represents the input capacitance of T_2 plus stray capacitances to the right of C_{c2}.

Analysis of the equivalent circuit shown in Fig. 9.29*b* is best carried out by considering it in three separate frequency ranges, as for the valve circuits. The simplified circuits for middle, low and high frequencies are shown in Figs. 9.30*a*, *b* and *c* respectively.

In these diagrams R_{IN}' represents the total resistance of R_{IN}, R_4 and R_5 in parallel. The current flowing into the base of T_2 will be equal to V_{IN}/R_{IN}, however, not V_{IN}/R_{IN}'.

At low frequencies the gain, voltage or current, is reduced from its value at middle frequencies because C_{c2} and R_{IN}' form a potential divider across the collector load resistor, R_L, and also because the decoupling of the emitter resistor is increasingly less efficient as the reactance of the decoupling capacitor increases. At high frequencies the gain is also reduced; there are two reasons for this, (a) the total shunt capacitance, $C_T = C_{s1} + C_{s2}$, has a reduced reactance and

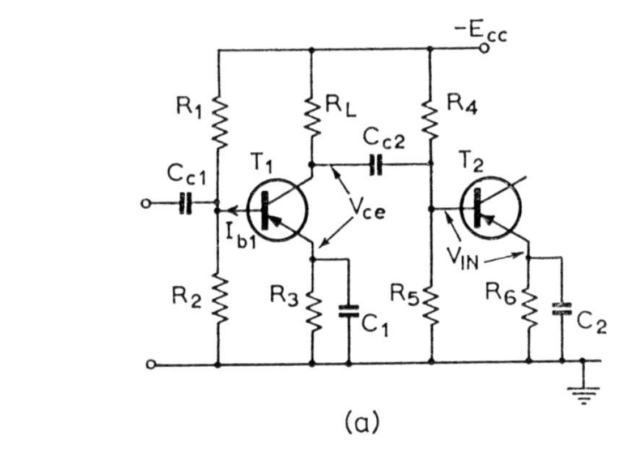

(a)

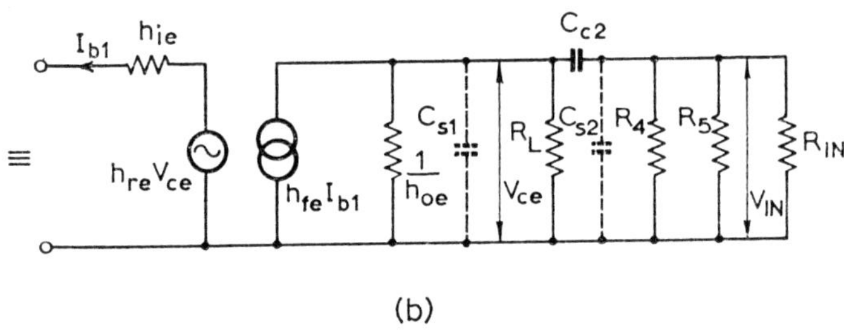

(b)

Fig. 9.29

An *R-C* coupled common-emitter transistor amplifier and its *h* parameter
equivalent circuit

shunts the effective collector load impedance, and (b) the actual current gain h_{fe} of the transistor falls with increase in frequency. For audio-frequency amplifiers, however, the loss of gain due to the transistor itself can be eliminated by the use of a suitable type of transistor in which the decrease in current gain does not occur until frequencies well beyond the audio range are reached. The gain/frequency characteristic of a transistor *R-C* coupled amplifier is similar to the characteristic of a pentode amplifier, Fig. 9.28.

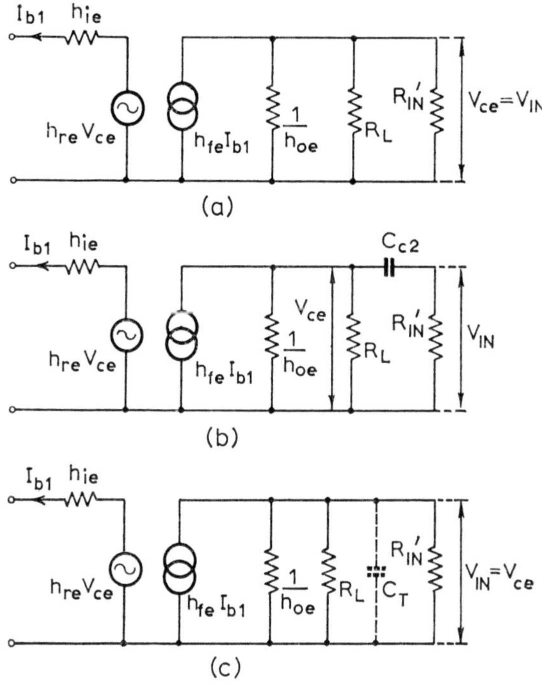

Fig. 9.30

Simplified equivalent circuits of an *R-C* coupled common-emitter amplifier at (*a*) middle, (*b*) low and (*c*) high frequencies

CHOKE-CAPACITANCE COUPLING

The circuit arrangement of a choke-capacitance coupled triode amplifier is shown in Fig. 9.31*a* and its equivalent circuit is shown in Fig. 9.31*b*.*

The input coupling components, C_{c1} and R_{g1}, have again been taken as part of the previous stage and are omitted from the equivalent circuit. The amplifier is similar to an *R-C* coupled amplifier, the difference lying in the use of an iron-cored inductor as the anode load instead of a resistor. The advantage of the choke-capacitance coupling arrangement is that the anode load has a low d.c. resistance; this results in only a small d.c. voltage drop across the anode load and permits the use of a lower h.t. supply voltage. The voltage gain

* The iron-cored inductor, or choke, in the anode circuit has merely been replaced in the equivalent circuit by an inductive reactance X_L. Note, however, that an iron-cored inductor itself has an equivalent circuit that represents its losses, etc. and in an exact analysis these would have to be included.

7—(T.1159)

of the circuit depends upon the reactance, $X_L = 2\pi fL$, of the anode inductor and is frequency dependent. The decrease in X_L with decrease in frequency is the main factor causing the gain to fall off at low frequencies since it occurs before the effect of coupling capacitor C_{c2} becomes noticeable. At high frequencies X_L increases with increase in frequency, but the gain decreases because of the increased effect of the shunt capacitances, C_{s1} and C_{s2}. The gain/frequency characteristic of a choke-capacitance coupled amplifier is not as flat as the characteristic of an *R-C* coupled amplifier, falling

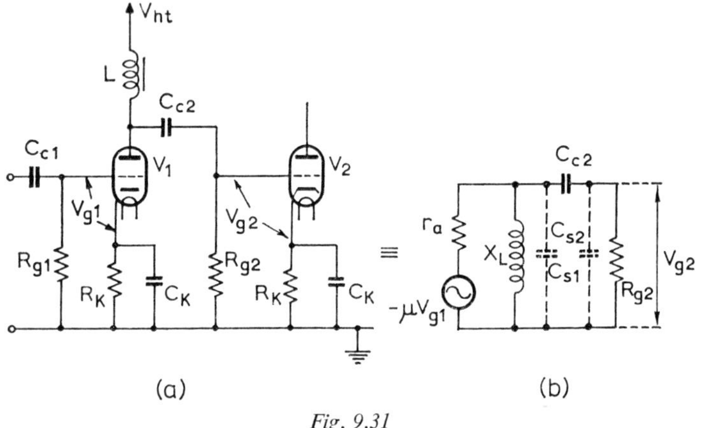

(a)   (b)

Fig. 9.31

A choke-capacitance coupled triode amplifier and its equivalent circuit

off more rapidly at low frequencies, and this is one of the disadvantages of choke-capacitance coupling. The other disadvantage is the physical size of the anode inductor.

TRANSFORMER COUPLING

The disadvantages of transformer coupling, which have already been mentioned, make its use in thermionic valve amplifiers rare for interstage coupling but, for reasons to be mentioned shortly, transformer-coupled transistor amplifier stages are widely employed. Fig. 9.32a shows a transformer-coupled triode amplifier and Fig. 9.32b shows its equivalent circuit.* The capacitance C_{s1} represents the output capacitance of V_1, the self-capacitance of the primary winding of TR_1 and any stray capacitances in the primary circuit; C_{s2} represents the input capacitance of V_2, the self-capacitance of the secondary winding of TR_1 and any stray capacitances in the

* For the full equivalent circuit the transformer should be replaced by its equivalent circuit.

secondary circuit. X_p is the reactance of the primary winding and is normally several henrys in value.

At middle frequencies neither X_p nor the shunt capacitances C_{s1}, C_{s2} have any effect on the performance of the amplifier and may be neglected. The secondary circuit of TR_1 is effectively open-circuited and so $V_{g2} \simeq V_{g1}n\mu$. The voltage gain A_v of a transformer-coupled triode amplifier at middle frequencies is

$$A_v = \frac{V_{g2}}{V_{g1}} = n\mu \qquad (9.15)$$

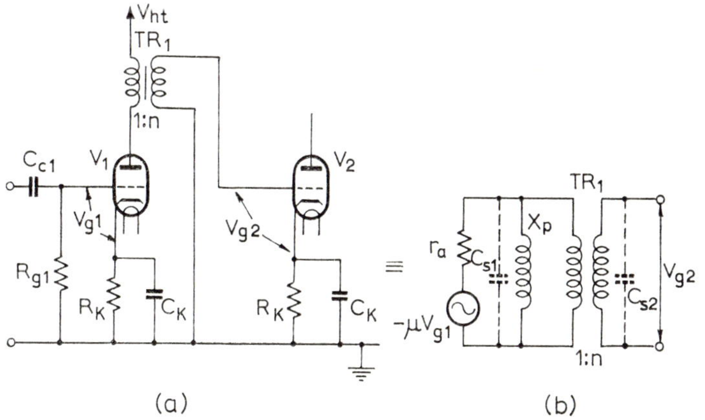

Fig. 9.32

A transformer-coupled triode amplifier and its equivalent circuit

Thus an advantage of transformer coupling is that it permits a stage gain greater than the amplification factor of the valve to be obtained. At low frequencies the inductive reactance X_p of the primary winding falls and has a shunting effect that increases with decrease in frequency. The voltage gain of a transformer-coupled triode amplifier therefore falls with decrease in frequency, the rate at which the gain falls being dependent upon the ratio r_a/L_p, where L_p is the primary inductance. It is to minimize this ratio that triodes and not pentodes are generally employed with transformer coupling. As the frequency is increased above the middle frequency range an initial *increase* in gain is obtained because of an interaction between the total shunt capacitance and the inductance of the transformer. Beyond this peak in the gain/frequency characteristic, Fig. 9.33, the gain decreases with increase in frequency at a much quicker rate than with *R-C* coupling. The peak in the characteristic is undesirable and careful design of the transformer is required to extend the region of flat gain and reduce the amplitude of the peak.

An alternative method of using transformer coupling between triode stages is shown in Fig. 9.34a and is known as parallel or shunt feed. The advantage of parallel-feed transformer coupling is that the d.c. component of the anode current does not flow in the primary

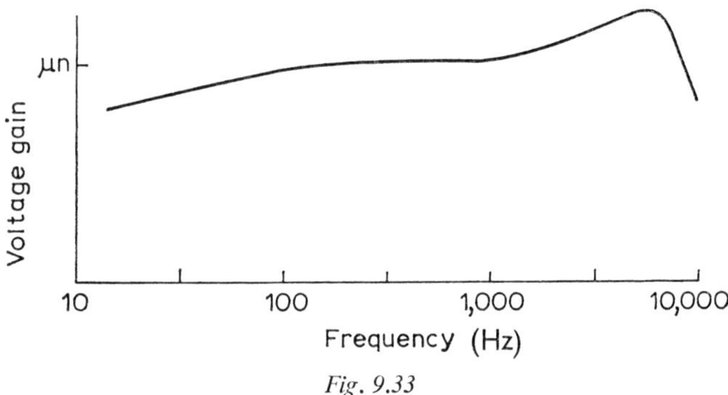

Fig. 9.33

Gain/frequency characteristic of a transformer-coupled triode amplifier

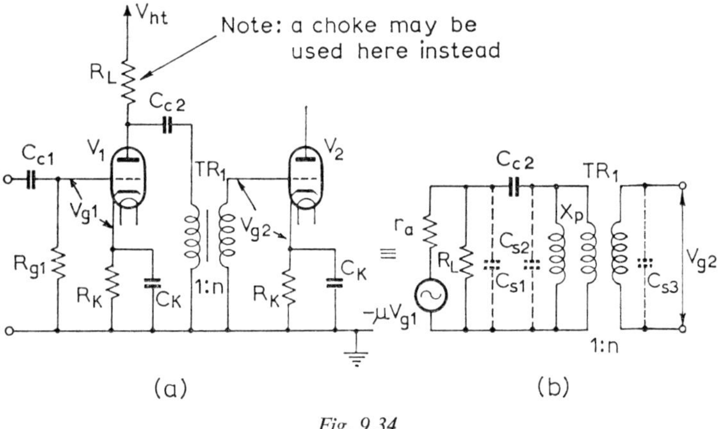

Fig. 9.34

A parallel-fed transformer-coupled triode amplifier and its equivalent circuit

winding of the transformer and so there is no risk of saturating the transformer core.

Fig. 9.34b shows the equivalent circuit of the amplifier and this is best analysed in three stages as before. At middle frequencies the reactance of the coupling capacitor, C_{c2}, is small and the reactances of the shunt capacitances, C_{s1}, C_{s2} and C_{s3} are large and so these capacitances may be neglected. The equivalent circuit then differs from the equivalent circuit of a series-fed transformer-coupled stage only in that the resistance R_L appears across the primary winding of TR_1. Normally R_L is several times the anode a.c. resistance of the triode and so it has little shunting effect on the signal path. At low frequencies the gain falls because of the increased reactance of the coupling capacitor C_{c2} and the reduced inductive reactance X_p of the primary winding of the transformer, but suitable choice of C_{c2} can minimize the rate of gain reduction. At high frequencies the gain varies with change in frequency in a similar way to the gain of a series-fed stage.

Transformer coupling of triode stages is more expensive, requires more space and has an inferior gain/frequency characteristic when compared to *R-C* coupling. Further, the gain per stage is less since pentodes cannot be used. For these reasons transformers are rarely employed for inter-stage coupling and their use is generally restricted to coupling the first and last stages of an amplifier to the input and output terminals, respectively, of the amplifier.

The input impedance of a transistor is, unlike that of a triode or pentode, fairly low and so an input current is taken. A transistor can, therefore, give a power gain. The overall power gain of a multi-stage transistor amplifier is the product of the power gains of the individual stages and has its maximum value when the stages are matched, i.e. the input resistance of one stage is equal to the output resistance of the preceding stage. The output resistance of one transistor is rarely equal to the input resistance of the next and matching can only be obtained if transformer inter-stage coupling is used. For matching, the transformer turns ratio required is given by

$$n = \sqrt{\frac{R_{OUT}}{R_{IN}}}$$

where R_{OUT} is the output resistance of one stage and R_{IN} is the input resistance of the next. In practice, correct matching is rarely achieved since a small amount of mismatch gives only a small loss of power gain but greatly reduces the distortion produced in the amplifier.

Normally the common-emitter configuration is chosen since it provides the greater gain, the common-base configuration being used only when considerations such as greater bandwidth, less distortion, and greater gain stability are of paramount importance. Fig. 9.35*a* shows the circuit of a transformer-coupled common-emitter transistor amplifier and Fig. 9.35*b* shows its T equivalent circuit. In the equivalent circuit C_{s1} represents the output capacitance of transistor T_1, the self-capacitance of the primary winding of TR_1 and any stray capacitances in the primary circuit; C_{s2} represents the input capacitance of transistor T_2, the self-capacitance of the secondary winding of TR_1 and any stray capacitances in the secondary circuit; X_p represents the inductive reactance of the primary winding and R_{IN}' is the total resistance of the input resistance of T_2 and resistors R_4 and R_5 all in parallel.

At middle frequencies the shunting effects of the primary reactance and the shunt capacitances can be neglected and the effective collector load for T_1 is equal to $n^2 R_{IN}'$. With such a collector load T_1 has particular values of input resistance and current gain and, therefore, a particular power gain.

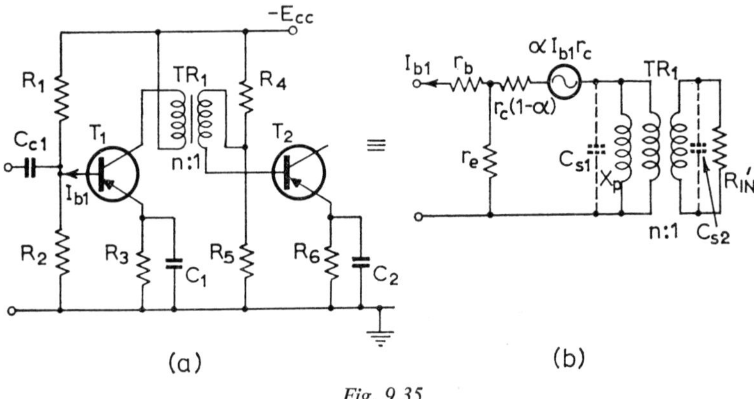

(a)   (b)

Fig. 9.35

A transformer-coupled common-emitter transistor amplifier and its T equivalent circuit

Remember that power gain

$$= \frac{(\text{current gain})^2 \times \text{collector load resistance}}{\text{input resistance}}$$

At low frequencies the primary reactance is reduced and shunts the effective collector load of T_1, reducing its value and reducing the power gain of the circuit. Similarly, the gain falls at high frequencies because of the shunting effect of the shunt capacitances.

DIRECT COUPLING

The high reactance of the coupling capacitor at very low frequencies prevents an *R-C* coupled amplifier from handling very low frequency and direct signals; such signals are also incapable of passing through a transformer-coupled stage. The low-frequency range of an amplifier can only be extended to zero frequency if the use of reactive coupling components is eliminated. Amplifiers in which stages are directly coupled are known as direct-coupled or d.c. amplifiers.

The anode of one valve in an amplifier cannot be directly coupled to the grid of the next valve because the anode voltage of the first valve would appear on the grid of the next and an excessive anode current would flow. A common method of overcoming this disadvantage is shown in Fig. 9.36 and involves the use of a negative

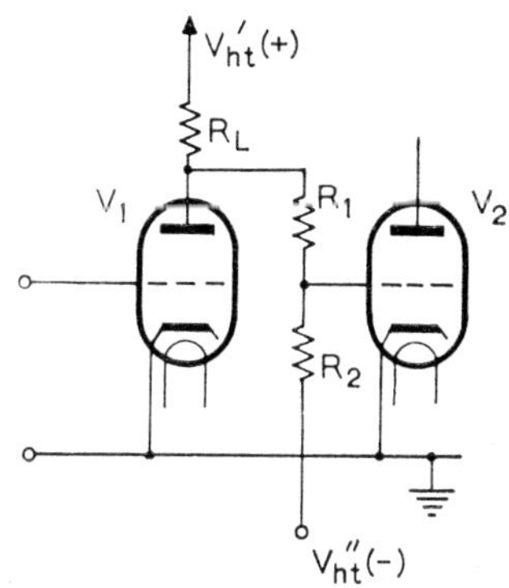

Fig. 9.36

A d.c. coupled triode amplifier

h.t. supply voltage. Usually the value of the potential divider, $R_1 + R_2$, is much greater than the value of the anode load resistor R_L and has negligible shunting effect on it. The steady anode voltage V_a of V_1 is equal to V_{ht}' minus the d.c. voltage dropped across R_L

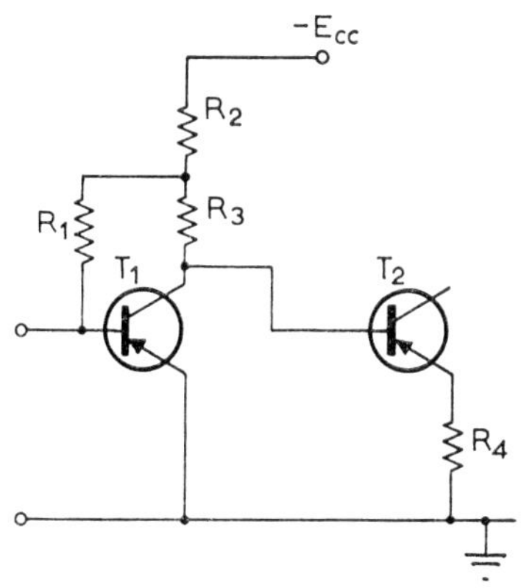

Fig. 9.37

A d.c. coupled transistor amplifier

by the steady anode current. The voltage across the potential divider, $R_1 + R_2$, is then equal to $V_a + V_{ht}''$ and by suitable choice of the values of R_1 and R_2 the grid bias voltage of V_2 can be given any desired value. This arrangement has the disadvantage that the signal voltage applied to the grid of V_2 is only a fraction, $R_2/(R_1 + R_2)$, of the voltage developed across R_L, i.e. the gain is reduced.

Transistor stages can be directly coupled without requiring an extra supply line, a possible arrangement being shown in Fig. 9.37.

The major difficulty with d.c. amplifiers is the reduction of drift. Any small fluctuations, or drifts, in the first stage in a *R-C* or transformer coupled amplifier are not passed on to the second stage by the coupling network. In a d.c. amplifier, however, such drifts act as spurious signals and are amplified by the following stages. A transistor amplifier, for example, is unable to distinguish between a signal of, say, 2 mV and a 2 mV change in the emitter/base bias voltage.

The causes of drift in a d.c. amplifier are changes in component values and valve or transistor characteristics with voltage and/or with time, and changes in the power supplies, and many circuits have been devised to reduce drift to a minimum.

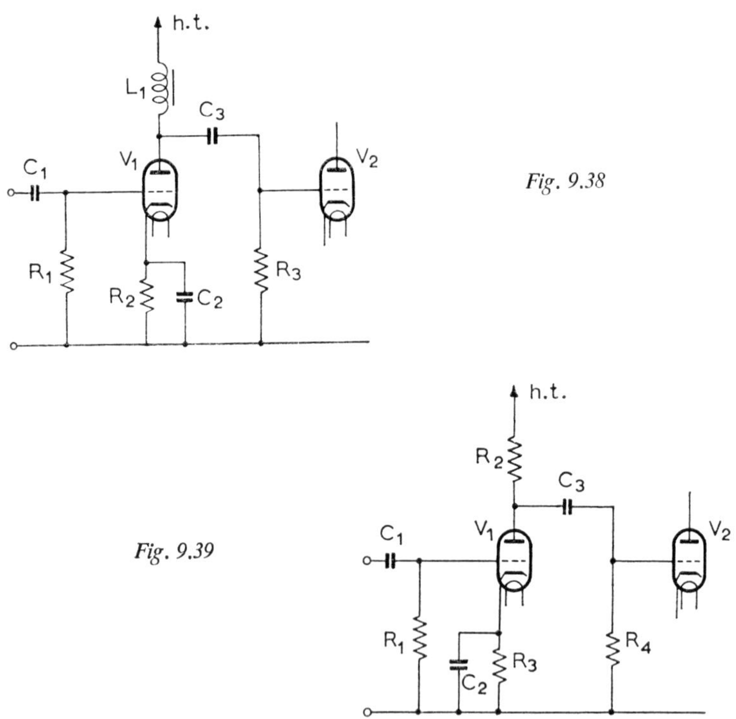

Fig. 9.38

Fig. 9.39

Exercises

1. Draw the equivalent circuit diagrams for the choke-capacitor coupled audio-frequency amplifier shown in Fig. 9.38 at low, middle and high frequencies.

By reference to gain/frequency response curves, briefly discuss the relative advantages and disadvantages of (*a*) resistor-capacitor, (*b*) transformer, and (*c*) choke-capacitance coupled audio-frequency thermionic valve amplifiers.
(A 1965)

2. Draw the equivalent circuit diagram for the resistance-capacitance coupled audio-frequency amplifier shown in Fig. 9.39 at low, middle and high frequencies.

   Use these diagrams to explain the gain/frequency characteristics of such an amplifier. Show that at middle frequencies the gain is dependent upon the parallel combination of R_2 and R_4. (A 1963)

3. Outline the functions of the electrodes in a triode valve when used as an amplifier and explain the process of amplification. Define the terms *mutual conductance* and *anode a.c. resistance*.

   Derive an expression for the amplification of a single-stage triode valve amplifier with resistance as anode load. (P1 1964)

4. Draw the circuit diagram of a two-stage *R-C* coupled audio-frequency amplifier using transistors. Show typical values for the components and suitable bias and stabilization arrangements.

   What factors influence the frequency response of the amplifier you have described? (A 1962)

5. Explain the following terms with reference to sketches of typical I_a/V_a and I_a/V_g curves for a thermionic triode valve—

   (*a*) the anode a.c. resistance,
   (*b*) the mutual conductance,
   (*c*) the amplification factor.

   Draw the circuit of a single valve amplifier with a resistance as anode load. Derive an expression for the voltage amplification of this stage, and hence show why this is less than the amplification factor of the valve. (P1 1962)

6. Draw diagrams of (*a*) resistor-capacitor and (*b*) transformer interstage coupling arrangements.

   Explain why neither method would be suitable for use in an amplifier required to transmit extremely low frequencies. Draw a diagram of a coupling arrangement that would be suitable for this purpose. (A 1961)

7. With the aid of sketches describe any one type of transistor, and explain briefly why it can act as an amplifier.

   Give a circuit of a single-stage audio-amplifier using a transistor.

   What effect has temperature change on the operation of such an amplifier?
(P1 1960)

8. Describe briefly, with reference to circuit diagrams, two methods of inter-stage coupling used in audio-frequency amplifiers. Sketch typical gain/frequency characteristics for each type, and explain why a logarithmic frequency scale is normally used for this purpose. (A 1959)

9. An audio-frequency amplifier uses a triode valve with an anode a.c. resistance of 10,000 Ω and an amplification factor of 50. What value of anode load resistor will be required to give a voltage amplification of 40 times?

   If the direct current taken by the valve is 5 mA, what should be the power rating of the anode load resistor? (I 1958)

10. Give the circuit diagram of a resistance-capacitance coupled audio-frequency amplifier and explain its action.

    What value of cathode bias resistor would be used in such an amplifier if the mean value of cathode current taken by the valve is 3 mA and the grid bias voltage is $-1 \cdot 5$ V? (I 1957)

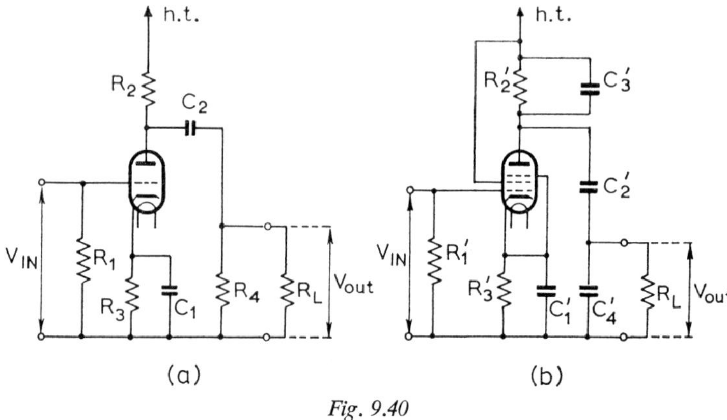

Fig. 9.40

11. For each of the two circuit diagrams given, draw an equivalent circuit diagram. In Fig. 9.40*a*, which single component could be modified in value to reduce the gain at low frequencies relative to the gain at higher frequencies?

(A 1960)

10 *Tuned Circuits*

A tuned circuit consists of an inductor and a capacitor connected either in series or in parallel. Some resistance is always present as well because of the unavoidable resistance of the inductor winding, but the resistance of the type of capacitor employed in a tuned circuit is so small that it may be neglected.

A tuned circuit has characteristics that are of the utmost use in electronic and radio circuitry; for example, at a particular frequency, known as the resonant frequency, a tuned circuit appears as a pure resistance that is either very low or very high in value. Often, one of the components of a tuned circuit is variable and this permits the circuit to be tuned to a particular resonant frequency. The tuning is possible because of the ways in which the reactances of an inductor and a capacitor vary with change in frequency.

The reactance of a pure (i.e. no resistance) inductor or capacitor is a measure of the opposition to the flow of a.c. current provided by the component. The reactance X of a component is given by the ratio

$$X = \frac{\text{voltage across component}}{\text{current through component}} \tag{10.1}$$

and it is measured in ohms.

It can be shown that

$$\text{Inductive reactance } X_L = 2\pi f L = \omega L \tag{10.2}$$

and

$$\text{Capacitive reactance } X_c = \frac{1}{2\pi f C} = \frac{1}{\omega C} \tag{10.3}$$

where $C =$ capacitance in farads and $L =$ inductance in henrys.

Clearly, the reactance of an inductor is proportional to frequency and the reactance of a capacitor is inversely proportional to frequency. The current through an inductor lags behind the voltage across it by $90°$ but the current through a capacitor leads the voltage across it by $90°$; that is, inductive reactance acts in the opposite direction to capacitive reactance. It is usual to consider inductive reactance to be positive and capacitive reactance to be negative. Graphs showing reactance plotted against frequency are given in Figs. 10.1a and b for an inductor and a capacitor respectively.

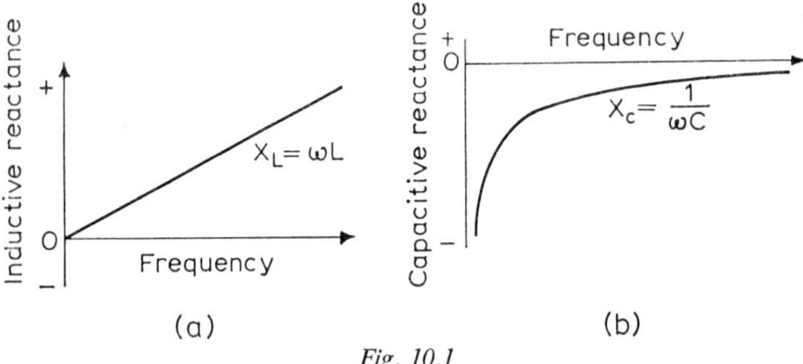

(a)   (b)

Fig. 10.1

Reactance/frequency curves for (*a*) an inductor and (*b*) a capacitor

The Series-tuned Circuit

A series-tuned circuit consists of an inductor and a capacitor connected in series as shown in Fig. 10.2. The resistance R is the total resistance of the circuit including the resistance of the inductor windings. At any given frequency f the effective reactance X_T of the circuit is the difference between the reactance of the inductor and the reactance of the capacitor, that is

$$X_T = 2\pi fL - \frac{1}{2\pi fC} \tag{10.4}$$

If $2\pi fL$ is greater than $1/(2\pi fC)$ the effective reactance of the tuned circuit is inductive and the circuit appears, electrically, to consist of an inductor in series with a resistor and the current I flowing in the circuit lags behind the applied voltage V. Conversely, if $1/(2\pi fC)$ is greater than $2\pi fL$ the circuit appears to consist of the series connection of a capacitor and a resistor and the current leads the voltage.

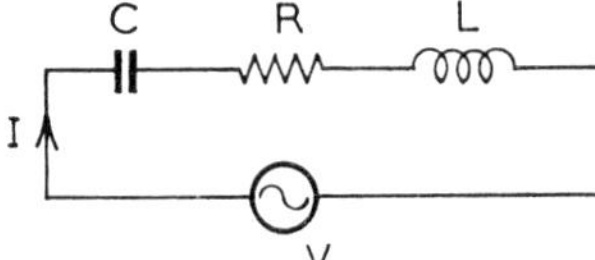

Fig. 10.2

The series-tuned circuit

The vector sum of the effective reactance X_T and the resistance R gives a quantity known as the impedance Z of the circuit, i.e.

$$Z = \sqrt{(R^2 + X_T^2)} \text{ ohms} \tag{10.5}$$

The impedance of a series-tuned circuit varies with change in frequency because the effective reactance X_T is frequency dependent, but Ohm's law can still be applied to find the current flowing.

Thus
$$I = \frac{V}{Z} \tag{10.6}$$

At a particular frequency, known as the resonant frequency, the reactance of the inductor is equal to the reactance of the capacitor and the effective reactance is zero. Substituting $X_T = 0$ in equation (10.5) shows that the impedance of a series-tuned circuit at resonance is a minimum and equal to R ohms. At resonance, the current flowing in a series-tuned circuit is given by

$$I = \frac{V}{R} \tag{10.7}$$

and it is in phase with the applied voltage.

Although at resonance the effective reactance of a series-tuned circuit is zero, the reactances of the inductor and the capacitor are not themselves zero, and voltages $V_L = IX_L$ and $V_C = IX_C$ are developed across the inductor and capacitor respectively. Because $I = V/R$, $V_L = VX_L/R$ and $V_C = VX_C/R$ and if, as is usual, R is very much smaller than X_L and X_C the voltages appearing across the inductor and the capacitor may be much greater than the voltage V, applied to the circuit. Care must be taken to ensure that the components used are capable of withstanding the high voltages that may appear across them at resonance.

RESONANT FREQUENCY

An expression may be obtained for the resonant frequency of a series-tuned circuit by noting that at the resonant frequency f_0 the inductive and capacitive reactances are equal.

i.e.
$$2\pi f_0 L = \frac{1}{2\pi f_0 C}$$

$$f_0^2 = \frac{1}{4\pi^2 LC}$$

or
$$f_0 = \frac{1}{2\pi\sqrt{(LC)}} \tag{10.8}$$

If L is in henrys and C is in farads the resonant frequency is in hertz.

Example 10.1
A tuned circuit consists of an inductor of 10 mH connected in series with a capacitor of 4 μF. Calculate the resonant frequency of the circuit.

Solution. From equation (10.8),
$$f_0 = 1/2\pi\sqrt{(10 \times 10^{-3} \times 4 \times 10^{-6})} = 796 \text{ Hz} \quad Ans.$$

Example 10.2
What value of inductance is required to tune a 0·2 μF capacitor to resonance at 2,520 Hz?

Solution. From equation (10·8),

$$L = 1/4\pi^2 f_0^2 C = 1/(4\pi^2 \times 2{,}520^2 \times 0{\cdot}2 \times 10^{-6}) = 20\ \text{mH} \qquad Ans.$$

SELECTIVITY

If the voltage of the signal applied to a series-tuned circuit is held at a constant value and its frequency is varied the current that flows varies with frequency in the manner shown in Fig. 10.3. At frequencies well below resonance the capacitive reactance is much

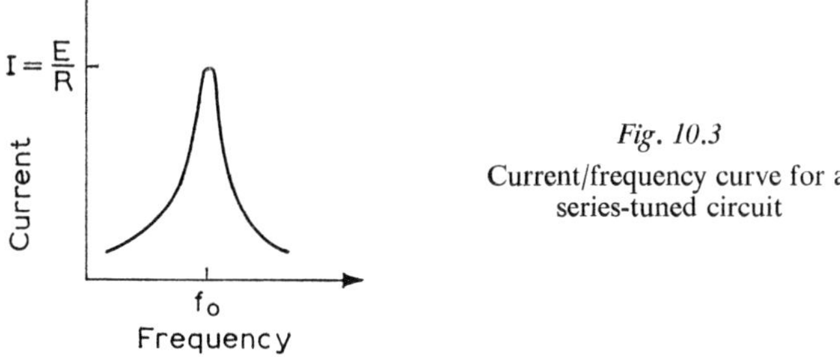

Fig. 10.3

Current/frequency curve for a series-tuned circuit

larger than the inductive reactance and a small, leading, current flows. With increase in frequency the capacitive reactance falls and the inductive reactance increases; the effective reactance of the circuit decreases and so the current increases, but it is still a leading current. Further increase in frequency brings about the resonant condition when the effective reactance is zero, and the current has

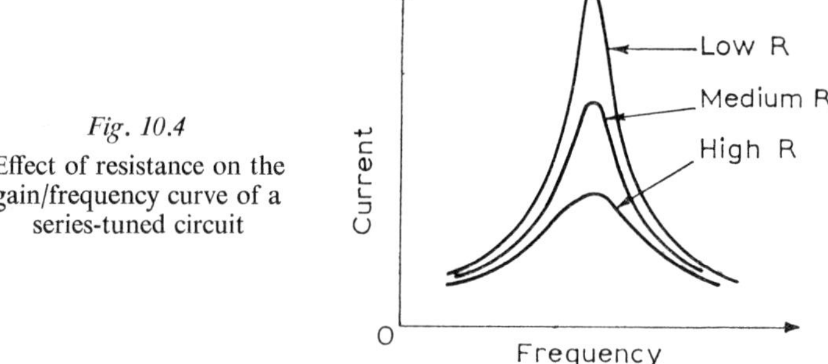

Fig. 10.4

Effect of resistance on the gain/frequency curve of a series-tuned circuit

its maximum value of V/R and is in phase with the voltage. Increasing the frequency above the resonant point makes the inductive reactance greater than the capacitive reactance and so the effective reactance is inductive. Now, further increase in frequency increases the effective inductive reactance and causes the current to fall. For particular values of L and C decreasing the circuit resistance gives a greater current at resonance and increases the slope of the curve either side of resonance, Fig. 10.4. Some control of the response curve of a series-tuned circuit can therefore be achieved by suitable choice of R.

If a complex signal containing components at a number of different frequencies is applied to a low-resistance series-tuned circuit and the output is taken from the capacitor, Fig. 10.5, only those components at, or near, the resonant frequency will be at a relatively high level in the output signal. The component V_0 at the resonant frequency of the circuit causes a current, $I_0 = V_0/R$, to flow and this current produces a voltage $\dfrac{V_0}{R} \times \dfrac{1}{\omega_0 C}$ across the capacitor.

Fig. 10.5

Use of a series-tuned circuit

A voltage V_1 at a frequency f_1 below resonance produces a capacitor voltage $\dfrac{V_1}{Z_1} \times \dfrac{1}{\omega_1 C}$, where Z_1 is the impedance of the circuit at f_1. Similarly components V_2 and V_3 at frequencies f_2 and f_3 above resonance produce capacitor voltages of $\dfrac{V_2}{Z_2} \times \dfrac{1}{\omega_2 C}$ and $\dfrac{V_3}{Z_3} \times \dfrac{1}{\omega_3 C}$ respectively. If a graph is plotted of capacitor voltage against frequency for constant values of input voltage, i.e. $V_0 = V_1 = V_2 = V_3$, etc., a curve similar to that shown in Fig. 10.3 will be obtained, but displaced slightly on the low-frequency side of resonance. For a tuned circuit having a small value of R this displacement is small and it is usual to assume that the capacitor voltage varies with frequency in exactly the same way as does the current.

The selectivity of a series-tuned circuit is its ability to discriminate between signals at different frequencies.

By the use of a variable capacitor, or a variable inductor, a series-tuned circuit can be made to resonate at any particular frequency within a given band and pass a band of frequencies centred on that resonant frequency. The bandwidth of a series-tuned circuit is defined in exactly the same way as the bandwidth of an amplifier; that is, the bandwidth is the band of frequencies over which the current is not more than 3 dB down on its maximum value. The bandwidth must be wide enough to pass all the important frequencies contained in the wanted signal but must not be so wide that unwanted signals are also passed.

Consider, for example, the current/frequency characteristic of a series-tuned circuit shown in Fig. 10.6. The curve has its maximum value at the resonant frequency of 600 kHz and has a bandwidth of 20 kHz. Suppose that three equal-amplitude signals are applied to the tuned circuit, the signals consisting, respectively, of the d.s.b. waves resulting from the amplitude modulation of 600, 570 and 620

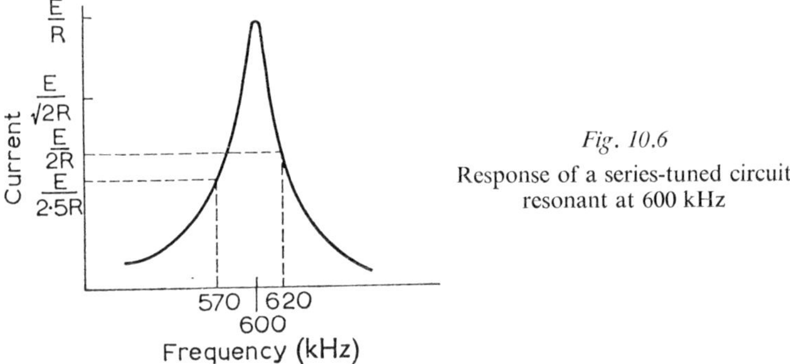

Fig. 10.6

Response of a series-tuned circuit resonant at 600 kHz

kHz carrier waves, Fig. 10.7a. Every component of the signals causes a current to flow in the tuned circuit and every current develops a voltage across the capacitor. The magnitudes of the voltages developed across the capacitor are not equal. It can be seen from Fig. 10.6 that the response of the tuned circuit to the signal centred on 570 kHz is only 1/2·5 times its response to the signal centred on

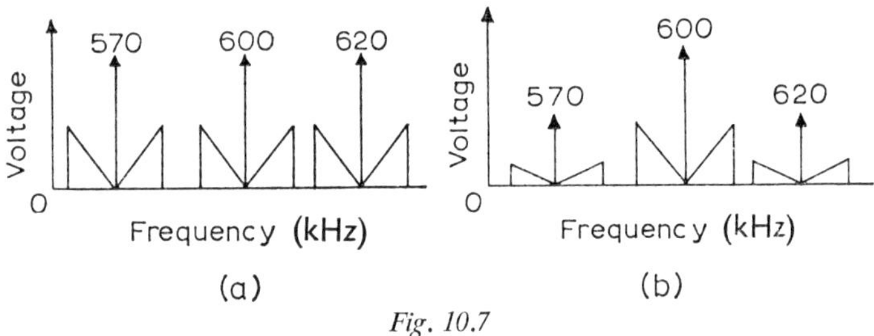

Fig. 10.7

Showing three amplitude-modulated signals (*a*) before and (*b*) after application to the series-tuned circuit of Fig. 10.6

600 kHz. Also, the response to the 620 kHz signal is only one-half of the maximum response. The tuned circuit has the effect of magnifying the 600 kHz signal two times relative to the 620 kHz signal and 2·5 times relative to the 570 kHz signal, Fig. 10.7b. Should it be desired to employ the tuned circuit to select, say, the 620 kHz signal instead of the 600 kHz signal, the circuit must be tuned to resonate at 620 kHz.

The Parallel-tuned Circuit

A parallel-tuned circuit consists of an inductor and a capacitor connected in parallel as shown in Fig. 10.8. Here, as before, R is the resistance of the inductor winding and the resistance of the capacitor is assumed to be zero. Since the two branches are in parallel the same voltage is applied across both. At low frequencies the inductive branch takes a large current I_L because the inductive reactance is low, and this current lags the applied voltage, V. The capacitive branch takes a small current I_c, because the reactance of the capacitor is large, that leads the applied voltage. The total current I is the vector sum of I_c and I_L and lags the applied voltage; this means that the parallel-tuned circuit below resonance appears, electrically, to be an inductor. At high frequencies the capacitive branch takes

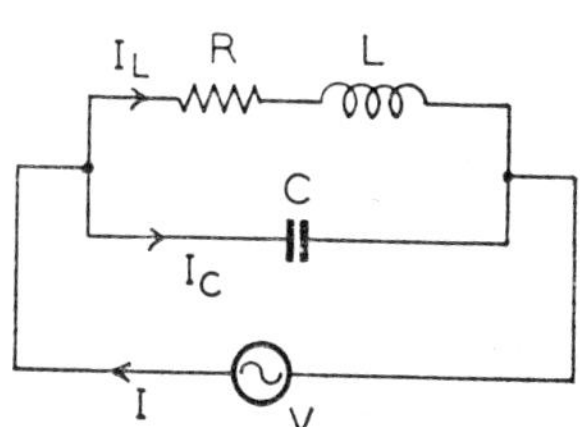

Fig. 10.8

The parallel-tuned circuit

the larger current and the total current I is capacitive, i.e. it leads the applied voltage. This means that the parallel-tuned circuit above resonance appears, electrically, to be a capacitor. At a certain frequency in between, the currents taken by the two branches are very nearly equal and the total current is very nearly zero. At this frequency, known as the parallel-resonant frequency, the impedance of the parallel-tuned circuit is very high and is purely resistive.

An analysis of the parallel-tuned circuit is more difficult than the analysis of a series-tuned circuit and it is beyond the scope of this book. It can be shown, however, that

(*a*) The resonant frequency f_0 of a parallel-resonant circuit depends not only the values of the inductance and capacitance but also on the value of the resistance R. If the resistance is small the resonant frequency is approximately equal to the frequency at which the same circuit is in series resonance, i.e.

$$f_0 \simeq \frac{1}{2\pi\sqrt{(LC)}} \tag{10.9}$$

If L and C are in henrys and farads respectively, the frequency is in hertz.

(b) The impedance of a parallel-tuned circuit at resonance is a pure resistance, known as the dynamic resistance R_d, and given by

$$R_d = \frac{L}{CR} \text{ ohms} \qquad (10.10)$$

Example 10.3
A tuned circuit is resonant at 600 kHz. What change in capacitance is required to tune the circuit to (a) 300 kHz and (b) 750 kHz?

Solution.

(a)
$$f_0 = \frac{1}{2\pi\sqrt{(LC)}}$$

$$\therefore \qquad 600 = \frac{k}{\sqrt{C}} \qquad (10.11)$$

where k is a constant equal to $\dfrac{1}{2\pi\sqrt{L}}$

and
$$300 = \frac{k}{\sqrt{C_1}} \qquad (10.12)$$

where C_1 is the new value of capacitance.
   Divide equation (10.11) by equation (10.12):

$$2 = \frac{\sqrt{C_1}}{\sqrt{C}}$$

$$\therefore \qquad \sqrt{C_1} = 2\sqrt{C}$$

or
$$C_1 = 4C$$

The capacitance must be increased to four times its original value. *Ans.*

(b)
$$750 = \frac{k}{\sqrt{C_2}} \qquad (10.13)$$

where C_2 is the other new value of capacitance.
   Divide equation (10.13) by equation (10.11):

$$\frac{75}{60} = \frac{\sqrt{C}}{\sqrt{C_2}}$$

or
$$C_2 = \frac{C}{1 \cdot 5625}$$

The capacitance must be decreased to 0·64 of its original value.   *Ans.*

SELECTIVITY
The selectivity of a parallel-tuned circuit is its ability to discriminate between a desired signal and signals at other frequencies. The voltage developed across a parallel tuned circuit is the product of the total current I flowing into the circuit and the impedance Z of the circuit. Assuming that the currents supplied by signals at different frequencies are of equal amplitude, the signal voltages developed across the

tuned circuit are directly proportional to the impedance of the circuit at each frequency. The impedance/frequency characteristic of a parallel-tuned circuit is shown in Fig. 10.9 and can be seen to be similar to the current/frequency characteristic of a series-tuned circuit. The voltage developed by a current I at the resonant frequency f_0 is IR_d and the voltages developed by currents at frequencies off resonance are less than this value, by an amount depending upon the actual frequency and the selectivity of the circuit.

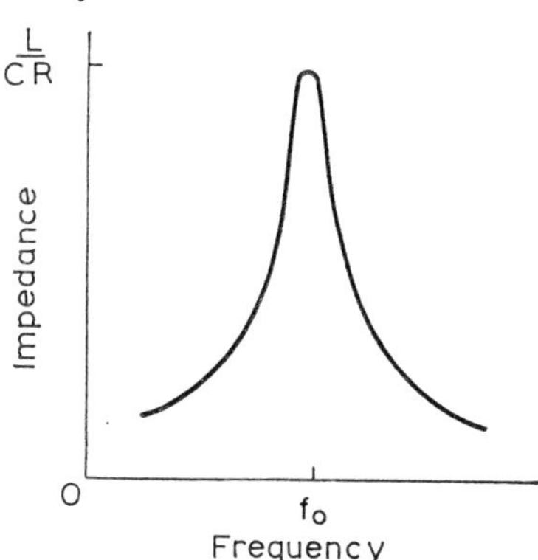

Fig. 10.9

Impedance/frequency curve for
a parallel-tuned circuit

By making either the capacitance or the inductance variable a parallel-tuned circuit can be made to resonate at any desired frequency and to select a band of frequencies centred on that resonant frequency.

Exercises

1. Explain what is meant by resonance in a tuned circuit. Briefly discuss the variation of impedance with frequency of a tuned circuit near resonance.

    An inductor is tuned to series resonance at 1,500 kHz by a capacitor of 1,600 pF. What is the new frequency of resonance if the capacitor is increased by 4,025 pF?
    (A 1965)

2. Describe the main differences in the constructional features of inductors used in the audio- and radio-frequency stages of a radio receiver.

    A tuned circuit comprises an inductor and a capacitor in series and it is tuned to a resonant frequency of 1 MHz.

    If the value of the inductor is increased by 21% what is then the frequency of resonance?
    (A 1964)

3. Sketch and explain the construction of one section of a two-gang variable tuning capacitor for use in a medium-wave broadcast receiver. What factor determines the shape of the plates of the capacitor?

    If one section of a two-gang variable tuning capacitor is used to tune the aerial circuit to the incoming signal, calculate the maximum and minimum values of the capacitor required to cover the frequency range 500 to 1,500 kHz when tuned with an inductance of 150 μH. Hence write down the range of the second section required to tune the frequency-changer oscillator over the range 1,000 kHz to 2,000 kHz if the same value of inductance is used to tune the oscillator circuit.
    (A 1963)

4. Explain what is meant by resonance in a tuned circuit. A tuned circuit comprises an inductor and a capacitor in series. The inductor has a value of 59 μH and the frequency of resonance is 1,500 kHz. If the capacitance is increased by 100 pF, at what frequency does the circuit resonate?

(A 1959)

5. The inductor in the tuned circuit of the first beating oscillator in a super-heterodyne receiver has an inductance of 250 μH and a self-capacitance of 10 pF. It is tuned by a variable capacitor of range 50–500 pF connected in series with a fixed capacitor of 1,000 pF. What are the highest and lowest oscillation frequencies?

(1 1955)

11 *Tuned Amplifiers*

A tuned amplifier is an amplifier that is required to handle a relatively narrow band of frequencies centred about a particular radio frequency. The anode, or collector, load of a tuned amplifier is provided by a parallel-tuned circuit that is adjusted to resonance at the desired operating frequency. The required LC product in the tuned circuit is fairly small and easily obtained; the valve, transistor, and stray capacitances that adversely affect the response of an untuned amplifier now provide a part of the tuned circuit capacitance. Thermionic-valve tuned amplifiers are employed as voltage amplifiers, when their function is to deliver the maximum voltage to the next stage, and as power amplifiers, when their function is to deliver the maximum power to the next stage or to the load, but in this book only voltage amplifiers are to be discussed. Pentode valves are always employed (except for a relatively few low-noise first stage amplifiers) because of their lower anode/grid capacitance and higher gain. Because a transistor amplifier is basically a power amplifier the collector tuned circuit must also provide matching, or near matching, between one stage and the next. The important characteristics of a tuned amplifier are its gain at the resonant frequency of its tuned circuit, its variation of gain at frequencies near resonance, and its discrimination against frequencies well removed from resonance.

Pentode Tuned Amplifiers

The circuit of a tuned-anode pentode r.f. amplifier is given in Fig. 11.1 and can be seen to differ from the circuit of an a.f. R-C coupled amplifier only in the use of a parallel-tuned circuit for the anode load. The parallel-tuned circuit has its maximum impedance of $R_d = L_1/C_1'r$ ohms, where r is the resistance of the inductor winding, at the resonant frequency of $1/(2\pi\sqrt{L_1C_1'})$ Hz. C_1' comprises a physical variable capacitor C_1 in parallel with the anode/cathode capacitance of V_1, the input capacitance of V_2 and the various stray capacitances existing in the circuit. At very high frequencies the required LC product is much smaller and the total valve and stray capacitance may well be large enough to obviate the need for a physical capacitor. The anode circuit then appears as shown in Fig. 11.2, permeability tuning of the inductor being employed.

The functions of all the remaining components in Fig. 11.1 are identical with the functions of the corresponding components in a R-C coupled amplifier. The reactances of the coupling capacitors C_{c1} and C_{c2} are negligible at the frequencies of operation.

The anode a.c. resistance of a pentode valve is very high and if (as is usual) the grid leak resistor R_{g2} of the following stage is made much greater than the dynamic resistance R_d of the tuned circuit, the approximate formula for the gain of a pentode stage can be used. Then the voltage gain A_v is given by

$$A_v = g_m Z_L \tag{11.1}$$

where Z_L is the impedance of the anode load.

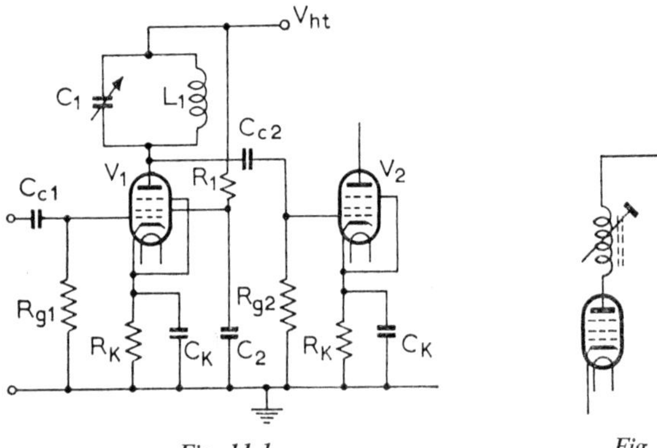
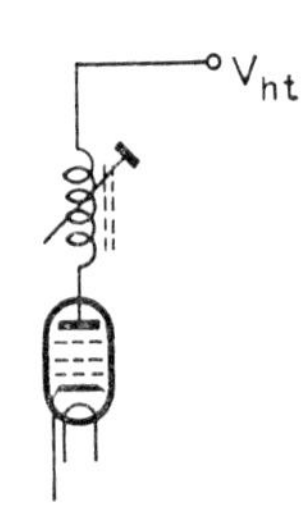

Fig. 11.1

The tuned-anode pentode amplifier

Fig. 11.2

Anode tuned by stray capacitance
and variable inductor

Since g_m is more or less constant the gain/frequency characteristic of the amplifier is substantially the same as the impedance/frequency characteristic of the anode tuned circuit. At the resonant frequency the amplifier gain is a maximum and is given by equation (11.2)

$$A_v = g_m R_d \tag{11.2}$$

$$= \frac{g_m L_1}{C_1' r} \tag{11.2a}$$

Since the resonant voltage gain depends upon the quotient L_1/C_1' the performance of the amplifier depends on whether tuning is accomplished by a variable capacitor or by a variable inductor. If the amplifier is to be tuned to a higher frequency then either the capacitance or the inductance in the tuned circuit must be reduced. If the capacitance is reduced the gain at resonance will increase, but if the inductance is reduced the resonant gain will decrease. Changing the L_1/C_1' ratio also affects the shape of the gain/frequency characteristic; increase in L_1/C_1' raises the whole curve but lifts the peak more than the sides; that is, an increase in the L_1/C_1' ratio increases both the resonant gain and the selectivity. Generally these effects are undesirable, since it is usually required for an amplifier to have the same gain irrespective of its operating frequency, and often complex coupling networks are employed to overcome these effects.

The bandwidth of a tuned amplifier is normally taken as the width of the band of frequencies over which the gain is not more than 3 dB down on the gain at the resonant frequency.

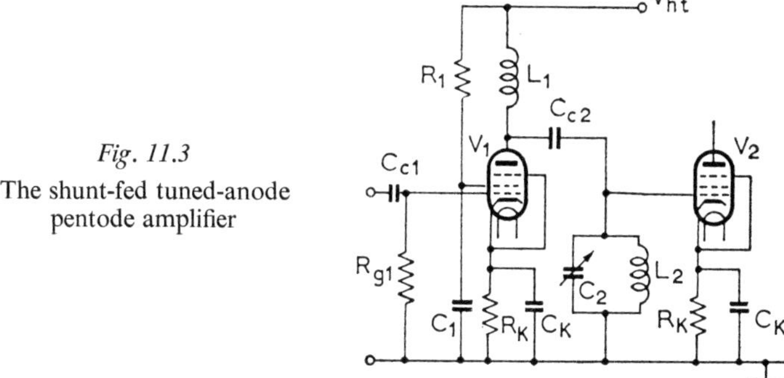

Fig. 11.3

The shunt-fed tuned-anode pentode amplifier

The tuned amplifier equivalent of choke-capacitance coupling is also possible and a typical circuit is shown in Fig. 11.3. The anode choke offers a high reactance to radio-frequency currents and prevents them entering the h.t. supply line, but it has a low d.c. resistance so that the steady anode voltage of V_1 is almost equal to the h.t. supply voltage. This arrangement has the advantage that one side

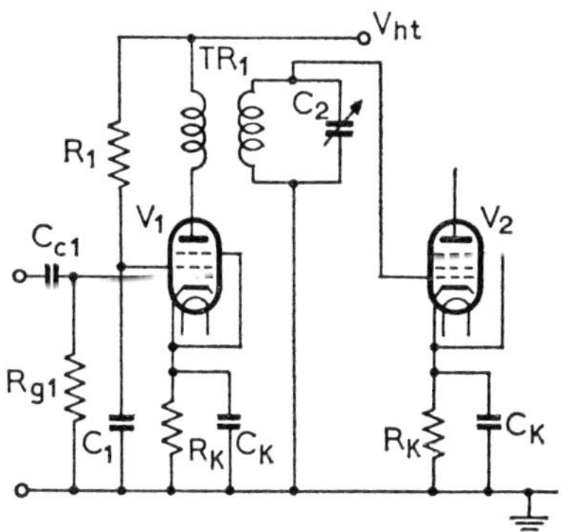

Fig. 11.4

The single-tuned transformer pentode amplifier

of the variable capacitor is at earth potential and not at h.t. potential as before.

A more commonly employed alternative method of coupling is the single-tuned transformer shown in Fig. 11.4. The primary winding of transformer TR_1 is connected in the anode circuit of V_1 and has a low d.c. resistance to minimize the d.c. voltage drop across it. The secondary winding is tuned to the required resonant frequency by capacitor C_2 and is connected in the grid-to-cathode circuit of V_2. A radio-frequency transformer such as TR_1 employs a very small coupling between its primary and secondary windings and this makes the concept of a turns ratio meaningless and analysis

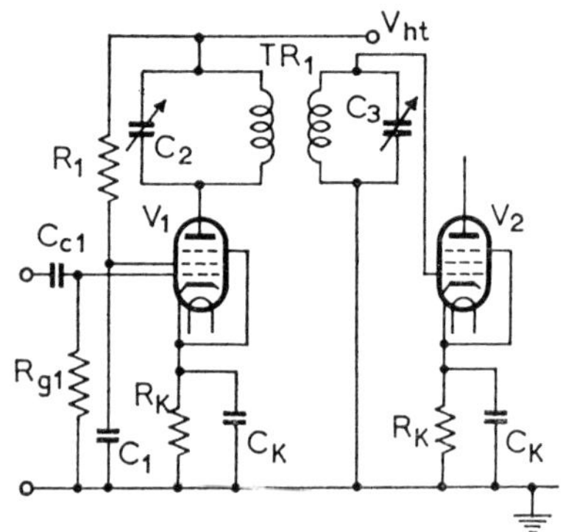

Fig. 11.5

The double-tuned transformer pentode amplifier

more difficult. However, the anode load of V_1 is the secondary impedance transformed to the primary circuit, and an a.c. voltage developed across the primary winding by a signal applied to the grid of V_1 is coupled into the secondary circuit and applied to the grid of V_2. The gain/frequency characteristic of the amplifier is determined by the impedance/frequency characteristic of the secondary parallel-tuned circuit.

The gain/frequency characteristics of the single-tuned amplifiers mentioned so far are rounded and fall away on either side of resonance. As a result a single-tuned amplifier cannot discriminate against unwanted frequencies near resonance without at the same time discriminating against some of the wanted frequencies. This disadvantage can be overcome in tuned amplifiers designed to work at a constant frequency, such as intermediate-frequency amplifiers in superheterodyne receivers, by the use of double-tuned transformer coupling.

A double-tuned amplifier, Fig. 11.5, employs transformer coupling in which both primary and secondary circuits are tuned to resonate at the desired operating frequency. If the coupling between the

windings is correctly adjusted a flat-topped gain/frequency characteristic can be obtained, as shown in Fig. 11.6. Such a characteristic is desirable for the amplification of a signal because it gives

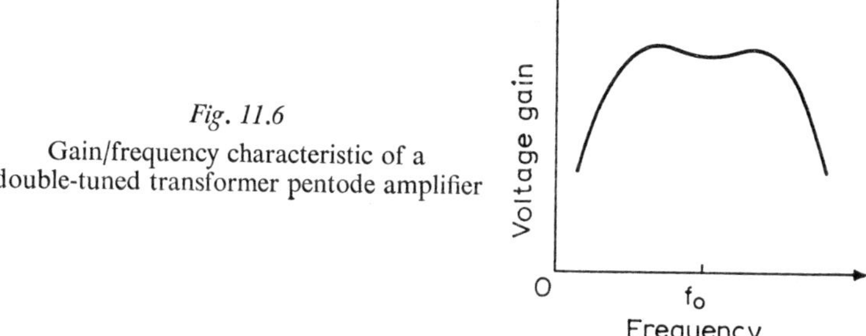

Fig. 11.6

Gain/frequency characteristic of a
double-tuned transformer pentode amplifier

substantially the same gain to all the frequency components of the
signal and discriminates sharply against unwanted signals. It is
rare to use double-tuned amplifiers over a range of frequencies
because correct alignment is difficult.

Transistor Tuned Amplifiers

A transistor amplifier stage takes an input current, and hence an
input power, and is essentially a power amplifier; for maximum
gain each stage must be matched, or nearly matched, to the following
stage. The interstage coupling network must determine the centre
frequency and bandwidth of the stage while, at the same time, trans-
fering sufficient r.f. power to the next stage and also providing matching. The tuned circuit arrangements employed in pentode r.f.
amplifiers may require some modification if the output impedance
of one transistor is to more nearly match the input impedance
of the next transistor.

The resonant frequency of a parallel-tuned circuit is not changed
if the inductor is tapped, Fig. 11.7a, or the capacitance is split
into two, Fig. 11.7b. The impedance seen looking into terminals 1
and 2 is much smaller than the impedance seen looking into terminals 1 and 3, the reduction depending upon the inductance (or
capacitance) ratio employed. If terminals 1 and 3 are connected to
the output terminals of one transistor and terminals 1 and 2 are
connected to the input terminals of the following transistor, suitable
choice of the inductance (or capacitance) ratio can give the required
match, or near match, condition, see Fig. 11.8a and b. If required, a
combination of both methods of matching can be used. The tuned
circuit is formed by L in parallel with the capacitors C_2 and C_3 in

Fig. 11.8*a* and C_2 alone in Fig. 11.8*b*, plus the transistor and stray capacitances. At higher frequencies the transistor and stray capacitances may be sufficiently large for a physical tuning capacitor to be unnecessary. Figs. 11.8*a* and *b* are completed by the resistors R_1 to R_6 and their associated decoupling capacitors; these provide bias and d.c. stabilization for the circuits. The only component not

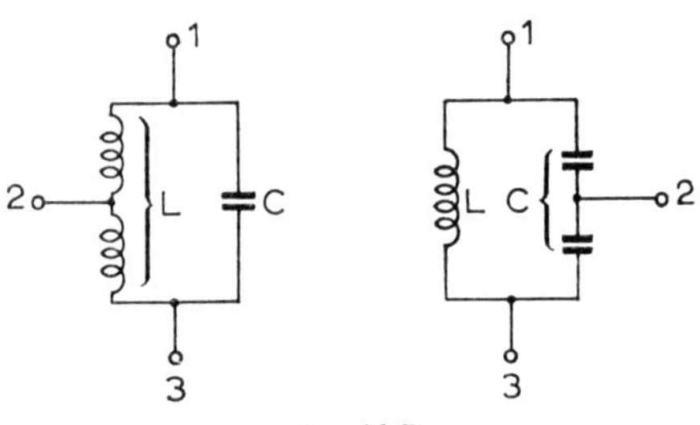

Fig. 11.7

Two methods of tapping a parallel-tuned circuit

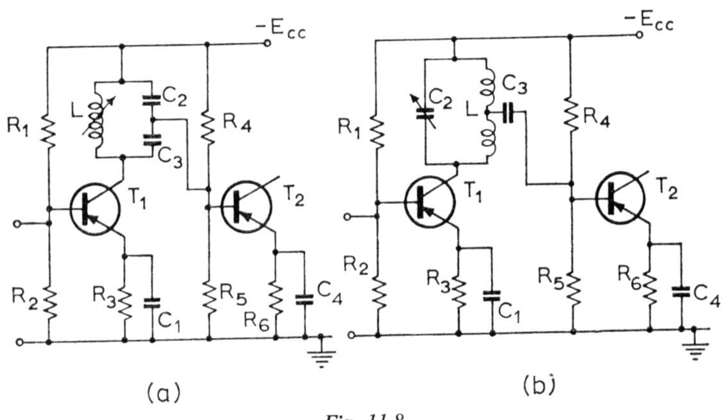

(a) (b)

Fig. 11.8

Two tuned-collector transistor amplifiers

mentioned is capacitor C_3 in Fig. 11.8*b*; this blocks d.c. and prevents the bias arrangement for transistor T_2 being upset.

The transistor versions of single-tuned and double-tuned transformer coupling are shown in Figs. 11.9 and 11.10 respectively. In the single-tuned circuit the primary of coupling transformer TR_1 is associated with various capacitances in the circuit and a physical capacitor is not employed. Tuning to the required frequency is effected by the movement of a screw-in dust core altering the inductance of the winding. Capacitor C_2 is provided to block the collector supply and ensure that the bias for T_2 is provided solely by the bias circuit. The remainder of the components have convential bias and d.c. stabilization functions.

Both the primary and the secondary windings of transformer TR_1 in Fig. 11.10 are tuned to the required centre frequency and the coupling between the windings is adjusted until a flat-topped response, as shown in Fig. 11.6, is obtained. When the inductance of

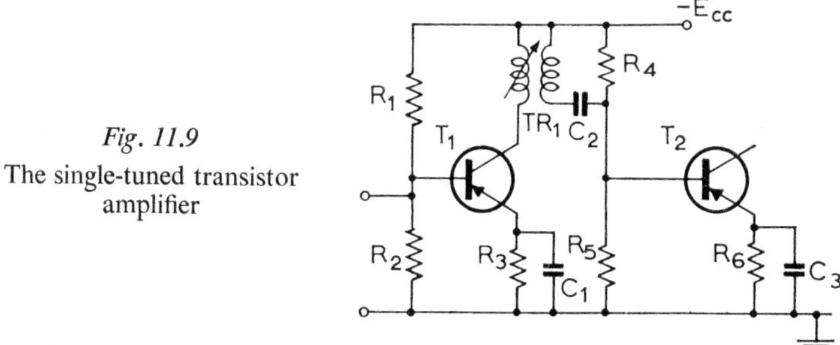

Fig. 11.9

The single-tuned transistor amplifier

the secondary winding is large the connection to the base of the following transistor is made via a tap as shown, but if the secondary inductance is small the tap is not necessary.

In conclusion it should be noted that at radio frequencies energy is fed internally from the output terminals of a transistor to its input

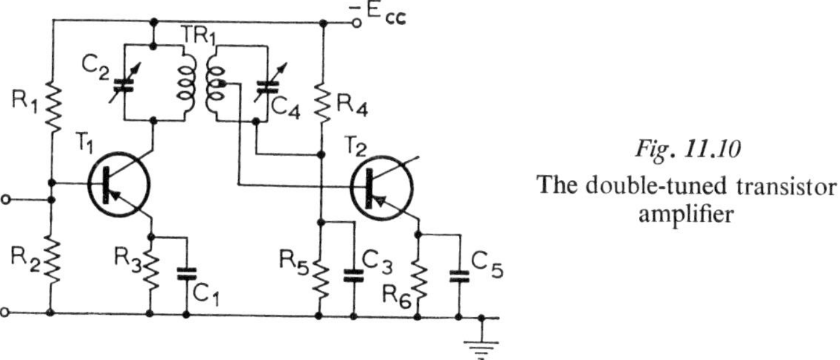

Fig. 11.10

The double-tuned transistor amplifier

terminals. This internal feedback is undesirable, because it causes signal distortion and may lead to oscillation, and it is generally reduced to tolerable proportions by the addition of circuitry external to the transistor.

Exercises

1. Sketch the basic circuit of a simple triode amplifier having a tuned anode load and derive an expression for the gain of the amplifier at the frequency of resonance of the load, neglecting stray capacitances.

In a single-stage amplifier, the anode slope resistance of the valve is 50,000 Ω and its amplification factor is 10. The anode load consists of a capacitor C of 300 pF in parallel with an inductor L of 400 μH having a resistance R of 20 Ω. Calculate the frequency of resonance.

Determine the voltage at the frequency of resonance that must be applied between the grid and cathode to give 1 V r.m.s. output across the tuned anode load.

The impedance of the anode load at resonance may be taken as given by the expression L/CR. (P2 1959)

2. The anode load of a radio-frequency amplifier consists of a parallel-tuned circuit, the inductor of which has a value of 150 μH. Neglecting circuit resistance, what value of capacitance will be required to tune the circuit to a frequency of 1 MHz?

Give a circuit diagram of such an amplifier with typical component values.

State briefly your reasons for the type of valve chosen. (1 1958)

12 *L-C Oscillators*

An oscillator is an electronic circuit that has been designed to produce
an alternating e.m.f. of known, periodic, waveform. A number of
different waveforms are obtained from different kinds of oscillator
but in this book only oscillators providing a sinusoidal output waveform are to be discussed. Further, discussion will be limited to the
simple types of oscillator in which the frequency of oscillation is
determined by a parallel-tuned *L-C* circuit.

A sinusoidal oscillator, valve or transistor, can be considered to be
an amplifier that provides its own input signal, the input signal being
derived from the output signal, see Fig. 12.1. This is possible because

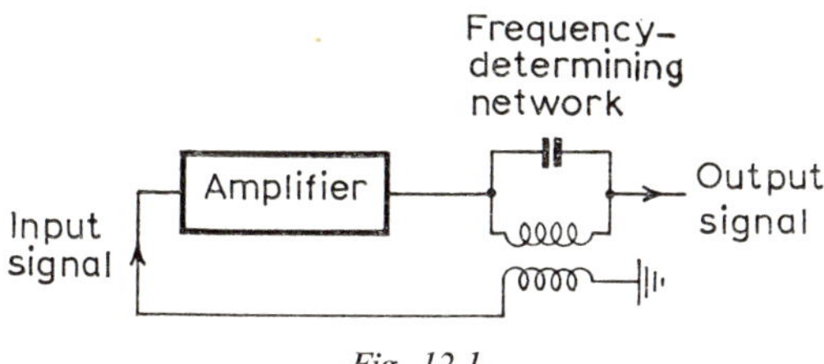

Fig. 12.1

The principle of an oscillator

the signal level required at the input terminals of an amplifier is
considerably less than the level of the amplified output signal. The
valve, or transistor, acts as a convertor of electrical energy, taking
d.c. power from the h.t. supply and converting a part of it into a.c.
power in the output signal. Either triodes or pentodes can be used
in the valve circuits to be described but only triode circuits will be
shown.

Oscillators are widely used in electronic, radio, and line communication equipment as sources of alternating e.m.f. and are
generally required to be extremely stable in frequency. Frequency
stability of the highest order cannot be obtained with the simple
oscillators to be described and other, high stability, oscillators are
widely employed. Other characteristics of importance are the purity
of the output waveform, and the constancy of the output level with
changes in frequency and/or in power supply voltage.

The Oscillatory Circuit

If a capacitor, *C* farads, is charged from a d.c. source a p.d. *V*

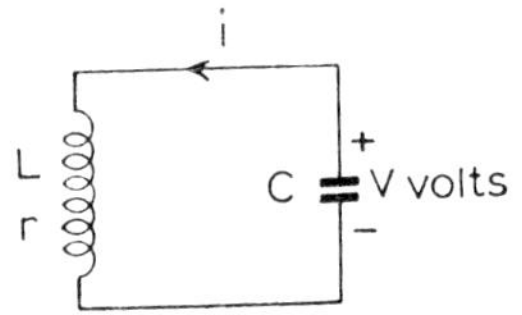

Fig. 12.2

The oscillatory circuit, initial conditions

volts will be developed across its terminals and an amount of electric energy, $\frac{1}{2}CV^2$ joules, will be stored in its dielectric. Consider such a charged capacitor to be connected across an inductor as shown in Fig. 12.2. A complete circuit exists and so the capacitor will discharge through the inductor and a current i will flow. This current commences to flow the instant the capacitor is connected across the inductor and it rises rapidly to a maximum value when the capacitor has fully discharged and there is zero voltage across its plates. The flow of a current in a conductor produces a magnetic field around that conductor; associated with the current flow in the inductor, therefore, is a magnetic field that reaches its maximum value at the same times as does the current. The energy stored in the magnetic field at this time is equal to $\frac{1}{2}LI^2$ joules, where L is the inductance of the inductor in henrys and I is the maximum value of the current in amperes. All the energy originally stored in the capacitor has now disappeared (since $V = 0$) and has been partly converted into magnetic energy and partly lost as power dissipation in the resistance r of the circuit. Since the p.d. across the capacitor terminals is now zero the current starts to fall and the magnetic field about the inductor starts to collapse. As the field collapses an e.m.f. is induced in each turn of the inductor that, according to Lenz's law, is in such a direction as to oppose the force creating it; that is, the total induced e.m.f. tends to keep the current flowing. Because the capacitor is completely discharged further current flow in this direction recharges it—but with the polarity opposite to what it was before.

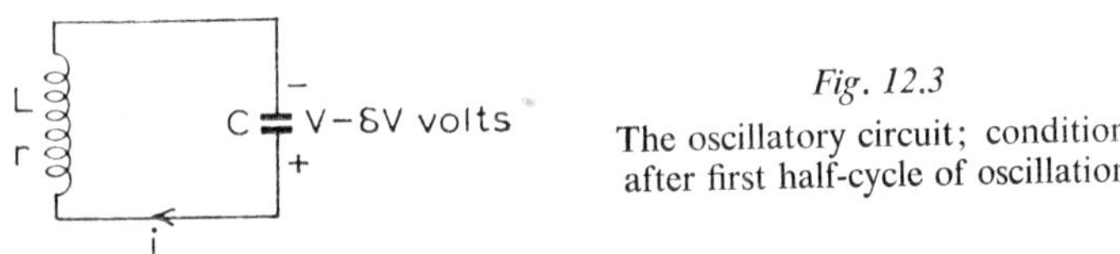

Fig. 12.3

The oscillatory circuit; conditions
after first half-cycle of oscillation

When the magnetic field has completely collapsed the current has fallen to zero and the capacitor is fully re-charged to a voltage somewhat less than before, say $(V - \delta V)$, where δV is a small voltage increment, Fig. 12.3. Almost all of the magnetic energy has been converted back to the form of electric energy stored in the dielectric of the capacitor, some energy having again been lost as i^2r dissipation in the circuit resistance. The capacitor now starts to discharge through the inductor again but this time the current flow is in a direction opposite to what it was before. A magnetic field is again set up around the inductor that increases with increase in the discharge

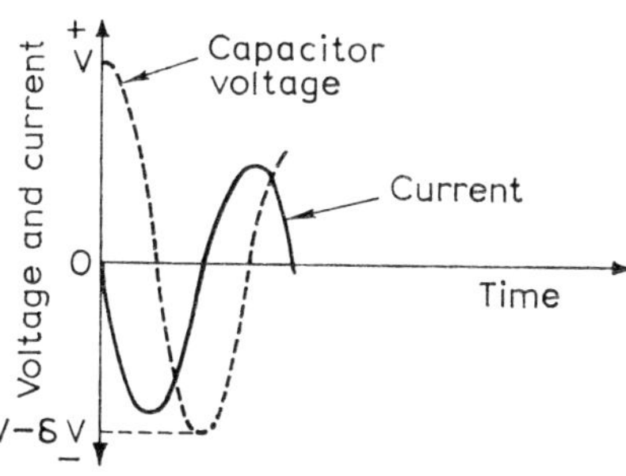

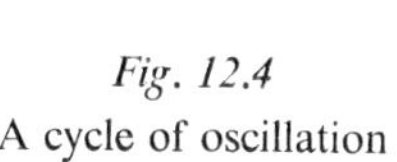

Fig. 12.4

A cycle of oscillation

current. When the capacitor is fully discharged the current starts to fall and the collapsing magnetic field induces an e.m.f. in the inductor windings that tends to maintain the current in its new direction. The capacitor is recharged with its original polarity by this current and when it is fully charged (to a voltage less than before) one cycle of the oscillatory current has been completed, see Fig. 12.4.

A continual interchange of energy between the capacitor and the

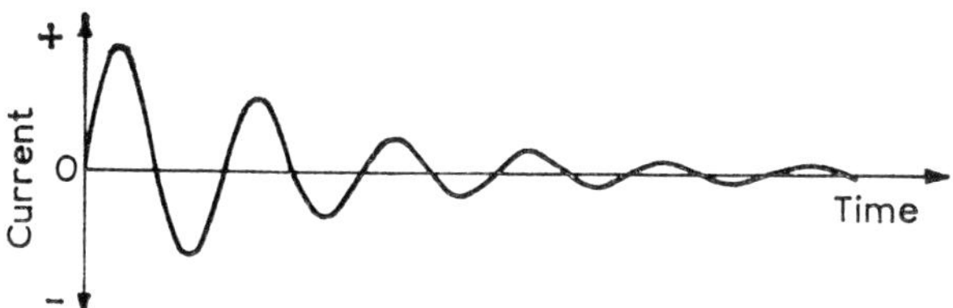

Fig. 12.5

A damped oscillation

inductor takes place at a constant frequency, but with the amplitude of the oscillatory current decreasing steadily until, eventually, the oscillation dies away. An oscillation of this type, shown in Fig. 12.5, is known as a damped oscillation. The rate at which the oscillation dies away depends upon the circuit resistance; the greater the resistance the sooner the oscillations disappear.

If energy can be supplied to the oscillatory circuit to replace the energy lost by i^2r dissipation an undamped oscillation can be obtained. An undamped oscillation is shown in Fig. 12.6, and it is clear that the current amplitude is constant and the oscillation does not die away but can be maintained indefinitely. The energy supplied to the oscillatory circuit must be sufficiently large to make good the losses and must be in phase with the oscillation.

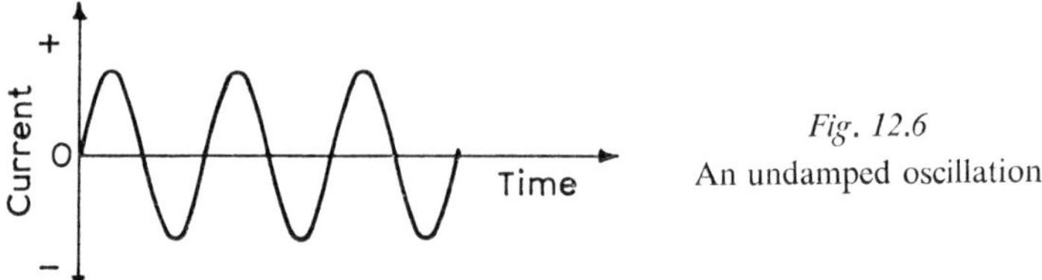

Fig. 12.6

An undamped oscillation

If the losses in the tuned circuit are low the frequency f_0 of oscillation is given approximately by the expression

$$f_0 \simeq \frac{1}{2\pi\sqrt{(LC)}} \text{ Hz} \qquad (12.1)$$

where L is in henrys and C is in farads.

Simple Types of L-C Oscillator

An oscillator must contain a frequency-determining section and a maintaining section, the former being provided by a parallel-tuned circuit and the latter by an amplifier.

The bias and d.c. stabilization requirements of a transistor oscillator are similar to those of a transistor amplifier and the same circuitry is used. Valve oscillators may employ either triodes or

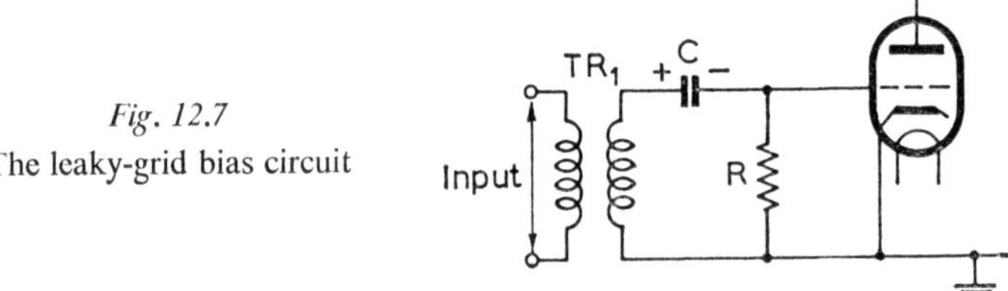

Fig. 12.7

The leaky-grid bias circuit

pentodes and generally incorporate leaky-grid bias, the circuit for which is given in Fig. 12.7.

The leaky-grid **bias circuit** consists of the capacitor C and the resistor R connected in the grid/cathode circuit of the valve. Initially the grid bias voltage is zero and the first positive half-cycle of the input signal takes the grid positive with respect to the cathode; grid current flows and charges the capacitor in the direction shown. During the following negative half-cycle of the input signal the grid is taken negative relative to the cathode and no grid current flows. The capacitor then starts to discharge through resistor R and the secondary winding of the input transformer TR_1. The rate at which the capacitor discharges depends upon the time constant, CR seconds, of the discharge circuit, and this time constant is chosen to ensure that the capacitor has not discharged completely before the next positive half-cycle occurs. During the next positive half-cycle of the input signal, grid current again flows and replaces the charge

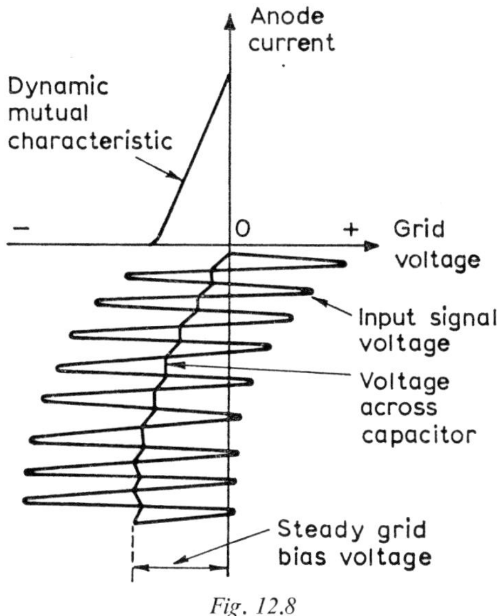

Fig. 12.8

Illustrating the build-up of bias voltage in a leaky grid circuit

lost by the capacitor, but during the next negative half-cycle some of this charge is again lost. For each complete cycle of the input signal more charge is put into the capacitor than is taken out of it and the charge stored gradually builds up. The p.d. across a capacitor is proportional to the charge stored in it and so the capacitor voltage— the grid bias voltage—gradually increases, see Fig. 12.8. Eventually the condition is reached where the grid current flowing during the positive half-cycles of the input signal is just sufficient to restore the charge lost during the negative half-cycles. The grid bias voltage has then attained a steady value that is proportional to the amplitude of the input signal. If the amplitude of the input signal changes, the grid bias voltage changes also and this action tends to keep the amplitude of the output signal at a constant level.

8—(T.1159)

Grid-leak bias is generally employed for valve oscillators because it gives improved frequency stability and makes an oscillator self-starting. When an oscillator using leaky-grid bias is first switched on there is no input signal and so the grid bias voltage is zero and a large anode current flows; any noise inherent in the circuit, or the disturbance caused by switching on, will produce a small oscillatory current in the tuned circuit that will cause oscillations to start.

THE TUNED-ANODE OSCILLATOR

A tuned-anode oscillator is one in which the frequency-determining tuned circuit is connected in the anode circuit of the valve (triode or pentode). The h.t. supply voltage may be fed to the anode either in series with the tuned circuit or in parallel with the tuned circuit. The

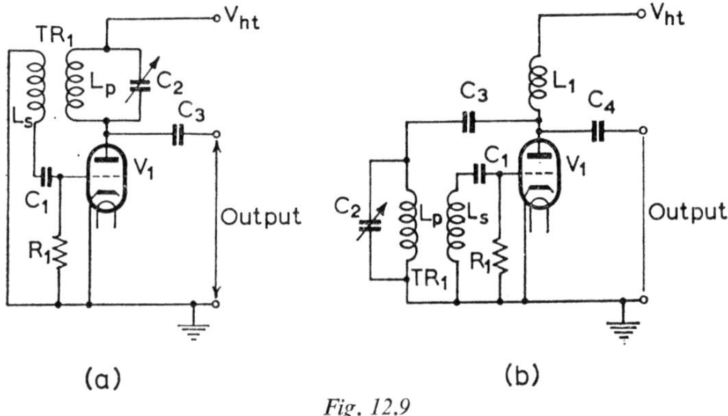

Fig. 12.9

Tuned-anode oscillators, (*a*) series-fed and (*b*) parallel-fed

first arrangement, known as series feed, is shown in Fig. 12.9*a* and the second arrangement, known as parallel feed, is shown in Fig. 12.9*b*. The parallel-feed circuit has the advantage that the tuned circuit is at earth potential and not h.t. potential. In both circuits the frequency of oscillation is determined by the product $L_p C_2$ and grid bias is provided by R_1 and C_1. In Fig. 12.9*b* the inductor L_1 is an r.f. choke whose function is to block r.f. currents from the h.t. line whilst dropping the minimum d.c. voltage. Capacitors C_3 and C_4 are d.c. blocking capacitors that have negligible reactance at the oscillation frequency and that block the h.t. supply voltage. The action of either of these circuits is as follows—when the h.t. supply voltage is switched on, any noise or small fluctuation in the grid/cathode circuit is amplified and causes an oscillatory current to be set up in the tuned circuit. The oscillatory current in the primary winding, L_p, of the transformer induces an e.m.f. at the same frequency into the secondary winding, L_s, and this voltage is applied to the grid of the valve. The valve introduces 180° phase shift and the transformer connections must be arranged to give a further 180° phase change to achieve an overall phase change around the oscillator loop of zero. Also, the coupling between the transformer windings must be tight enough to ensure that the loop gain is greater than unity; otherwise the oscillations will die away.

THE TUNED-COLLECTOR OSCILLATOR

The circuit of a tuned-collector oscillator is given in Fig. 12.10. R_1, R_2, R_3, C_1, and C_2 are bias and d.c. stabilization components, TR_1 is an r.f. transformer, C_3 a variable tuning capacitor, and C_4 a

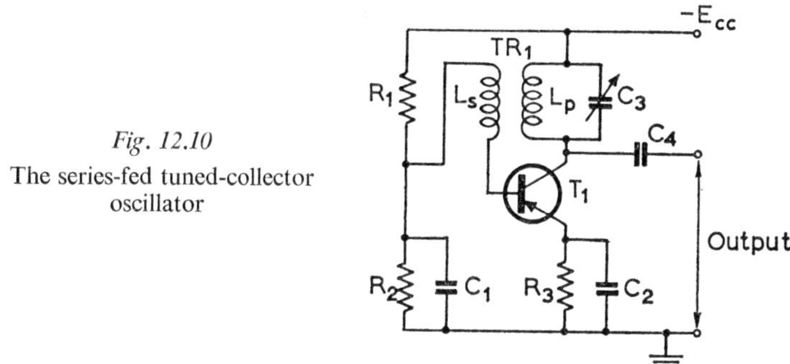

Fig. 12.10

The series-fed tuned-collector oscillator

blocking capacitor. The operation of the circuit is similar to the operation of the tuned-anode oscillator; differences exist in the

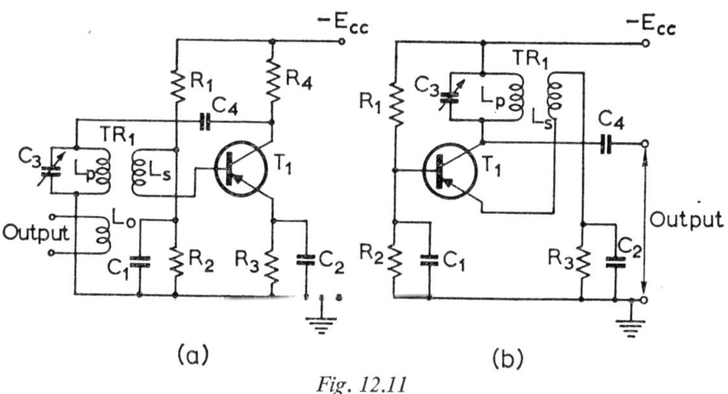

(a)   (b)

Fig. 12.11

Tuned-collector oscillators (*a*) parallel-fed and (*b*) employing emitter coupling

methods of design and analysis and the method of reduction of oscillation amplitude, but these matters are beyond the scope

of this book. The frequency of oscillation f_0 is given approximately
by

$$f_0 \simeq \frac{1}{2\pi\sqrt{(L_pC_3)}} \text{ Hz} \qquad (12.2)$$

The output signal can be taken off as shown in the diagram, or,
alternatively, taken from a third winding, L_o coupled to L_p and L_s as
shown in the parallel-feed circuit of Fig. 12.11a. Finally, the oscillatory voltage may be fed into the emitter circuit instead of the base,
Fig. 12.11b.

THE TUNED-GRID OSCILLATOR

A tuned grid oscillator has the frequency-determining tuned circuit
connected in the grid/cathode circuit of the valve as shown in Fig.
12.12. The frequency of oscillation is approximately given by

$$f_0 \simeq \frac{1}{2\pi\sqrt{(L_sC_2)}} \text{ Hz} \qquad (12.3)$$

and the coils must be connected to give the required 360° phase
shift around the oscillatory loop. The circuit has the output taken

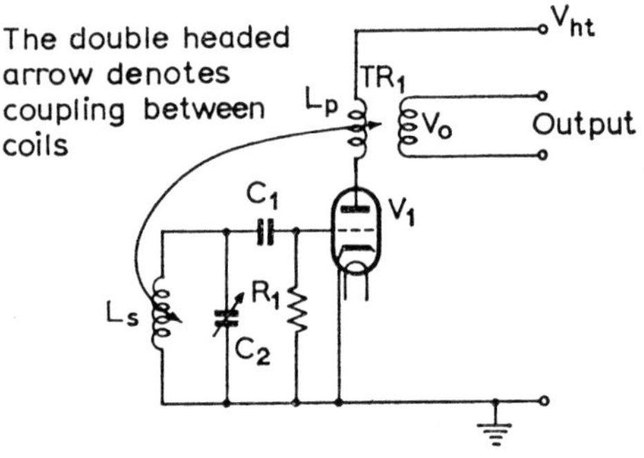

Fig. 12.12

The tuned-grid oscillator

off via a third winding on the r.f. transformer TR_1 but capacitor
coupling of the output is equally possible. Leaky-grid bias is provided by components C_1 and R_1 and series-feeding of the h.t. supply
is employed.

THE TUNED-BASE OSCILLATOR

The circuit of a tuned base transistor oscillator is shown in Fig.
12.13 and is similar to the circuit of a parallel-feed tuned-grid oscillator. Conventional bias and d.c. stabilization circuitry is used and

the frequency of oscillation is determined by the parallel-tuned circuit of L_s and C_3. Feedback from output to input is via capacitor C_1 and the r.f. transformer TR_1. Capacitor C_4 is necessary to prevent the bias for transistor T_1 being determined primarily by the low d.c. resistance of winding L_s. Parallel feed of the collector supply

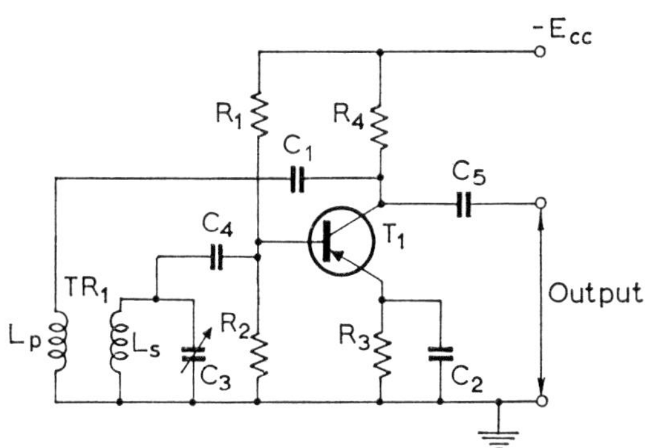

Fig. 12.13

The tuned-base oscillator

voltage is shown and has the usual advantage over series feeding in that no d.c. current flows in the windings of the transformer. The operation of the circuit is similar to the operation of the preceding circuits.

THE TUNED-ANODE TUNED-GRID OSCILLATOR

A tuned-anode tuned-grid oscillator, Fig. 12.14, employs two parallel-tuned circuits, one connected in the anode circuit of the

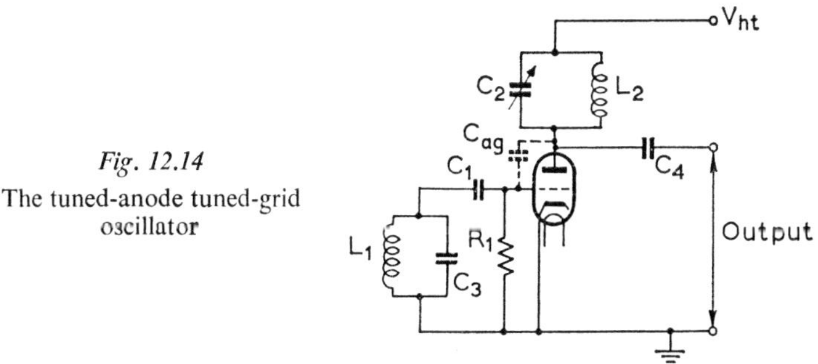

Fig. 12.14

The tuned-anode tuned-grid
oscillator

valve and the other connected in the grid/cathode circuit. The two coils are *not* coupled together, coupling between the anode and grid/cathode circuits being via the anode/grid capacitance C_{ag} of the valve (shown dotted). For the circuit to oscillate it is necessary for both of the tuned circuits to be inductive; this means that they must be tuned to a frequency somewhat higher than the required oscillation frequency.

Frequency Stability

The frequency of oscillation of an oscillator is primarily dependent upon the resonant frequency of its tuned circuit but it is also a function of various other parameters. The main causes of frequency instability are changes in the value of the tuned-circuit capacitance and inductance because of changes in temperature, changes in valve

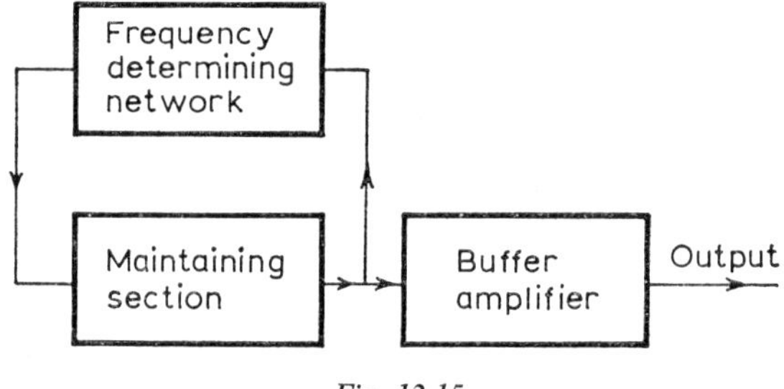

Fig. 12.15

The use of a buffer amplifier

or transistor parameters due to fluctuations in power supplies, and variations in the external load.

The first cause of frequency instability can be minimized by using inductors and capacitors in the tuned circuit that have small temperature coefficients, and keeping the tuned circuit at as constant a temperature as possible. Fluctuations in the power supplies can be reduced by the use of adequate power stabilization circuitry, and variations in the external load can effectively be removed by feeding the load via a buffer amplifier. Fig. 12.15 shows the use of a buffer amplifier in an oscillator. The amplifier is an ordinary audio-frequency or radio-frequency amplifier—depending upon the oscillation frequency—whose function is to isolate the oscillator from any changes in the load and also to increase the output power level.

When very good frequency stability is required none of these measures are adequate and crystal oscillators are used. A crystal oscillator is an oscillator whose frequency is determined by a piezo-electric crystal such as quartz.

Amplitude Modulation of an Oscillator

An oscillator may be amplitude-modulated by injecting a modulating signal into the grid or anode circuit of a valve oscillator, or into the base, emitter or collector circuit of a transistor oscillator.

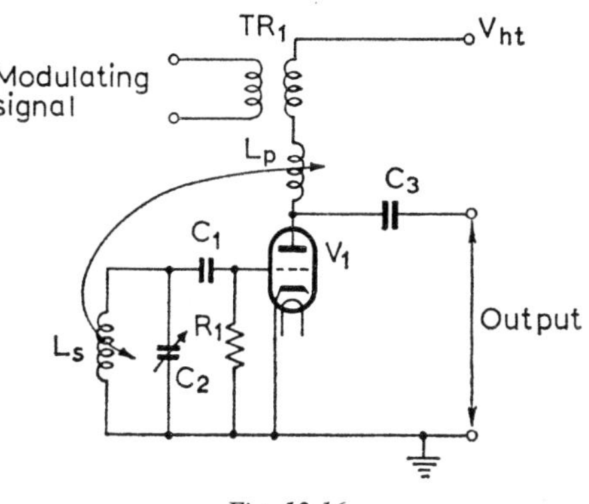

Fig. 12.16

The anode-modulated valve oscillator

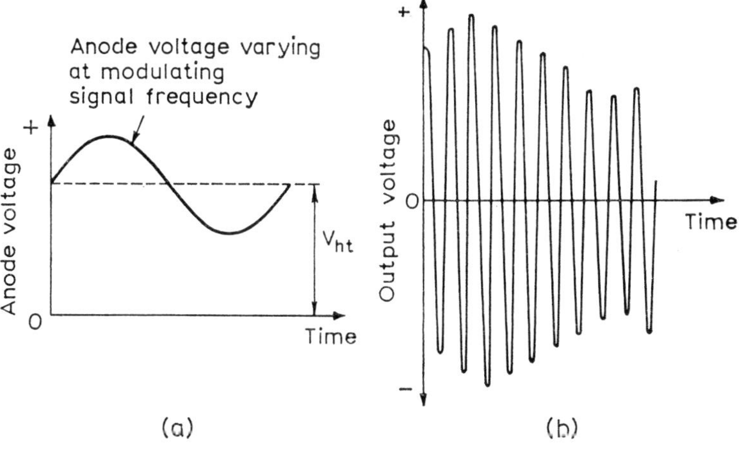

Fig. 12.17

Waveforms in an anode-modulated valve oscillator

The most common form of amplitude-modulated valve oscillator has the modulating signal injected into its anode circuit, and the circuit of an anode-modulated tuned-grid oscillator is given in Fig. 12.16. The oscillation frequency is fixed by the product L_sC_2, and can be varied by adjustment of C_2. Coils L_p and L_s comprise an r.f. transformer and provide feedback at the required level and phase for the maintenance of oscillations. C_1 and R_1 provide leaky-grid bias and C_3 couples the output signal to the load. The modulating signal is coupled into the anode circuit of the valve via the transformer TR_1 and causes the anode voltage of V_1 to vary in accordance with the waveform of the modulating signal. If the circuit is correctly adjusted the amplitude of the oscillations is proportional to the anode voltage and so the oscillations are amplitude modulated.

Fig. 12.17 shows the waveforms of the anode voltage of V_1 and of the modulated output signal for the particular case when the modulating signal is sinusoidal. The depth of modulation of the output waveform can be adjusted by altering the amplitude of the modulating signal, or by varying the h.t. supply voltage. A disadvantage of the anode-modulated oscillator is that its oscillation frequency is not very stable.

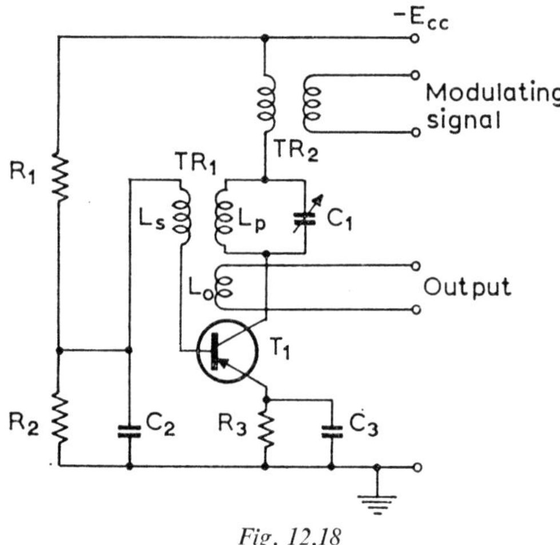

Fig. 12.18

The collector-modulated transistor oscillator

Transistor oscillators can also be amplitude-modulated and Fig. 12.18 shows a possible circuit. A tuned-collector oscillator with base coupling is used with the modulating signal introduced into the collector circuit. If required the modulating signal can, instead, be injected into the base circuit, or into the emitter circuit, by suitable positioning of transformer TR_2. In all cases the injected modulating signal changes the operating point of the transistor and this results in changes in the transistor's current gain. If the current gain varies in accordance with the modulating signal waveform the amplitude of the oscillations will also vary and an amplitude-modulated wave will be produced.

Exercises

1. Explain, with the aid of a diagram, how an oscillator may be anode modulated by an external audio-frequency signal.

   The amplitude of a 205 kHz wave is modulated sinusoidally at a frequency of 3 kHz between 0·5 V and 1·5 V.

   Determine the depth of modulation and state the frequencies present in the modulated wave. (A 1964)

2. Draw the circuit of a simple medium-frequency *L-C* oscillator using either a transistor or a triode valve, and explain how its output may be amplitude modulated.

   If such an oscillator has a fixed inductance of 60 μH and it is required to tune over the band 1–2 MHz, calculate the range of the variable capacitor to be used. (A 1962)

3. With the aid of a circuit diagram, describe how an oscillator may be anode modulated by an external audio-frequency source.

   Briefly explain what you understand by the sidebands of an amplitude-modulated wave, illustrating your answer with reference to a carrier modulated by a single-frequency tone. (A 1960)

4. Draw the circuit of an inductance-capacitance valve oscillator, indicating suitable values for the components.

   Explain the purpose of the various components and the factors which determine their values. (A 1959)

5. Describe briefly, with the aid of a circuit diagram, a method of producing an amplitude-modulated wave.

   An amplitude-modulated wave has a maximum value of 3 V and a minimum value of 1 V. What is the depth of modulation? (1 1958)

6. Describe the operation of a valve oscillator and indicate the way the component values of the tuned circuit affect the frequency of oscillation. (1 1957)

7. Explain how oscillations are maintained in a simple tuned-anode oscillator; where does the oscillatory power come from? An oscillator is to be designed to cover the frequency range 10 kHz to 100 kHz. If the single tuning capacitor used has a maximum-to-minimum capacitance range of 4:1, how many coils will be needed? (1 1956)

13 *Detection*

When intelligence is passed over a telecommunication system the intelligence signal is often not directly transmitted but, instead, is used to modulate a radio-frequency carrier wave and the modulated carrier wave is transmitted. At the receiving end the intelligence signal must be extracted from the modulated wave before use can be made of it, because if the modulated wave were applied directly to a loudspeaker or telephone receiver the device would not be able to vibrate quickly enough, and even if it could the resulting sounds would be above the range of frequencies the human ear is capable of hearing. The extraction of the intelligence signal from a modulated wave is known as detection or demodulation and the circuit employed is known as a detector or demodulator. Generally, the terms demodulation and demodulator are restricted to applications of s.s.b. amplitude-modulated signals. In this chapter only the detection of d.s.b. amplitude-modulated waves is to be discussed.

With amplitude modulation the intelligence signal is carried in the form of variations in the amplitude of a carrier wave, and the process of detection involves the production of a voltage that varies in the same way as the envelope of the modulated carrier wave. The output signal of an ideal detector is an exact replica of the intelligence signal modulated on to the carrier applied to the detector. If the output signal is not identical with the intelligence signal some distortion has been introduced by the detector.

A number of different kinds of detector circuit exist but most of them can be placed into one of two categories, (*a*) non-linear detectors and (*b*) linear detectors. A non-linear detector is a detector that has a non-linear current/voltage characteristic, as in Fig. 13.1*a*, and that detects by distorting the input waveform and extracting the mean value of the resulting output current. A linear detector, on the other hand, has a linear current/voltage characteristic, Fig. 13.1*b*, and performs the operation of detection by rectifying the input waveform and extracting the mean value of the resulting output current. In both cases the mean value, or d.c. component, fluctuates at the frequency of the modulating intelligence signal. Distortion, or rectification, of the input waveform before the mean value of the output current is extracted is necessary because the mean value of an undistorted amplitude-modulated wave over a large number of carrier cycles is zero.

A non-linear detector is the more sensitive because the required output voltage is proportional to the square of the input carrier voltage, whereas the output voltage of a linear detector is only proportional to the input carrier voltage itself.

The main difference between the design of detector circuits in valve radio receivers and in transistor receivers is that the detector

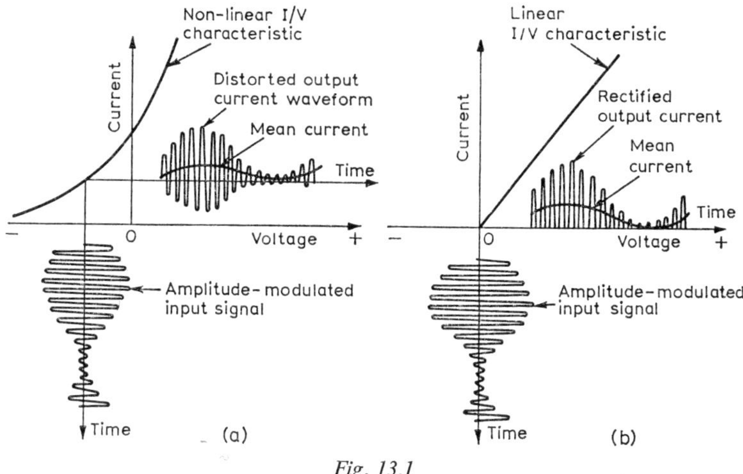

Fig. 13.1

The principle of (*a*) a non-linear detector and (*b*) a linear detector

in a valve receiver is required to provide a voltage for feeding the following stage (an audio-frequency amplifier), while in the transistor receiver the detector must supply power.

The Anode-bend Detector

The circuit of a pentode anode-bend detector is given in Fig. 13.2 and can be seen to be identical with the circuit of an R-C coupled pentode amplifier stage. The circuit acts primarily as a detector, and not as an amplifier, because of the operating voltages chosen. The valve is biased to operate either at the lower bend of its dynamic mutual characteristic or at the upper bend, see Fig. 13.3. The amplitude-modulated wave to be detected is applied across the input terminals of the circuit, which operates on a markedly non-linear portion of the dynamic mutual characteristic. The waveform of the anode current is distorted and its mean or d.c. component varies in the same way as does the modulation envelope and develops a voltage across the anode load resistor R_L. The detected voltage developed

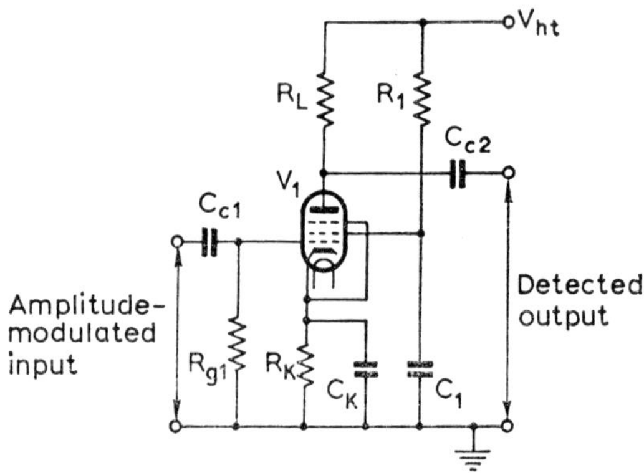

Fig. 13.2

The anode-bend detector

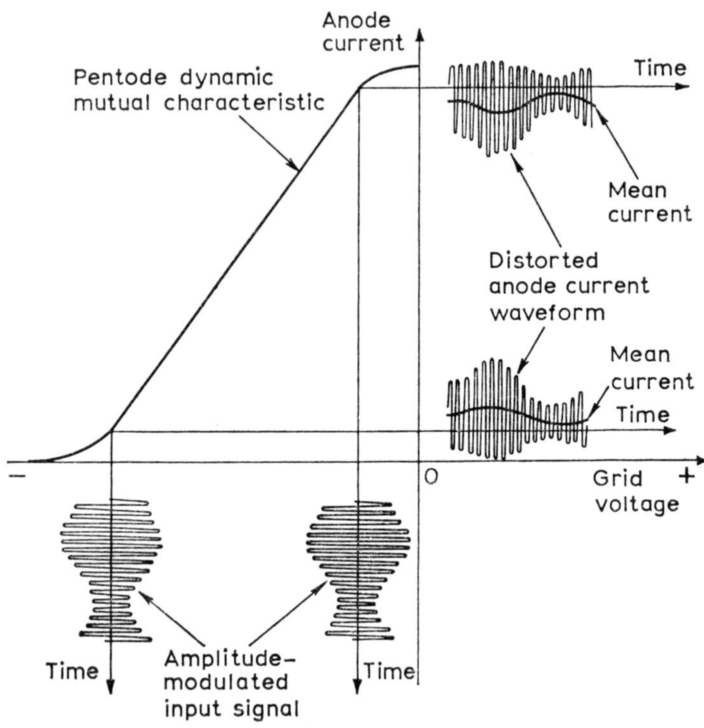

Fig. 13.3

The operation of an anode-bend detector

across R_L is coupled to the load via capacitor C_{c2}. A simple filter circuit is provided to remove a number of unwanted components at other frequencies that accompany the detected intelligence signal.

Anode-bend detection is rarely used nowadays because considerable distortion of the detected intelligence signal takes place, but it does have the advantages of a high input impedance and of providing amplification.

The Diode Detector

The diode detector is the most commonly used type of detector because it is capable of detection with the least distortion and it is also the cheapest circuit. A diode detector may give either a voltage output or a current output, depending on its circuit arrangement. For example, Fig. 13.4a shows the circuit of a voltage-output diode detector that consists essentially of a diode connected in series with a parallel resistor-capacitor network, and Fig. 13.4b shows a circuit

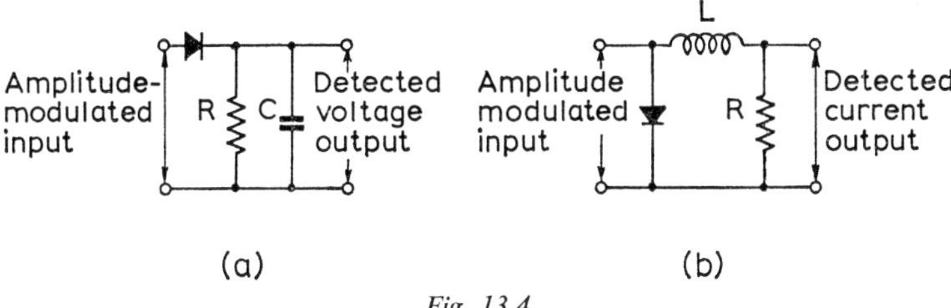

Fig. 13.4

Diode detector circuits providing (a) a voltage output and (b) a current output

that provides a current output and comprises a diode connected in parallel with a series inductor-resistor network.

Detectors giving a voltage output are always used in valve radio receivers because valves are voltage amplifying devices, but they are also widely employed in transistor receivers. In a valve receiver the diode may be either a thermionic valve or a semiconductor, but only semiconductor diodes are used in transistor receivers. The circuit giving a current output can be used in transistor receivers where the following stage (an audio-frequency amplifier) requires input power for its operation, a semiconductor diode always being used.

THE VOLTAGE-OUTPUT DIODE DETECTOR

Consider an amplitude-modulated wave to be applied to the input terminals of the circuit shown in Fig. 13.4a and suppose the capacitor to be disconnected. The diode will conduct during those half-cycles of the input signal that make the diode anode positive with respect to its cathode or, in the case of a semiconductor diode, make its p-type region positive with respect to its n-type region. Current will flow in a series of pulses whose amplitude is proportional to the amplitude of the voltage applied to the diode. The current pulses flow in the diode load resistor R and develop a voltage across it that varies in the manner shown in Fig. 13.1b. The mean value of the load voltage varies at the wanted modulation frequency and so detection has taken place. This simple circuit is an inefficient detector because only $1/\pi$ times the input voltage is available at the output terminals as the detected output.

A number of other components, at d.c., at the carrier frequency, and at harmonics of the carrier frequency, are also present.

If the capacitor is re-connected in the position shown in Fig. 13.4a the detection efficiency η (η is defined as the ratio of peak

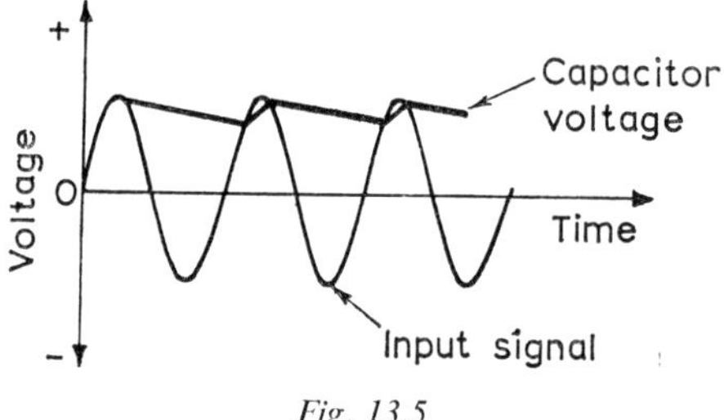

Fig. 13.5

Output voltage of a diode detector handling a signal of constant amplitude

output voltage at modulating frequency to peak input voltage), is improved approximately π times. The action of the capacitor in effecting this improvement is as follows. If an unmodulated carrier wave of constant amplitude is applied to the detector, the first positive half-cycle of the wave will cause the diode to conduct. The diode current will charge the capacitor to a voltage that is slightly less than the peak value of the input signal (slightly less because of a small voltage drop in the diode itself). At the end of this first half-cycle the diode ceases to conduct and the capacitor starts to discharge through the load resistor, R, at a rate determined by the time constant, CR seconds, of the discharge circuit. The time constant is chosen to ensure that the capacitor has not discharged very much before the next positive half-cycle of the input signal arrives to recharge the capacitor, see Fig. 13.5. The time constant for the charging of the capacitor is equal to Cr seconds, where r is the forward resistance of the diode and is much less than R. A nearly constant d.c. voltage is developed across the load resistor R; the fluctuations that exist are small and take place at the frequency of the input carrier signal. If, now, the input signal is amplitude-modulated the voltage across the diode load will vary in sympathy with the wave envelope, provided the time constant is small enough. The capacitor must be able to discharge rapidly enough for the voltage across it to follow those parts of the modulation cycle when the modulation envelope is decreasing in amplitude, see Fig. 13.6. The capacitor voltage falls until a positive half-cycle of the input signal makes the diode conduct and recharge the capacitor. When the modulation envelope is decreasing one positive half-cycle is of lower peak value than the preceding positive half-cycle and the

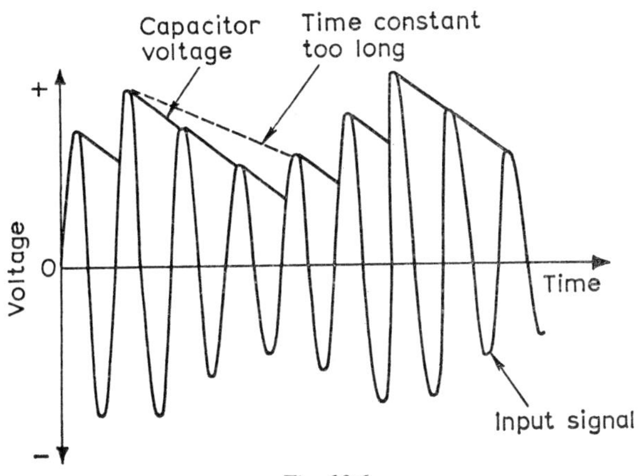

Fig. 13.6

Output voltage of a diode detector handling an amplitude-modulated signal

capacitor is recharged to a smaller voltage. If the time constant of the discharge path is too long, relative to the periodic time of the modulating signal, the capacitor voltage will not be able to follow the troughs of the modulation envelope; that is, the decay curve passes right over the top of one or more input voltage peaks, as shown by the dotted line in Fig. 13.6, and waveform distortion takes place. The time constant must not be too short, however, or the voltage across the load resistor will not be as large as it could be, because insufficient charge will be stored between successive pulses of diode current. The time constant determines the rapidity with which the detected voltage can change and must be long compared with the periodic time of the carrier wave and short compared with the periodic time of the modulating signal.

The voltage developed across the diode load resistor has three components: (*a*) a component at the wanted modulating signal frequency, (*b*) a d.c. component that is proportional to the peak value of the unmodulated wave (this component is not wanted for detection and must be prevented from reaching the following audio-frequency amplifier stage) and (*c*) components at the carrier frequency and harmonics of the carrier frequency that must also be prevented from reaching the audio-frequency amplifier. To eliminate the unwanted components the detector output is fed into a resistance-capacitance filter network before application to the audio-frequency amplifier.

THE CURRENT-OUTPUT DIODE DETECTOR

The circuit of a diode detector giving a current output is shown in Fig. 13.4*b*. During the first half-cycle of the input signal voltage that makes the semiconductor diode conduct, resistor R is shunted by the low forward resistance of the diode and little current flows in the inductor and the resistor. During the next half-cycle of the input signal the diode is non-conducting and current flows in the inductor and the resistor, setting up a magnetic field around the inductor. At the beginning of the next half-cycle, when the diode is again conducting, the magnetic field collapses and induces an e.m.f. in the inductor. This e.m.f. is in such a direction as to tend to maintain the flow of current for a time dependent upon the time constant, L/R seconds, of the circuit. If the time constant is chosen correctly a detected current output wave is obtained with minimum distortion of the waveform of the modulating signal.

THE RELATIVE MERITS OF VALVE AND SEMICONDUCTOR DIODES IN DETECTOR CIRCUITS

A thermionic-valve diode detector is capable of handling large signals with the minimum distortion of the modulating signal waveform but small input signals, less than about 1 volt peak, are distorted, because of the non-linear shape of a diode characteristic at low anode voltages. A thermionic valve requires a heater supply and must be mounted in a valve base. A semiconductor diode detector tends to introduce slightly more distortion of strong signals and has a lower input resistance, but can work into a lower-resistance load. The semiconductor diode has the advantages of small size, lower weight, not requiring a heater supply, being directly solderable into a circuit, and greater reliability.

The Leaky-grid Detector

With a leaky—or cumulative—grid detector, detection takes place in the grid/cathode circuit of a triode, or pentode, valve. The grid and cathode form, respectively, the anode and cathode of an

.equivalent diode valve and an *R-C* parallel network in the grid circuit provides the diode load. The detected signal appears across this load and is amplified by the valve acting as a triode, or pentode, in the usual way.

The circuit of a triode leaky-grid detector is shown in Fig. 13.7 and consists of a transformer-fed *R-C* coupled amplifier employing a form of leaky-grid bias. Consider, initially, that an unmodulated carrier wave is applied to the input terminals of the circuit. The peak of each positive half-cycle of the wave will take the grid positive with respect to the cathode; grid current will flow and charge capacitor C_1 to a voltage nearly equal to the peak value of the input signal with the polarity shown. During the remainder of the positive

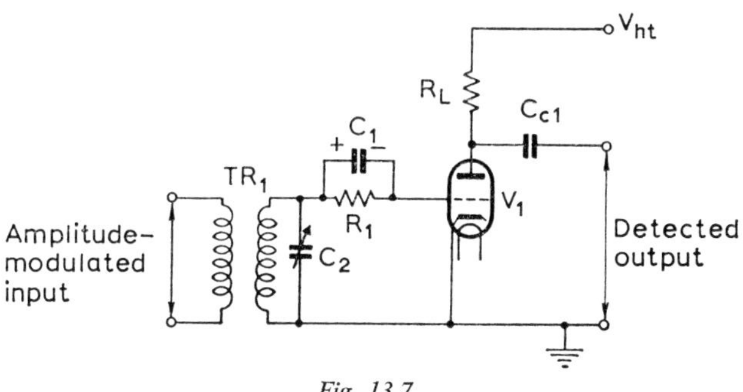

Fig. 13.7

The leaky-grid detector

half-cycles and all of the negative half-cycles the grid is negative relative to the cathode and no grid current flows. When C_1 is not being charged by the flow of grid current it starts to discharge through resistor R_1 at a rate determined by the time constant, $C_1 R_1$ seconds, of the discharge circuit. Provided the time constant is long compared with the periodic time of the input signal the capacitor voltage remains approximately constant. The grid of the valve is held negative relative to the cathode and a steady anode current flows. If the input signal is now amplitude-modulated the capacitor voltage will follow the modulation envelope provided that the capacitor is able to discharge rapidly enough. A voltage varying at the modulating signal frequency is thus applied to the grid of the valve and is amplified to appear across the anode load resistor R_L. The output voltage, which is taken off via coupling capacitor C_{c1}, also contains various unwanted radio frequencies and these are filtered off.

The leaky-grid detector is rarely used nowadays because it suffers from the disadvantage that it can only handle small signals without the introduction of an undue amount of distortion.

The Transistor Detector

The circuit of a common-emitter transistor detector is shown in Fig. 13.8. The operation of the circuit is very similar to the operation of the leaky-grid detector; rectification takes place in the emitter/base circuit and the rectified signal is amplified by the transistor in the usual way.

Components R_1, R_2, R_4, C_2 and C_3 provide bias and d.c. stabilization and R_3 is the collector load resistor. C_4 is a by-pass capacitor that is provided to prevent voltages at the carrier frequency appearing across R_3 and being fed, via C_5, to the output terminals of the circuit. The input amplitude-modulated signal is fed into the base/emitter circuit via r.f. transformer TR_1, the primary winding of which is

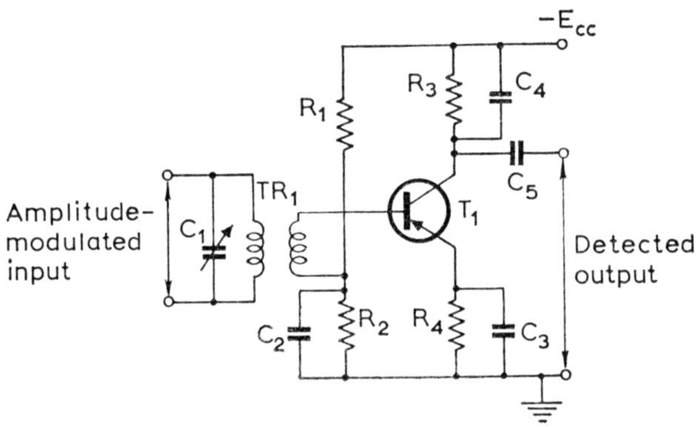

Fig. 13.8

The transistor detector

tuned to the carrier frequency. The base/emitter junction of transistor T_1 acts as a semiconductor diode and together with R_2 and C_2 forms a diode detector. The detected voltage appears across R_2 and varies the emitter/base bias voltage of the transistor and this variation causes the collector current to vary in accordance with the modulation envelope. A voltage at the modulating signal frequency appears across the collector load resistor, R_3, and this is coupled to the load by capacitor C_5.

A detector circuit using a transistor connected in the common-base configuration is also possible, detection again taking place in the emitter/base circuit.

Exercises

1. Explain the need for a detector in a radio receiver.

   Describe, with the aid of a diagram, the operation of a semiconductor diode in the detection of an amplitude-modulated wave.

   What are the relative merits of thermionic and semiconductor diodes for this purpose?   (A 1964)

2. Explain the rectifying action of a semiconductor diode.

   Briefly discuss the relative advantages and disadvantages of thermionic and semiconductor diodes for the detection of amplitude-modulated waves.
   (A 1963)

3. Briefly explain why it is necessary to include a detector stage in a receiver used for the reception of amplitude-modulated signals.

   Describe, with the aid of a diagram, the operation of a semiconductor diode in the detection of an amplitude-modulated wave. (A 1962)

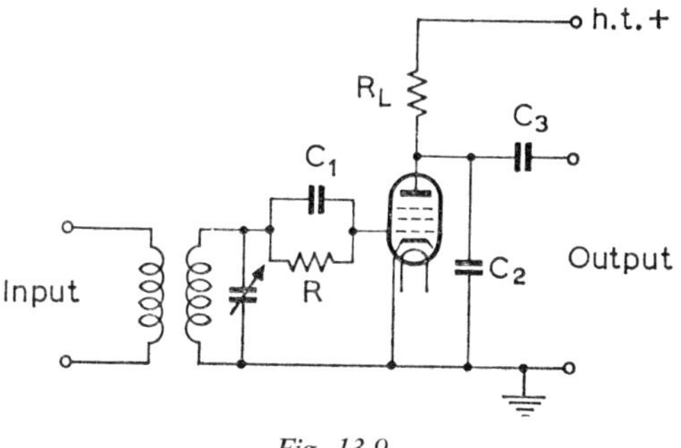

Fig. 13.9

4. Fig. 13.9 shows the circuit of a leaky-grid detector. By reference to an amplitude-modulated signal applied to the input of the circuit, explain the operation of the detector, mentioning particularly the function of each of the lettered components. (A 1961)

5. Briefly explain why it is necessary to include a detector stage in a receiver for amplitude-modulated signals.

   Discuss the characteristics of thermionic and semiconductor diodes which make them suitable for the detection of a.m. signals. (A 1960)

6. Explain, with the aid of diagrams, the action of either a leaky-grid detector or an anode-bend detector when used for the detection of amplitude-modulated waves. (1 1957)

14 *Receiving Aerials*

In a radio system, be it for point-to-point communication or for sound and/or television broadcasting, the intelligence signal is modulated on to a radio-frequency carrier wave and is then radiated into the atmosphere, in the form of an electromagnetic wave, by a transmitting aerial. For the signal to be received a receiving aerial is required that is capable of intercepting the electromagnetic wave and extracting a small amount of energy from it. This energy can then be applied to the receiving equipment for amplification and detection. The power radiated by a transmitting aerial may be very large, perhaps several kilowatts, but the power absorbed by a receiving aerial is small, being only a microwatt or so.

This chapter will be concerned only with the more common types of aerial that are used for the reception of sound and television broadcasts but, first, an outline of the radiation of energy from an aerial and the nature of an electromagnetic wave will be given.

Radiation from an Aerial

If an a.c. generator is connected to the middle of a length of conductor, as shown in Fig. 14.1, a current will flow because a complete circuit is formed by the capacitance between the ends of the conductor. Since the conductor must have some self-inductance a series-resonant circuit exists, and the current will be at a maximum if the frequency

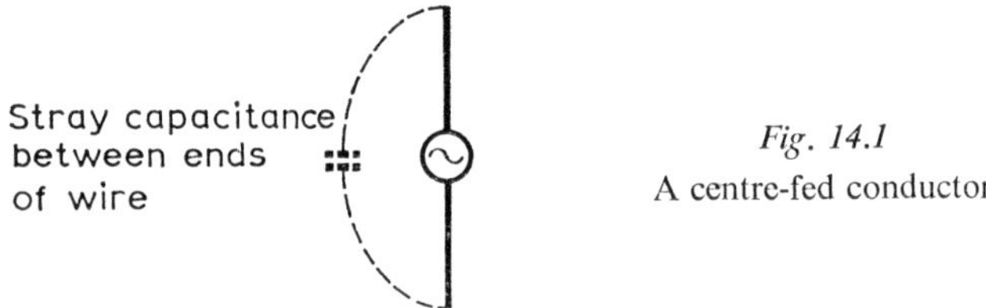

Fig. 14.1
A centre-fed conductor

of the generator corresponds to the resonant frequency of the circuit. Whenever a conductor carries a current it is surrounded by a magnetic field and if the current changes the magnetic field changes also. Now a varying magnetic field *always* produces an electric field, in the form of closed loops of electric force, that *only* exists while the magnetic field is changing. When the magnetic field is steady the electric field disappears. The direction of the electric field depends on whether the magnetic field is building up or is collapsing and can always be deduced by the application of Lenz's law. Similarly, a changing electric field *always* produces a magnetic field, and this means that a conductor carrying an a.c. current is surrounded by continually changing magnetic and electric fields that are completely dependent upon one another. Although a stationary electric field can exist without the presence of a magnetic field, and a stationary magnetic field does not require an associated electric field, it is impossible for either field to exist separately when changing.

If the current in a conductor is varying sinusoidally the electric

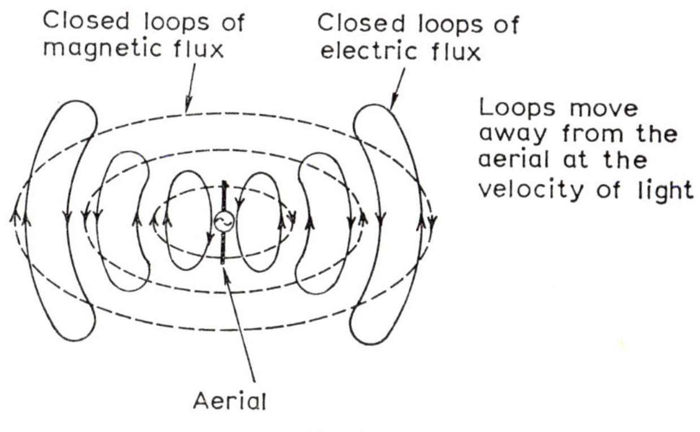

Fig. 14.2

Radiation from an aerial

and magnetic fields around the conductor will also attempt to vary sinusoidally. Thus when the current has its maximum value in one direction the two fields are also at a maximum in a particular direction: when the current decreases and then rises to its maximum value in the reverse direction the two fields attempt to follow suit. For this to happen successfully, all the energy contained in the magnetic and electric fields when they are at their maximum values must either return to the conductor *or* be radiated away from the conductor. At power and audio frequencies all the energy succeeds in returning to the conductor and no radiation takes place. A finite time is required for a magnetic field, and its associated electric field, to collapse, and at radio frequencies not all the energy has returned to the conductor before the current has started to increase in the opposite direction and to create new electric and magnetic fields. The energy left outside the conductor cannot return to it and, instead, is propagated away from the conductor at the velocity of light (approximately 3×10^8 metres/second), Fig. 14.2. The amount of energy radiated from the conductor increases with increase in frequency, since more and more energy is unable to return to the conductor.

The energy radiated from the conductor, or *aerial*, is transmitted in the form of an electromagnetic wave in which there is a continual interchange of energy between the electric and magnetic fields. The radiated field strength is proportional to the frequency of the wave and inversely proportional to the distance from the aerial. In the radiated field the electric and magnetic fields are at right angles to one another, and are also at right angles to the direction of propagation, as shown in Fig. 14.3 for a particular instant in

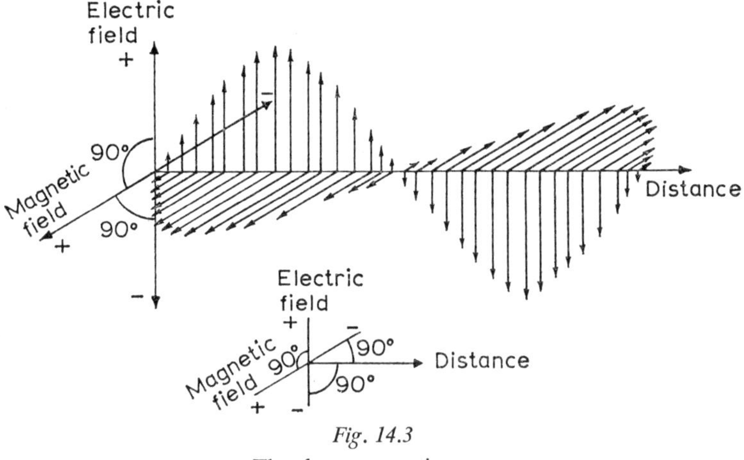

Fig. 14.3

The electromagnetic wave

time. The plane containing the electric field and the direction of propagation of the electromagnetic wave is known as the plane of polarization of the wave. For example, if the electric field is in the vertical plane and the magnetic field is at $90°$ to this, in the horizontal plane, the wave is said to be vertically polarized.

In the immediate vicinity of an aerial the electric and magnetic fields are of greater magnitude and different relative phase than in the radiated field. This is because there is, in addition to the radiation field, an induction field near the aerial. The induction field represents energy that is not radiated away from the aerial and corresponds to the fields that would surround the aerial if the aerial current was constant and not alternating. The magnitude of the induction field is independent of frequency and decays inversely as the square of the distance from the aerial. Near to the aerial the induction field is larger than the radiation field but, since the latter only falls off as the reciprocal of distance from the aerial, this situation changes. The radiation field becomes stronger than the induction field at distances greater than $\lambda/2\pi$, where λ is the wavelength of the signal.

For a simple one-conductor aerial, such as that shown in Fig. 14.1, the radiated electromagnetic wave moves away from the aerial in all directions normal to the aerial. For example, for a vertical aerial the wave travels in the horizontal plane as shown in Fig. 14.4. It should be noted that a vertical conductor produces a

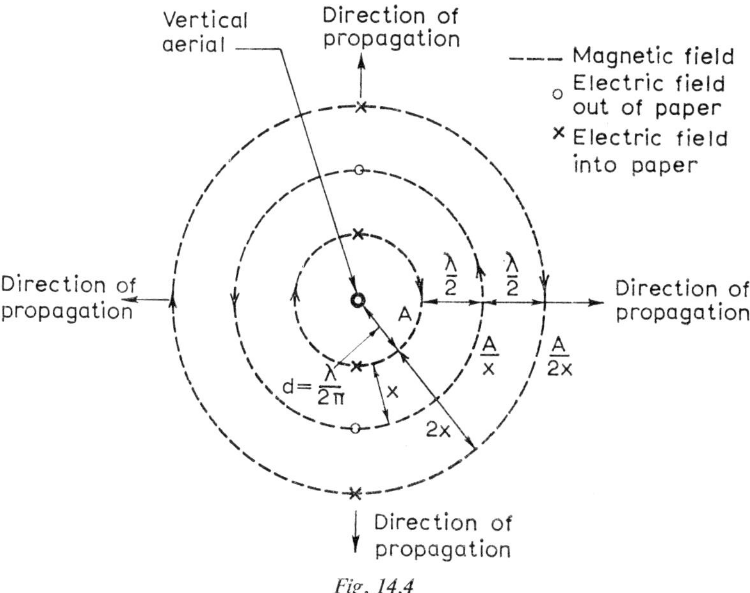

Fig. 14.4

Radiation fields about an omnidirectional vertical aerial

vertically polarized wave, i.e. the electric field is in the vertical plane. At a distance $d = \lambda/2\pi$ from the aerial the radiated field predominates and the electric field strength is equal to A volts/metre. At a distance $x = \lambda/2$ more distant from the aerial the polarities of the electric and magnetic fields are reversed and the field strength has fallen to a new value A/x volts/metre. At a distance $(\lambda/2\pi) + \lambda$ from the aerial the fields have again reversed polarity to their original directions at $d = \lambda/2\pi$, and the field strength has still further fallen to $A/2x$ volts/metre, and so on.

The amplitudes of the electric E and magnetic H fields in an electromagnetic wave bear a constant relationship to one another. This relationship is known as the impedance of free space and is the ratio of the electric field strength to the magnetic field strength, that is

$$\text{Impedance of free space} = \frac{E \text{ volts/metre}}{H \text{ ampere turns/metre}}$$

$$= 120\pi \text{ ohms} \qquad (14.1)$$

$$= 377 \text{ ohms} \qquad (14.1a)$$

Example 14.1
The magnetic field strength at a point 5 miles from a transmitting aerial is 0.0265×10^{-3} AT/m. Calculate the electric field strength at a point 10 miles from the aerial in the same direction.

Solution. From equation (14.1*a*),

$$\frac{E}{H} = 377$$

or $$E = 377\,H$$
∴ $$E_5 = 377 \times 0.0265 \times 10^{-3} = 10 \text{ mV/m}$$

∴ At a point 10 miles distant from the aerial the electric field strength E_{10} is,

$$E_{10} = \frac{10}{2} = 5 \text{ mV/m} \qquad Ans.$$

Radiation Patterns, Directivity and Aerial Gain

Aerials have the property of being able to transmit power better in some directions than in others, and this directional characteristic is very useful because it allows most of the transmitted power to be sent in a wanted direction and very little in unwanted directions. The directional characteristic of a particular type of aerial is expressed by its radiation pattern (or polar diagram as it used to be called).

The radiation pattern of an aerial is a graphical representation of the way in which the power transmitted by the aerial varies at equal distances from the aerial. A radiation pattern is normally drawn in either the vertical plane or the horizontal plane, and refers to the performance of the aerial itself, i.e. when mounted well away from any objects, such as buildings, that might have some effect on the field strength produced by it. Since aerials are generally mounted near to some object or other, a radiation pattern does not give a true indication of the performance to be expected from a particular aerial installation, but it does give a method of comparing different types of aerial. Since any aerial is equally good at receiving signals as transmitting them, the radiation pattern of an aerial also gives an indication of the aerial's receiving capabilities.

Directivity in a receiving aerial is useful because it enables the aerial to distinguish, to some extent, between wanted and unwanted signals. A simple vertical dipole aerial will transmit, or receive, equally well in all directions in the horizontal plane and so its horizontal-plane radiation pattern is a circle, Fig. 14.5. Further radiation patterns are given later.

A map of the field strength actually obtained in the neighbourhood of an aerial installation is given by a field strength diagram. A

Fig. 14.5

The horizontal-plane radiation
pattern of a vertical dipole

field strength diagram shows a number of plots of points of equal field strength around an aerial, a typical diagram being given in Fig. 14.6. All lines in the diagram join points of equal field strength, e.g. 5 mV/m, and are clearly not at constant distances from the aerial. Factors such as tall buildings and hills can cause a drastic reduction

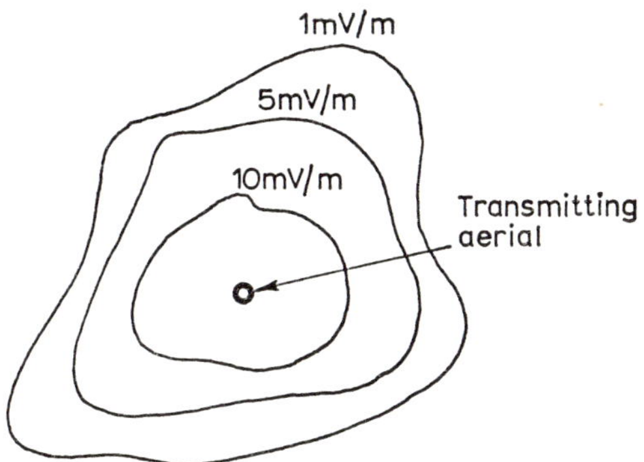

Fig. 14.6

A field strength diagram

in field strength and this is shown in the diagram by a line moving nearer to the aerial.

The gain of an aerial is a measure of its directional properties and it indicates either the extent to which radiation is concentrated in a particular direction, or the extent to which the aerial receives signals better from one direction than from others. Aerial gain is defined relative to a reference aerial; the reference aerial is usually a single straight conductor that is half a wavelength long and is known as the $\lambda/2$ dipole. The gain of an aerial is the same whether it is used for transmitting or receiving but may be defined in terms of either. The gain of a transmitting aerial is the ratio G of the field strength produced at a point in the direction of maximum radiation from the aerial to the field strength produced at the same point by the reference aerial, both aerials being fed with the same power. The gain of a receiving aerial is the ratio G of the voltage appearing at the open-circuited terminals of the aerial to the voltage appearing at the open-circuited terminals of the reference aerial, the field strengths at the locations of the aerials being the same. In decibels, the gain of an aerial is given by $20 \log_{10} G$.

Receiving Aerials

A receiving aerial is able to abstract energy from an incident electromagnetic wave because the magnetic field of the wave cuts the aerial conductor and induces an e.m.f. in it. This means that the aerial must not be in the same plane as the magnetic field, i.e. a vertical aerial is required to receive a vertically polarized wave. The magnitude of the induced e.m.f. is proportional to the density of the magnetic flux cutting the aerial which, in turn, is proportional to the field strength and the length of the aerial. If the aerial terminals are closed in a suitable impedance a current will flow in the aerial and this current will be a maximum when the aerial is resonant (remember that an aerial is similar to a series-tuned circuit). For the aerial to resonate it must be a whole number of quarter-wavelengths long but, for mechanical and economic reasons, this is not possible at the lower radio frequencies.

THE DIPOLE AERIAL

A dipole aerial consists of a straight conductor, of resonant length at the signal frequency to be received, that is mounted, horizontally or vertically, in the air and that has a pair of output terminals across which a load, such as the input impedance of a radio receiver, can be connected. An incident electromagnetic wave will induce an e.m.f. into the dipole and, if its terminals are closed, an aerial current will flow. The current and voltage distributions in a dipole aerial depend upon the length of the aerial and three typical distributions are given in Fig. 14.7. The magnitude of the aerial current varies along the length of the dipole but it is always zero at the ends. Dipoles longer than $\lambda/2$ have currents in antiphase in different parts of the aerial and, since this adversely affects aerial performance, the $\lambda/2$ length is by far the most common. The ratio of voltage to current at the input terminals of the aerial is the aerial input impedance and for a $\lambda/2$ dipole is approximately 73 ohms pure resistance.

The radiation pattern of a vertical $\lambda/2$ dipole is a circle in the horizontal plane, Fig. 14.5, and a figure-of-eight shape in the vertical plane, Fig. 14.8. The circular pattern shows that in the horizontal plane a vertical $\lambda/2$ dipole receives signals from all directions

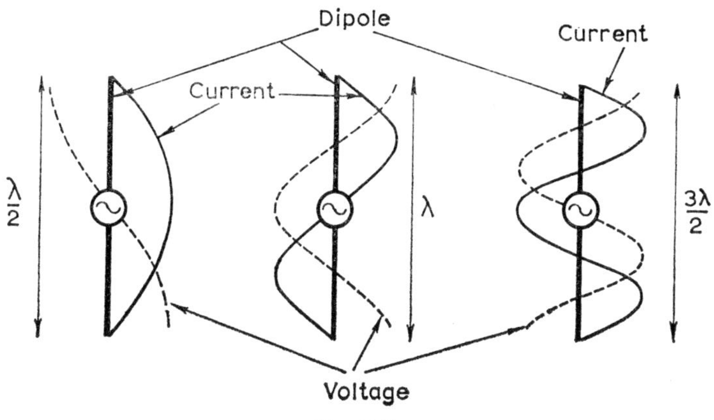

Fig. 14.7

Current and voltage distributions in vertical dipoles of **different lengths**

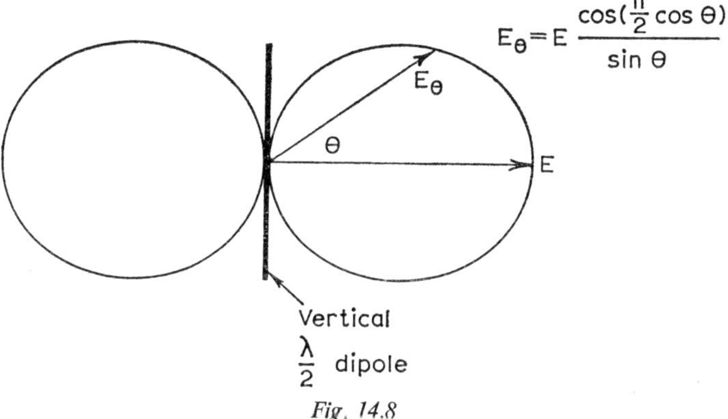

Fig. 14.8

The vertical plane radiation pattern of a vertical dipole

with equal efficiency, the vertical plane pattern shows that a vertical dipole does not receive at all in the line of the aerial axis.

THE UNIPOLE AERIAL

Fig. 14.9 shows a $\lambda/4$, or unipole, aerial that is mounted immediately above earth and whose input terminals arc between the base of the aerial and earth. The e.m.f. induced in such an aerial is directly proportional to the total magnetic field that cuts the aerial, and this field consists of waves that reach the aerial directly and waves that reach the aerial after reflection from the earth. The induced e.m.f. is exactly the same as the e.m.f. that would be induced in an aerial of twice the length, mounted half above the ground and half below, if no reflection took place but, instead, the reflected wave took the

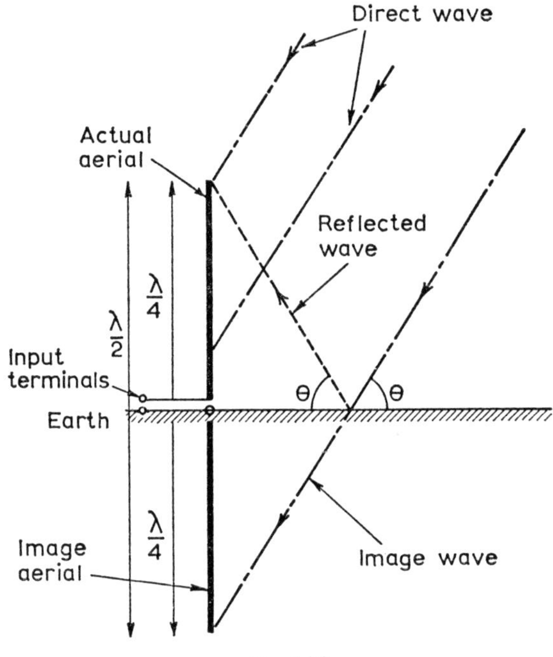

Fig. 14.9

The unipole, $\lambda/4$, aerial

path marked "image wave". The imaginary half-aerial beneath the ground is known as the image aerial.

At frequencies in the long and medium wavebands it is impracticable to use either $\lambda/2$ dipoles or $\lambda/4$ unipoles as receiving aerials because of the excessively large structures that would be required. For example, consider an aerial that is to receive the B.B.C. Home Programme on 908 kHz. The wavelength of this transmission is $(3 \times 10^8)/(908 \times 10^3) = 330$ metres and hence even a $\lambda/4$ aerial would need to be 82·5 metres high. At the lower radio frequencies, therefore, the receiving aerials used are much less than $\lambda/4$ in length.

THE INVERTED-L AERIAL

The inverted-L aerial is designed to receive low and medium frequency, vertically polarized signals and its construction is shown in Fig. 14.10. The aerial proper in which e.m.f.s are induced is the downlead and it is much less than $\lambda/2$ in length. The downlead must be cut by the horizontal magnetic field of the incident electromagnetic wave and it should be as near vertical as possible. No e.m.f. is induced

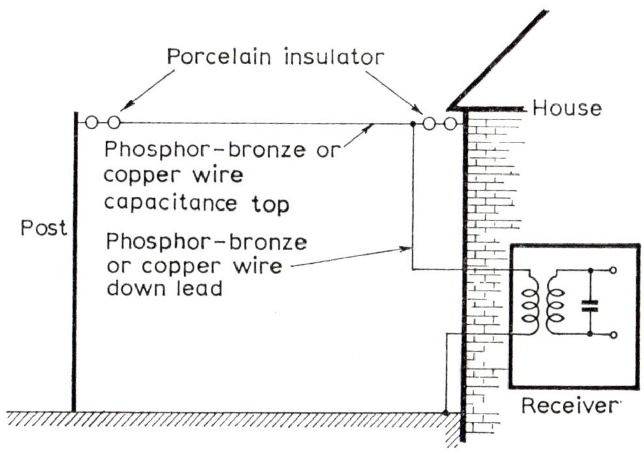

Fig. 14.10

The inverted-L aerial

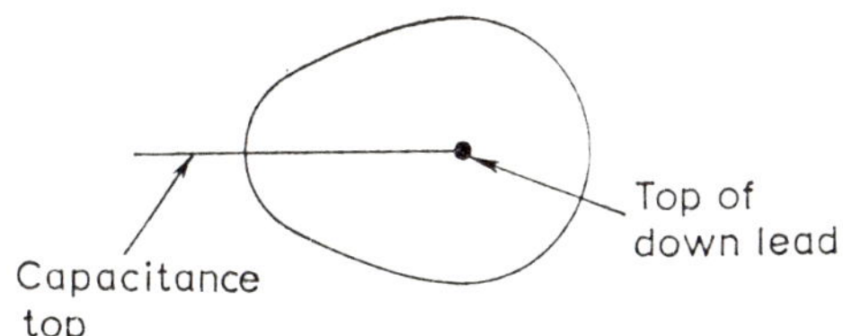

Fig. 14.11

The horizontal plane radiation pattern of an inverted-L aerial

in the long horizontal section of the aerial, which is provided to increase the effective length of the aerial and thereby increase the aerial current. Increase in the aerial current makes the aerial more efficient. The horizontal section is known as the capacitance top.

The radiation pattern of an inverted-L aerial is shown in Fig. 14.11 and it can be seen that the aerial exhibits slight directivity, receiving somewhat better in the direction from capacitance top to downlead.

For the best results the aerial should be mounted as high as possible and be well clear of buildings that might reduce the field strength of the signals to be received. This type of aerial can be used for domestic radio installations and is ideal when reception of distant stations is required.

THE T AERIAL

For some installations it may be more convenient to have the downlead connected to the centre of the capacitance top, as shown in Fig. 14.12. Such an arrangement, known as a T aerial because of its

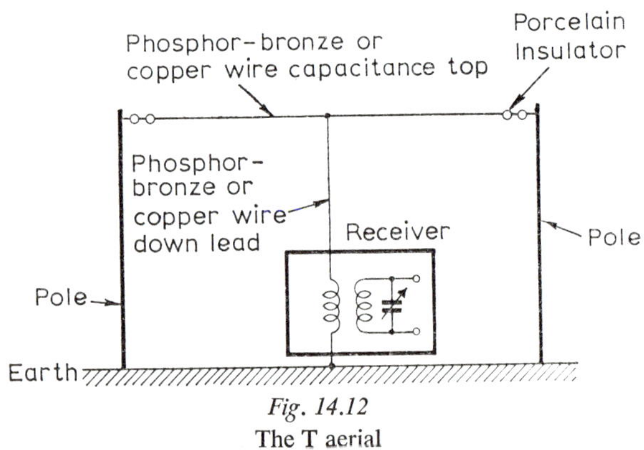

Fig. 14.12

The T aerial

appearance, would not be used in conjunction with a receiver in a house but it is employed, for example, on board ship. The T aerial receives equally well from all directions and so its radiation pattern is a circle.

THE ROD AERIAL

For the best results an aerial should not only have the maximum response in the required direction(s) but it should also pick up the minimum noise and interference. Noise signals can originate from a number of different sources and may be either directly radiated or propagated along the mains supply wiring and then radiated from it. There is a risk that an inverted-L aerial may pick up unwanted noise signals, and if the level of such signals is considerable it may be necessary to employ a rod aerial instead. A rod aerial consists merely of a short length of conductor, mounted on a chimney or a gutter, that employs a screened, balanced, feeder to connect it to the receiver, see Fig. 14.13. The aerial is connected to the feeder via a screened transformer, the turns ratio of which is chosen to match the impedance of the feeder to the aerial impedance. At its other end the

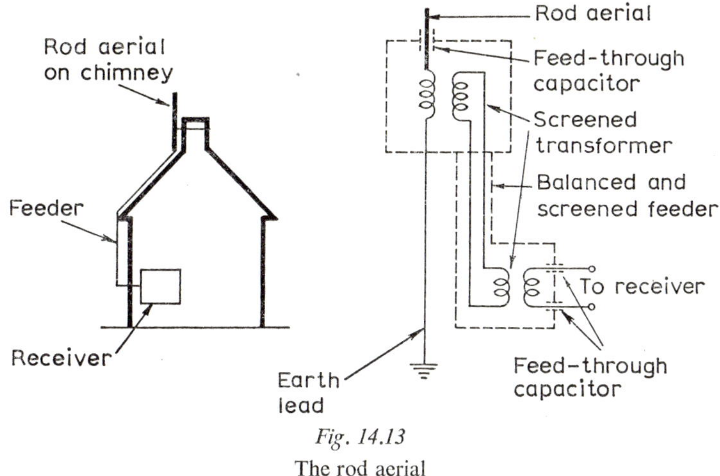

Fig. 14.13

The rod aerial

feeder is connected, via another screened transformer, to the input terminals of the receiver. An earth lead is connected to the primary winding of the aerial transformer to complete a circuit for the aerial current to flow. The screening arrangements and the use of a balanced feeder reduce the interference picked up by the aerial to a low level.

INDOOR AERIALS

If a receiver is only to be used for the reception of strong signals, or where a portable receiver is concerned, a small aerial that will fit inside the receiver cabinet is required. Two types of indoor aerial are in common use, the loop, or frame, aerial and the ferrite-rod aerial.

A loop aerial consists of one, or more, turns of wire that are connected to the input terminals of the receiver. Fig. 14.14 shows a rectangular loop aerial that has four turns and is tuned to the signal frequency by a variable capacitor. The voltage delivered to a receiver by an aerial of this type depends upon the area of the loop and the number of turns of wire employed; the former is limited by the need to fit the aerial within a confined space. A loop aerial has a figure-of-eight radiation pattern in the horizontal plane, see Fig. 14.15.

Modern radio receivers, particularly transistorized models, usually employ a ferrite-rod aerial. A ferrite-rod aerial consists of one or more tuning coils mounted on a rod of ferrite that is some $\frac{1}{4}$ in. to $\frac{1}{2}$ in. in diameter. This type of aerial has the advantage of small physical size enabling it to be mounted horizontally on the chassis of a radio receiver. Fig. 14.16 shows the construction of a typical ferrite-rod aerial. A single coil giving coverage of one waveband is shown wound on the rod; if coverage of two bands, e.g. long and medium, is required another coil must be wound. Ferrite is a material that combines high permeability with high resistivity and the action of the ferrite rod is to concentrate the magnetic field of an incident electromagnetic wave. This action increases the flux density of the

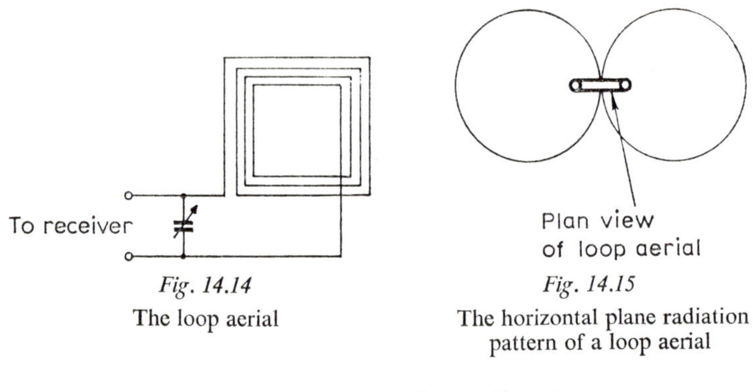

<table>
<tr><td>

To receiver

Fig. 14.14

The loop aerial

</td><td>

Fig. 14.15

The horizontal plane radiation
pattern of a loop aerial

</td></tr>
</table>

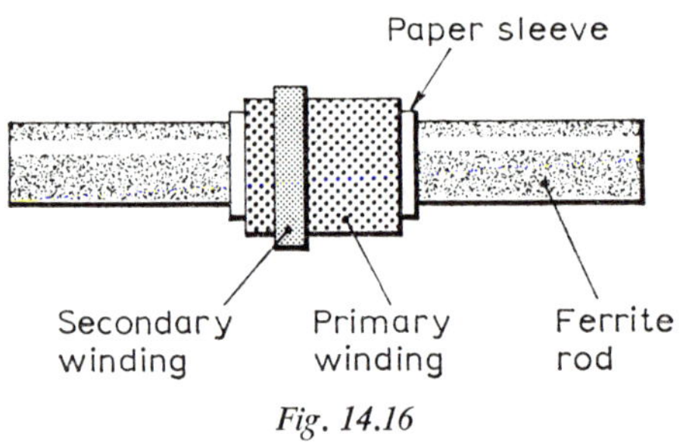

Fig. 14.16

The ferrite-rod aerial

field that cuts any coils wound on the ferrite rod and thereby increases the e.m.f. induced in them. The radiation pattern of a ferrite-rod aerial is given in Fig. 14.17, clearly for maximum response the aerial should be orientated to lie 90° to the direction of the incident electromagnetic wave. The ferrite-rod aerial always forms an integral part of the radio-frequency section of a radio receiver and a typical circuit arrangement is shown in Fig. 14.18.

DIPOLES FOR THE RECEPTION OF V.H.F. SIGNALS
At v.h.f. the wavelength of a signal to be received is fairly short, e.g. at 50 MHz $\lambda = 6$ metres, at 100 MHz $\lambda = 3$ metres, and it is practical to employ resonant $\lambda/2$ dipole aerials. The $\lambda/2$ dipole

consists of two rods of aluminium tubing, each $\lambda/4$ in length, which are mounted one above the other and terminated by a coaxial feeder at their centre point, see Fig. 14.19a. If a two-wire feeder is employed this method of connection will probably not be possible because

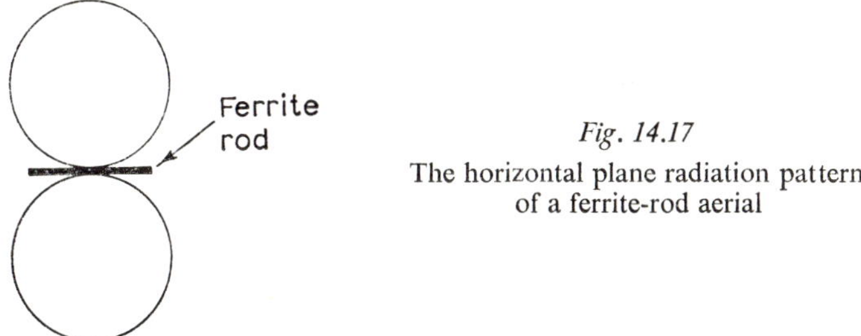

Fig. 14.17

The horizontal plane radiation pattern
of a ferrite-rod aerial

of the mismatch caused by the much higher impedance of the feeder. Connection between feeder and aerial can then be made using the method of Fig. 14.19b; in this method the tapping distance x is selected to give a reasonable match between the aerial and feeder

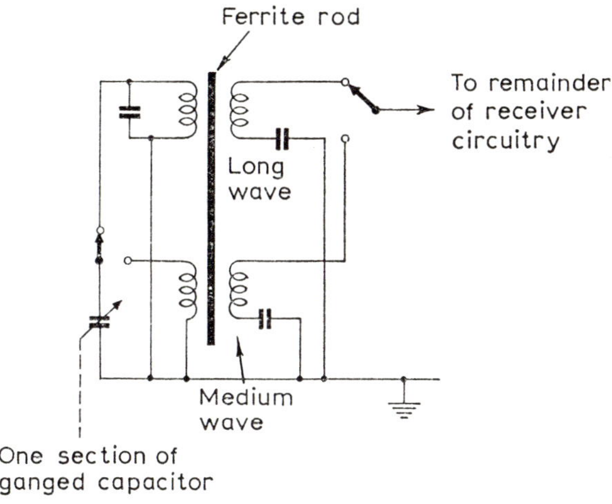

Fig. 14.18

Coupling the ferrite-rod aerial to a radio receiver

impedances (remember, from Fig. 14.7, the current and voltage distributions, and hence the aerial impedance, vary along the length of the aerial).

It is often necessary to raise the input impedance of a $\lambda/2$ dipole to facilitate matching to the feeder and this can be achieved by folding the dipole. A $\lambda/2$ dipole can be folded in a number of different ways to give an increase in input impedance of some 3 to 12 times. The appearance of one type of folded $\lambda/2$ dipole is shown in Fig. 14.20; such an aerial gives an increase in input impedance of about four times.

The bandwidth of a $\lambda/2$ dipole, defined in the same way as for

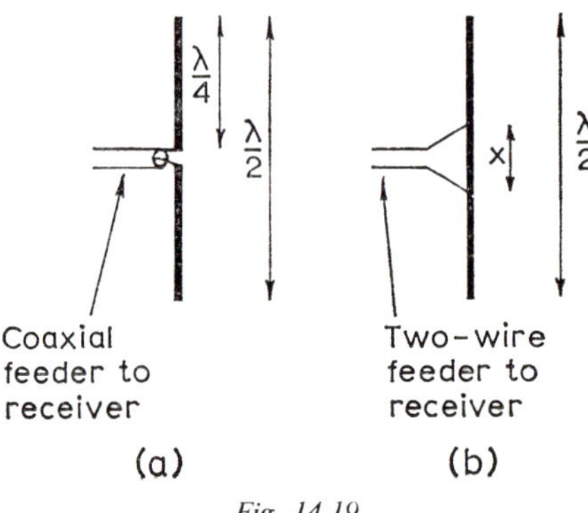

Fig. 14.19

$\lambda/2$ dipole fed by (*a*) coaxial feeder and (*b*) two-wire feeder

amplifiers, is a function of the diameter of the tubing from which it is constructed. The bandwidth for a particular centre frequency can be fixed by the suitable choice of tube diameter; the wider the bandwidth required the greater must be the tube diameter employed. Folding a dipole increases its bandwidth.

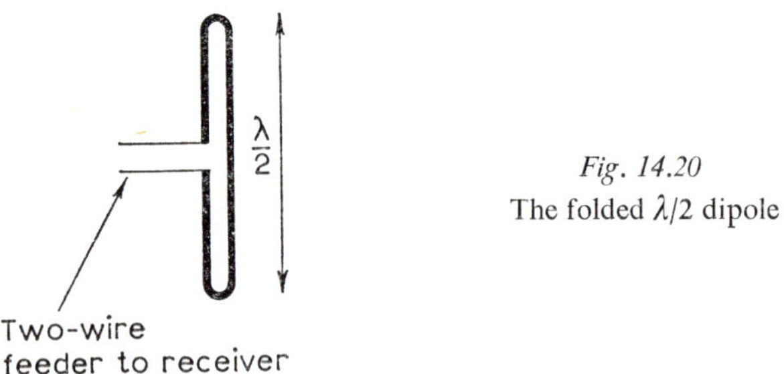

Fig. 14.20

The folded $\lambda/2$ dipole

In the horizontal plane the radiation pattern of a vertical $\lambda/2$ dipole is a circle, which means that the aerial will receive vertically polarized signals equally well from all directions. The performance of such an aerial can be greatly improved if it is given some measure of directivity. A $\lambda/2$ dipole aerial can be made directive by the addition of a reflector and one or more directors.

Both reflectors and directors are passive elements made of straight lengths of aluminium rod mounted at their centre on a metal support rod. The length of these elements is such that they are near, but not at, resonance at the signal frequency. A reflector is always

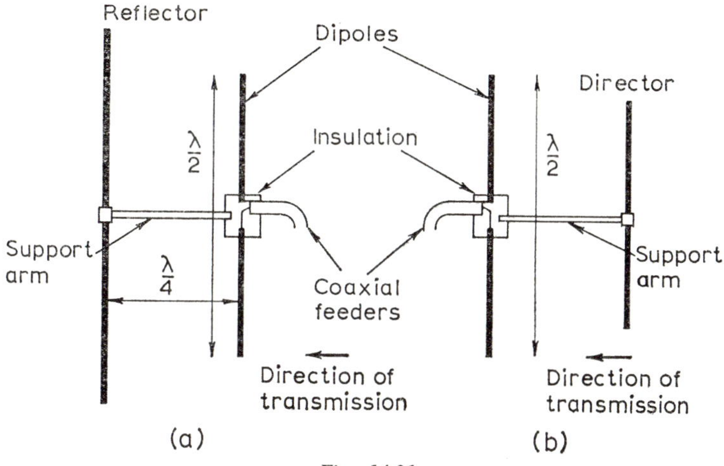

Fig. 14.21

$\lambda/2$ dipole with (*a*) a reflector and (*b*) a director

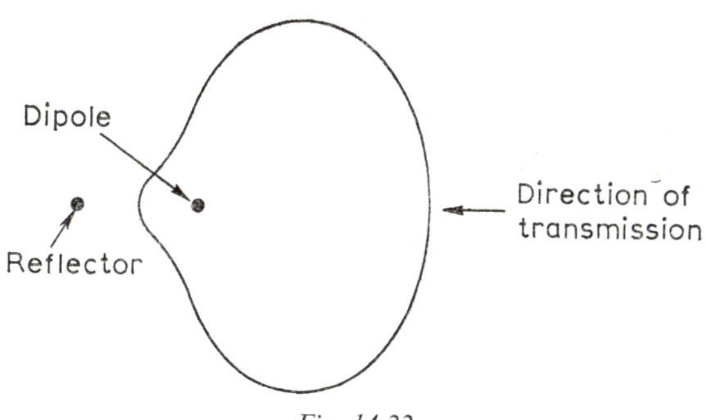

Fig. 14.22

Radiation pattern of a $\lambda/2$ dipole and reflector

slightly longer than $\lambda/2$ in length and is mounted approximately $\lambda/4$ behind the dipole, see Fig. 14.21*a*. A director is always somewhat shorter than $\lambda/2$ in length and is mounted between the dipole and the transmitter, Fig. 14.21*b*. The action of a reflector element is rather complex but, briefly, e.m.f.s are induced into the reflector and their energy is re-radiated with such a phase that, after travelling to the dipole, it aids signals arriving from the wanted direction and opposes signals arriving from the unwanted directions. The overall effect of a reflector is to make a $\lambda/2$ dipole more directive, as shown by the radiation pattern of a $\lambda/2$ dipole with a reflector given in Fig. 14.22. It is evident that while the aerial gain is now greater in the wanted direction considerable gain still exists in other directions, and it is the function of director elements to still further increase the aerial directivity. A progressive increase in gain can be achieved by adding director elements, each added director being slightly shorter than the previous one. The addition of directors to a dipole

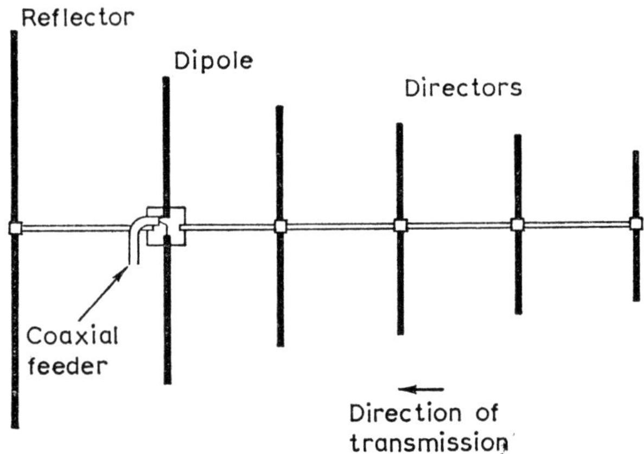

Fig. 14.23

A multi-element $\lambda/2$ dipole aerial

aerial reduces the impedance of the aerial, and it often becomes necessary to increase the impedance by the use of a folded dipole. A typical multi-element dipole aerial is shown in Fig. 14.23.

For the best performance a dipole aerial should be mounted as high as possible and be well clear of surrounding buildings, and it should then be orientated to give the best results.

Exercises

1. Sketch the polar diagrams, in the horizontal plane, of the following types of aerial—

    (*a*) a horizontal dipole,
    (*b*) a half-wave vertical dipole with reflector,
    (*c*) an inverted-L,
    (*d*) a ferrite rod as used in a portable broadcast receiver.

Sketch, and explain the construction, including the lead-in to the receiver, of either—

    (i) an inverted-L aerial for the reception of medium-wave signals, or

    (ii) a dipole and reflector type of aerial commonly used for the reception of television signals. (A 1965)

2. Sketch and explain the construction of (*a*) an inverted-L aerial and (*b*) a T aerial for the reception of medium-wave signals.

   Indicate the manner of coupling these aerials to the receiver. Briefly discuss the directional properties of each of these aerials.

   What type of aerial (other than a frame aerial) is now often used in portable broadcast receivers. Briefly discuss the construction of such an aerial and its orientation for the best reception. (A 1964)

3. Sketch and describe a type of aerial commonly used for the reception of television signals. Give approximate dimensions.

   Explain what is meant by the radiation pattern (polar diagram) of an aerial. Sketch the radiation pattern, in the horizontal plane, of the aerial described. (A 1963)

4. Why is it desirable to use an outdoor aerial for the reception of medium-wave broadcast signals near the edge of the service area?

   What form might such an aerial take? (1 1958)

5. Describe a type of aerial commonly used for the reception of television signals.

   Indicate the precautions necessary when installing such an aerial to ensure that it is used to best advantage. (1 1957)

6. A radio amateur wishes to erect a horizontal half-wave dipole aerial for communication in the high-frequency band with amateurs overseas. Describe a method of connecting the aerial to his equipment and indicate from which directions he might expect to receive useful signals if the aerial is orientated with its axis along a north-south line. (1 1956)

15 *Power Supplies*

For the correct operation of electronic and radio equipment suitable power supplies must be available; d.c. power is required for the anode and screen grid supplies of thermionic valve equipment and the collector supplies of transistor equipment, and a.c. power is required for the heater supplies of thermionic valve equipment. Rarely are the correct supplies available and power supply equipment is necessary to convert an available supply into the wanted supply. The d.c. requirements of a small transistor equipment can be supplied directly by a small dry battery, but for larger transistor and thermionic valve equipments more complicated arrangements are necessary. The power conversions that may be required, either singly or in combination, to convert a given power source into the power supply required for a particular application are (*a*) the conversion of a.c. from one voltage to another, (*b*) the conversion of a.c. into d.c., (*c*) the conversion of d.c. into a.c. and (*d*) the conversion of d.c. from one voltage to another. The first of these conversions can be easily achieved by the use of a transformer of the appropriate turns ratio; methods of achieving the remaining conversions form the subject of this chapter. Only relatively low power supplies, such as those required for radio receivers, tape recorders and other small equipments, are to be discussed.

The Metal Rectifier

Three kinds of rectifying elements are used in power supply units, thermionic valve diodes, semiconductor diodes and metal rectifiers. The first two of these have been previously discussed in this book and do not require further mention.

The metal rectifier is a device that conducts an electric current more readily in one direction than in the other and consists of a number of elements mounted on a spindle. Two types of metal rectifier are in common use, namely the copper-oxide rectifier and the selenium rectifier. An element in a copper-oxide rectifier is shown in Fig. 15.1*a*; it consists of a copper disc that has a layer of cuprous oxide formed on one of its faces; the layer is coated with a thin film of graphite and in tight contact with this is a lead plate. The construction of a selenium rectifier element is shown in Fig. 15.1*b*; the element consists of a nickel-plated iron disc that has a

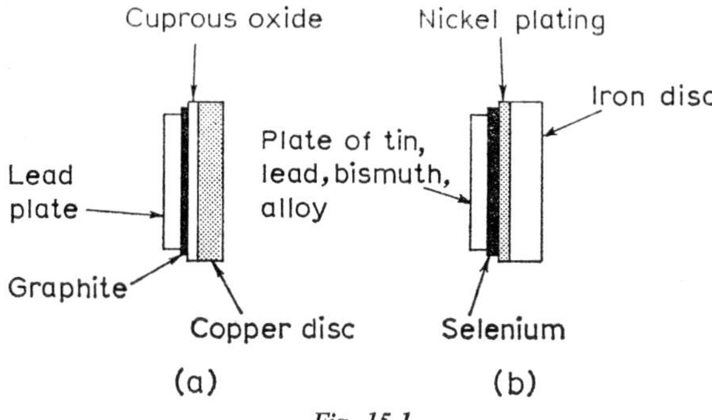

Fig. 15.1

Metal rectifier elements, (*a*) copper-oxide, (*b*) selenium

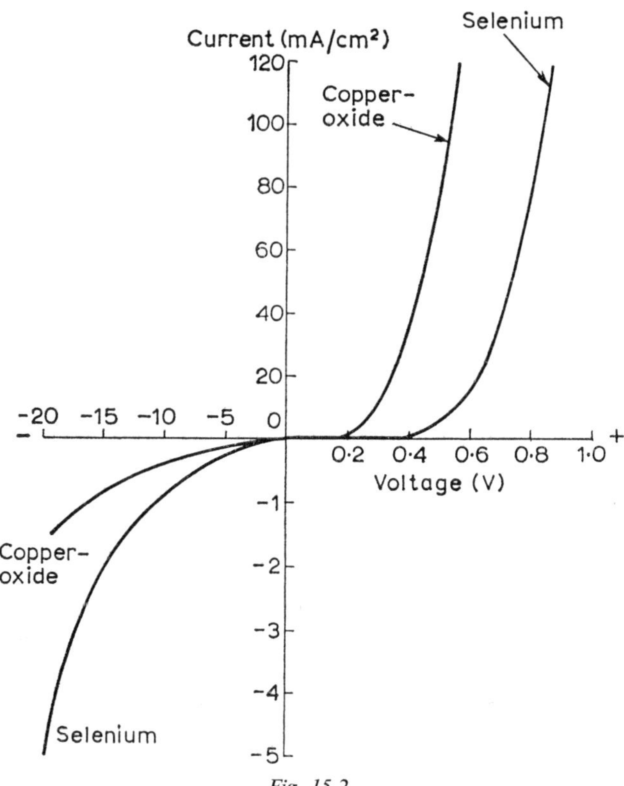

Fig. 15.2

Current/voltage characteristics of copper-oxide and selenium metal rectifiers

layer of selenium on one of its faces and pressed tightly against this is a plate made of an alloy of lead, tin and bismuth.

The current/voltage characteristics of copper oxide and selenium rectifier elements are shown in Fig. 15.2 and may be obtained in the same way as the characteristic of a semiconductor diode, see Fig. 6.21. In the forward direction (oxide to copper, or iron to selenium) the current per unit of surface area increases with increase in voltage, slowly at first but then more rapidly. In the reverse direction the current density is small at first and then increases fairly rapidly with increase in the reverse voltage. Operation as a rectifier requires a low forward resistance and a high reverse resistance and therefore

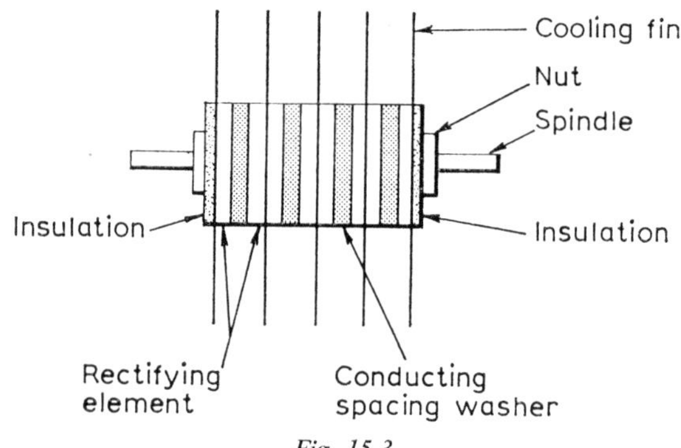

Fig. 15.3

The construction of a metal rectifier

the voltage that can be applied across an element is limited to some 10 volts or so. When higher voltages than this are to be handled a number of elements must be connected in series on a single spindle. The current-carrying capacity of an element depends upon its surface area and can always be increased by enlarging the area. In the larger rectifiers considerable heat may be produced, and it may be necessary to include cooling fins and conducting spacing washers at intervals along the spindle to assist in dissipating the heat. A typical assembly of a multi-element metal rectifier is shown in Fig. 15.3. The spindle is provided with screw threads at each end so that the assembly can be bolted up tightly to ensure good contact between the elements and plates. Selenium rectifiers have the advantage of being able to work at higher temperatures but the copper-oxide rectifier can be made in smaller sizes.

Metal rectifiers were widely used in the past, not only in power supply units, but are nowadays being increasingly superseded by the more efficient semiconductor diodes.

Rectifier Circuits

A number of different circuits exist that are capable of converting an a.c. supply into a pulsating d.c. current and they may be broadly divided into one of two classes, half-wave rectifiers and full-wave rectifiers. In the circuits that follow, diodes, thermionic valve or semiconductor, or metal rectifiers may be used as the rectifying device, but of course the valve diodes require a heater supply.

HALF-WAVE RECTIFICATION

In its simplest form half-wave rectification consists merely in the connection of a rectifying device in series with the a.c. supply and the load, as shown in Fig. 15.4*a*.

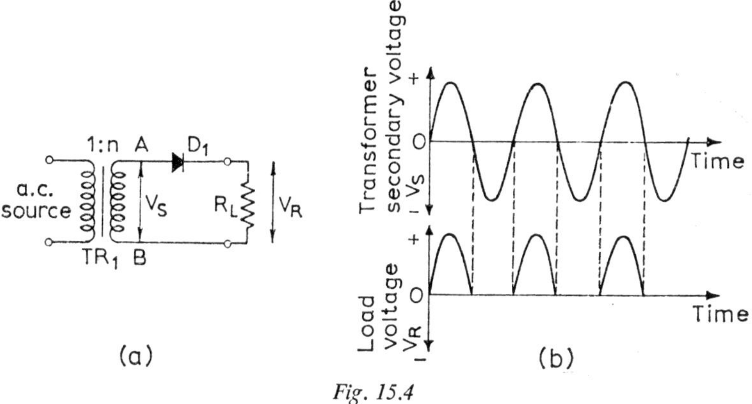

Fig. 15.4

The half-wave rectifier with resistance load

The rectifier conducts only during those alternate half-cycles of the a.c. supply voltage V_s that make point A positive relative to point B, and so the load current consists of a series of half sine-wave pulses. The voltage V_R developed across the load is the product of the load current and the load resistance and has the same waveform as the load current, Fig. 15.4*b*. The disadvantage of this simple rectifying circuit is very clear: the load voltage, although unidirectional, varies considerably and is, indeed, zero for half the time. Such a waveform is only suitable for simple applications, such as battery charging, since the variations will appear as noise at the output of any equipment fed by the supply. The d.c. output of a rectifier circuit is required to be as steady as possible and a great step towards this goal could be achieved if the load voltage could be prevented from falling to zero during alternate half-cycles. One way of achieving this is to connect a capacitor C in parallel with the load as shown in Fig. 15.5*a*. Each time the rectifier conducts, the current that flows charges the capacitor and the voltage across the capacitor builds up. During the intervals of time when the rectifier is non-conducting the capacitor discharges via the load resistance and prevents the load voltage falling to zero, Fig. 15.5*b*. The capacitor continues to discharge, at a rate determined by the time constant CR_L seconds, until the point A is taken more positive than the capacitor voltage by a positive half-cycle of the input voltage V_s. The rectifier then conducts, the capacitor is recharged, and the capacitor voltage rises again. If the load current is fairly small the capacitor

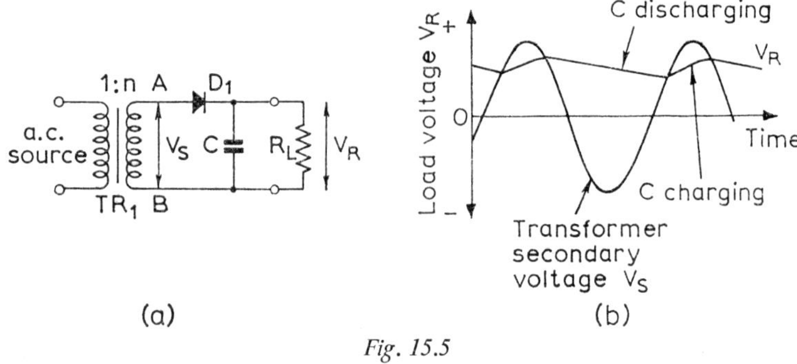

Fig. 15.5

The half-wave rectifier with resistance-capacitance load

does not discharge very much between charging pulses and the average load voltage V_R is only slightly less than the applied voltage V_s.

The value of the load voltage is adjustable, within limits, by suitable choice of the turns ratio n of the input transformer. An increase in the load, i.e. in the load current, means that the load resistance is less; this, in turn, means that the time constant of the discharge path is smaller.

Capacitor C then discharges more rapidly and the load voltage is not as constant, see Fig. 15.6.

A completely steady load voltage cannot be obtained in this way since too large a value of capacitance would be required. The maximum value of capacitance that can be employed is limited, because the larger the capacitance value the greater the current required to charge the capacitor to a given voltage, and the current that can be handled by a rectifier is limited to a figure quoted by the manufacturer. The fluctuating, unidirectional, voltage appearing across the load may be regarded as a d.c. voltage having an a.c. voltage superimposed upon it. This a.c. voltage is known as the ripple voltage and is at the frequency of the supply voltage, usually 50 Hz. The ripple voltage is undesirable, since the object of rectification is to provide a steady d.c. voltage, and can be removed by a smoothing, or filter, circuit connected between the rectifier and the load.

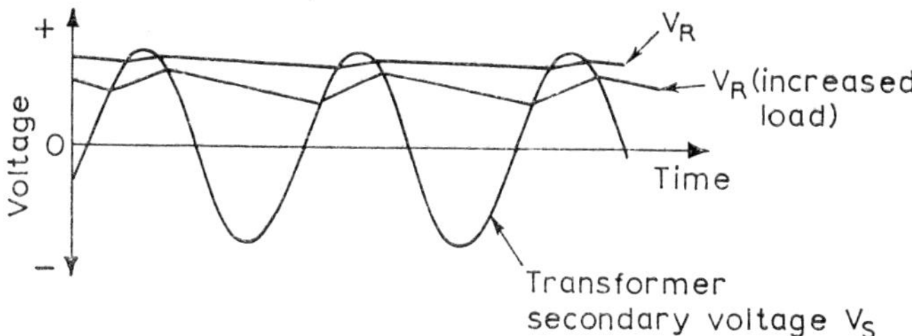

Fig. 15.6

Showing the effect of load changes on a half-wave rectifier with resistance-capacitance load

FULL-WAVE RECTIFICATION

With full-wave rectification of an a.c. source both half-cycles of the input waveform are utilized, alternate half-cycles being inverted to give a unidirectional load current. The circuit of a full-wave rectifier is shown in Fig. 15.7*a* and can be seen to require two

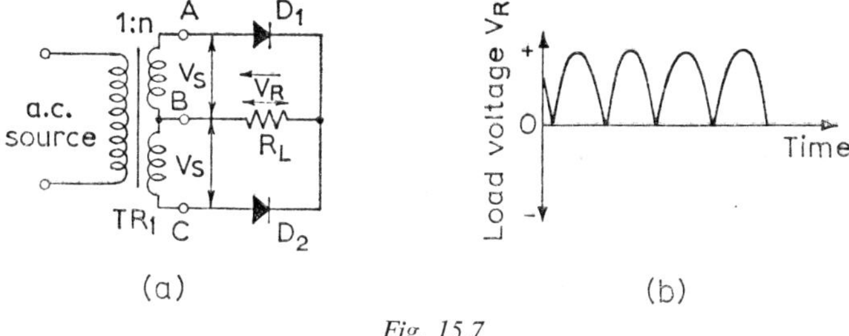

Fig. 15.7

The full-wave rectifier with resistance load

rectifiers. The secondary winding of the input transformer, TR_1, is accurately centre-tapped so that equal voltages are applied across the two rectifiers D_1 and D_2. During those half-cycles of the input waveform that make point A positive with respect to point B and point C negative relative to point B, D_1 conducts and D_2 does not and current flows in the load in the direction indicated by the arrow. When the point C is positive with respect to point B and point A is negative relative to point B, D_2 is conducting and D_1 non-conducting and current flows in the load in the same direction as before. The waveform of the current, and hence of the load voltage V_R, is shown in Fig. 15.7*b*.

A more constant value of load voltage can be obtained by the connection of a capacitor across the load as shown in Fig. 15.8a. The action of the reservoir capacitor is exactly the same as in the half-wave circuit but now the capacitor is re-charged twice per input cycle instead of only once. Between charging pulses the capacitor commences to discharge through the load but, provided the time constant is not too short, the load voltage has not fallen by much

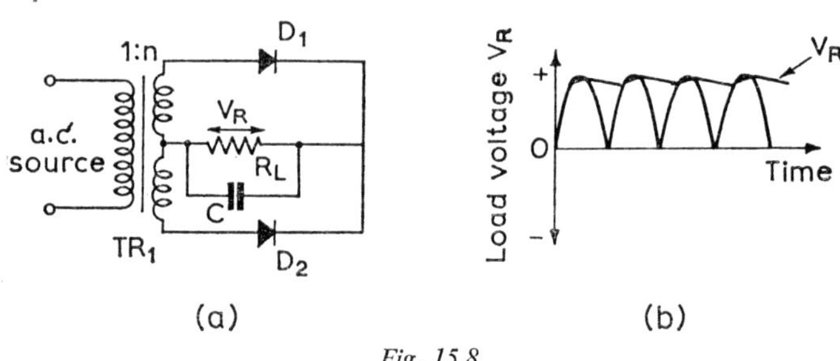

(a)   (b)

Fig. 15.8

The full-wave rectifier with resistance-capacitance load

before the next charging pulse occurs, Fig. 15.8b. The load voltage attains a mean value only slightly less than the peak voltage appearing across one half of the input transformer secondary winding. As before, the ripple content of the load voltage increases with increase in load current and can be reduced by the use of a suitable filter network.

An alternative method of full-wave rectification is the use of a bridge network, see Fig. 15.9. The bridge rectifier circuit requires

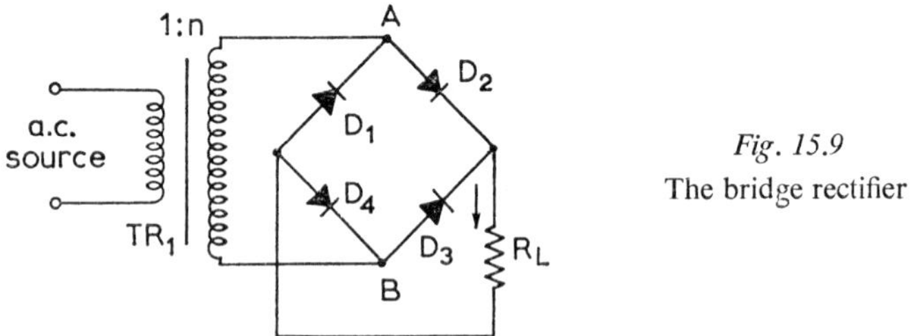

Fig. 15.9

The bridge rectifier

four rectifiers instead of two but avoids the need for a centre-tapped input transformer. Further, the arrangement gives a load voltage that is nearly twice as great as can be obtained from the circuit of Fig. 15.8a—assuming, of course, the same transformer secondary voltage. During those half-cycles of the input that make point A positive with respect to point B, rectifiers D_2 and D_4 are conducting and rectifiers D_1 and D_3 are non-conducting; current therefore flows from point A to point B via D_2, the load R_L and D_4. When point A is negative relative to point B, D_1 and D_3 are conducting and D_2 and D_4 are non-conducting; current then flows from point B to point A via D_3, the load R_L, and D_1. Both currents pass through the load in the same direction and so a fluctuating, unidirectional, voltage is developed across the load having the waveform of Fig. 15.7*b*. The variations in load voltage can be reduced

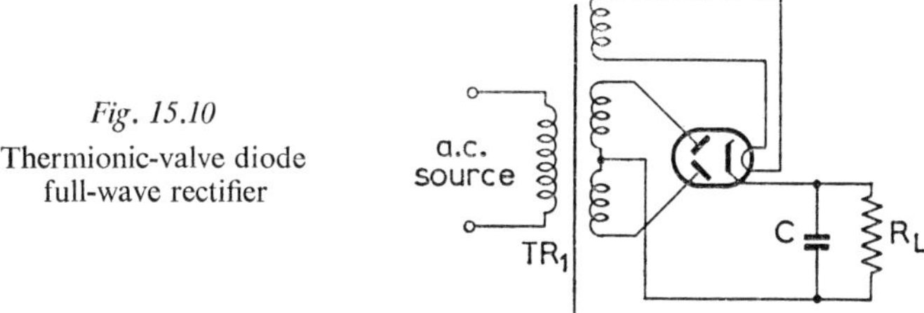

Fig. 15.10

Thermionic-valve diode
full-wave rectifier

by the connection of a capacitor across the load; the load voltage waveform is then that of Fig. 15.8*b*. The bridge rectifier circuit is rarely used in conjunction with thermionic valve diodes because the heater supply problem is rather difficult.

When thermionic-valve diodes are employed in a rectifier circuit the diodes must be supplied with heater current and this is usually provided from a tertiary winding on the input transformer. The arrangement is shown in Fig. 15.10 for a full-wave circuit that employs a double-diode valve.

Filter Circuits

A power supply unit for electronic equipment must provide a d.c. voltage of minimum ripple content. To reduce the ripple voltage to a tolerable level it is generally necessary to include some kind of filter circuit between a rectifier and its load. The simple capacitor connected in shunt across the load reduces ripple and is, therefore, a simple filter that may be adequate for some applications. Further smoothing of the output voltage can be achieved if the capacitor is followed by either an *L-C* or an *R-C* network, or, alternatively, the rectifier may feed directly into a series inductor. The effectiveness of a filter can be judged in terms of the reduction in ripple voltage it gives, but another, equally important (usually), criterion is its voltage regulation. The voltage regulation of a rectifier circuit is a measure of how its output voltage changes as the current taken from it is varied. Ideally, of course, the output voltage should remain constant, so good regulation implies that the voltage changes very little as the load current is varied from minimum to maximum.

CAPACITOR-INPUT FILTERS

A capacitor-input filter consists of a shunt capacitor, connected across the output terminals of the rectifier, followed by a basic low-pass filter. The low-pass filter may consist of a series inductor and a shunt capacitor as in Fig. 15.11, or a series resistor and a

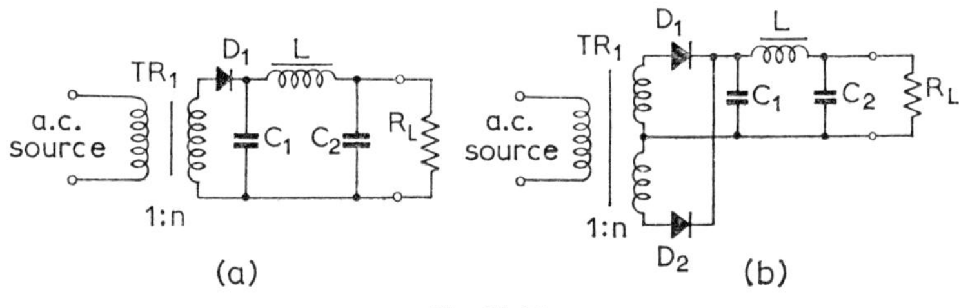

(a)   (b)

Fig. 15.11

The capacitor-input *L-C* filter

shunt capacitor as shown in Fig. 15.12. In the *L-C* filter the value of capacitance C_1 is chosen to give a reasonably smooth output voltage from the rectifier proper, and the values of L and C_2 are chosen to give adequate ripple suppression. Typical values for C_1 and C_2 for a half-wave rectifier are 32 μF each and for L, 30 H, and at 50 Hz, the half-wave rectifier ripple frequency, these components have reactances of $1/(2\pi \times 50 \times 32 \times 10^{-6})$, or approximately 100 Ω, and $2\pi \times 50 \times 30$, or 9,426 Ω, respectively. The reactances of C_2 and L act as a potential divider across the rectifier output and reduce the ripple voltage to approximately 100/9,426 times its original value. In the full-wave rectifier the ripple frequency is 100 Hz and this means that a filter using the same component values would be approximately four times as efficient. For a given amount of ripple smaller components can be used and typical values are $C_1 = C_2 = 8$ μF and $L = 15$ H. Both filters have negligible effect on the wanted d.c. component of the rectified output voltage. In order to obtain the fairly large capacitance values required cheaply and in the minimum of space, electrolytic capacitors are normally used.

The disadvantages of the capacitor-input *L-C* filter are (*a*) the cost, weight, size and external fields of the series inductor, (*b*) the high peak currents that must be handled by the rectifier (because C_1 must charge up in a very short time), and (*c*) the relatively poor voltage regulation. The first of these disadvantages can be overcome by replacing the series inductor with a series resistor, although this has the obvious disadvantage of increasing the d.c. voltage drop in the filter. The *R-C* filter, therefore, has poor regulation and requires adequate ventilation to conduct away the heat produced in the resistor. As a result it is only used to supply equipments taking only a

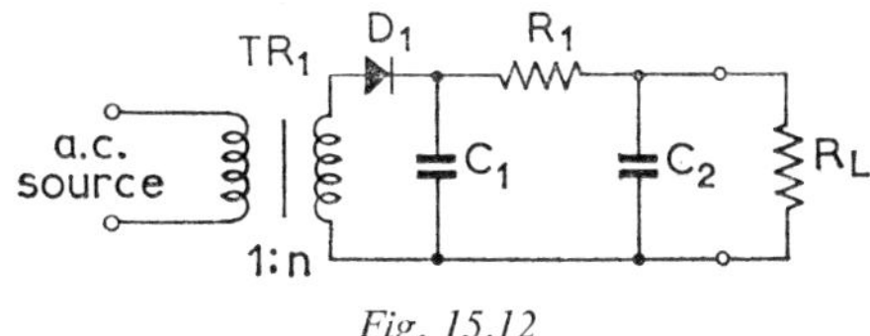

Fig. *15.12*

The capacitor-input *R-C* filter

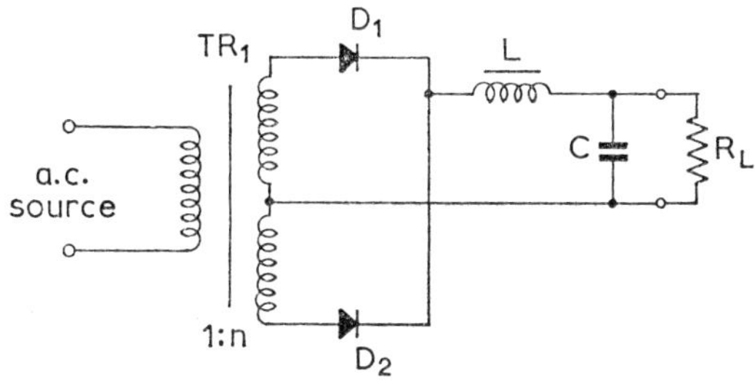

Fig. *15.13*

The choke-input filter

small current and/or whose performance would be adversely affected by the magnetic field of an *L-C* filter. Disadvantage (*b*) is particularly serious in thermionic-valve diode rectifiers where the ratio (peak current/average current) is fairly small; this disadvantage can be overcome by using either semiconductor diodes or selenium metal rectifiers, both of which are capable of handling high values of peak current.

The circuit of a capacitor-input *R-C* filter is shown in Fig. 15.12; such filters are used in the power supply units of some television receivers and cathode-ray oscilloscopes. Typical values are $R = 100$ to $200\,\Omega$, $C_1 = 100\,\mu\text{F}$ and $C_2 = 150$ to $200\,\mu\text{F}$.

CHOKE-INPUT FILTERS

When a choke-input filter is used there is no reservoir capacitor and the rectifier feeds directly into the filter, Fig. 15.13. Inductor *L* and capacitor C form a potential divider across the output of the rectifier and reduce the ripple voltage to a low value. The choke-input filter can only be used in conjunction with a full-wave rectifier since it requires current to flow at all times, the current being provided first by D_1 and then by D_2. The fact that current flows continuously, instead of in a series of pulse as in the capacitor-input filter circuits, means that the input transformer is used more efficiently. A further advantage is that the ripple content at the output of the filter is independent of the load current (compare with Fig. 15.6). Typical values for L are 5 to 30 H and for C, 5 to 40 μF.

Fig. 15.14 gives a comparison of the voltage regulation of the two types of filter. It can be seen that although the output voltage of the

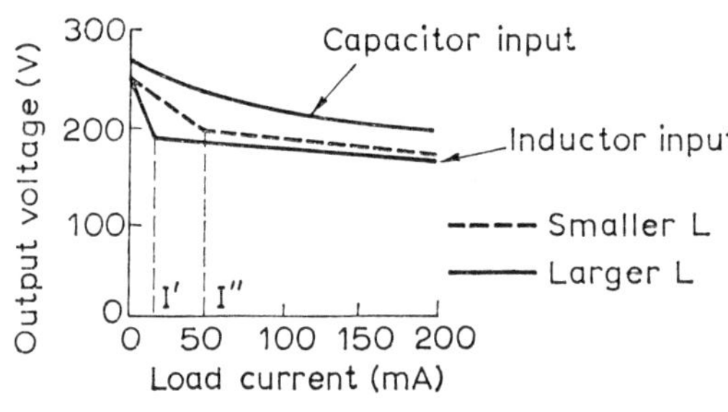

Fig. 15.14

Regulation curves for capacitor- and choke-input filters

choke-input filter is the smaller, its regulation is better. The regulation can be improved by the use of a larger value of inductance as shown by the two lower curves. If the load current falls below a certain critical value I' or I'' (this could happen each time the equipment was first switched on) the output voltage will rise abruptly. To prevent this happening it is usual to connect a resistor, known as a "bleeder", across the output terminals of the filter to ensure that a current greater than the critical value is always taken.

Choke-input filters were widely employed in the past for power supplies in excess of about 50 W because of the need to protect thermionic valve diodes from excessive peak currents. Good regulation can be obtained from the capacitor-input filter if a large value of reservoir capacitance is employed, and nowadays the use of silicon diodes, in conjunction with electrolytic capacitors connected in series, permits the capacitor-input filter to be used for quite large powers. This has the advantage of obviating the need for a large and expensive choke, and also a bleeder resistor with its accompanying power loss.

Voltage Multiplying

The use of a suitable combination of rectifiers and capacitors can give a d.c. output voltage that is several times greater than the peak voltage appearing across the secondary winding of the input transformer. Consider, for example, Fig. 15.15 which shows a voltage-doubling circuit. During the half-cycles of the input when point A is positive with respect to point B, rectifier D_1 conducts and capacitor C_1 is charged to the peak voltage V_s appearing across the secondary winding of transformer TR_1. When point B is positive relative to point A, rectifier D_2 conducts and capacitor C_2 is charged to the same voltage. Capacitors C_1 and C_2 are connected in series across the output terminals and so the voltage appearing across these terminals

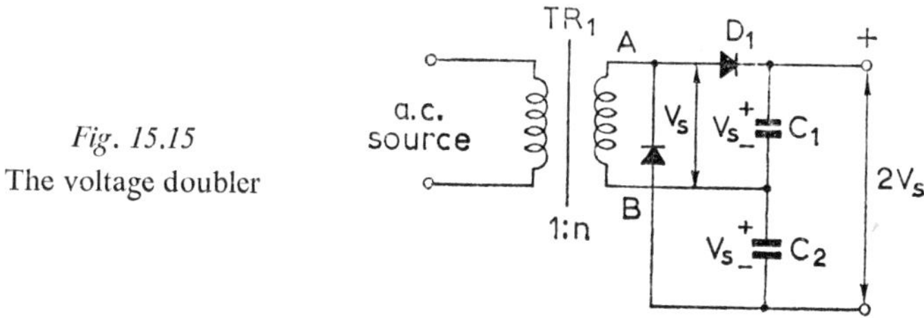

Fig. 15.15

The voltage doubler

is equal to twice the peak secondary voltage, i.e. $2V_s$. If large values of capacitance are used and the load current is fairly small the output voltage has small ripple content and good regulation. Since the capacitors are charged during alternate half-cycles the ripple frequency is at twice the supply frequency, that is, 100 Hz for 50 Hz mains.

The principle of the voltage doubler can be extended to voltage tripling, quadrupling or even higher. Fig. 15.16 shows possible arrangements for (*a*) a voltage tripler and (*b*) a voltage quadrupler. Consider the voltage tripler. During half-cycles when point A is positive with respect to point B, rectifier D_1 conducts and capacitor C_2 is charged to V_s volts. During the half-cycles when point B is positive relative to point A, D_2 conducts and C_1 is charged to $2V_s$ volts—because the voltage applied across it is the sum of the transformer secondary voltage V_s and the voltage V_s across C_2. Also, when point A is positive relative to point B, D_3 conducts and C_3 is charged to $3V_s$, because the voltage applied across it is the sum of the transformer secondary voltage V_s and the voltage $2V_s$ across C_1.

Typical Rectifier Circuits for Valve Radio Receivers

The power supply arrangements for radio receivers working off the mains supply fall into one of two categories. Firstly, power supply units capable of producing the required d.c. supplies from an a.c. mains supply, and, secondly, units that will operate from either an a.c. or a d.c. supply.

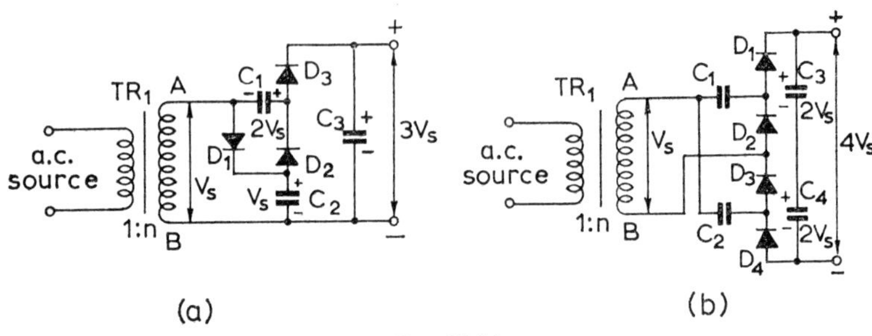

(a)   (b)

Fig. 15.16

(*a*) The voltage tripler and (*b*) the voltage quadrupler

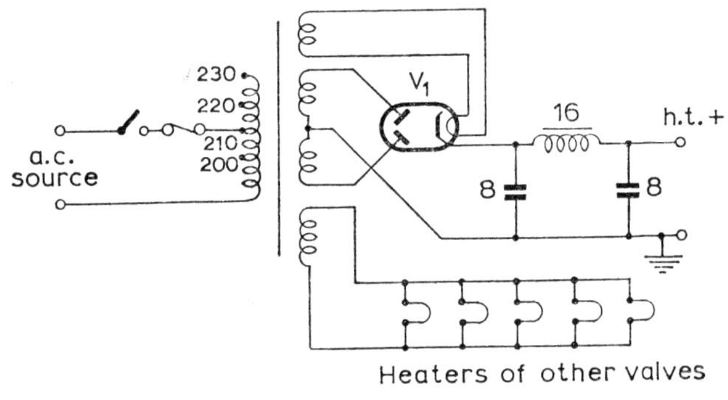

Fig. 15.17

Power supply unit for an a.c. thermionic-valve radio receiver

Domestic valve radio receivers, and other small power equipments, require d.c. voltages of some 150 to 300 V for the h.t. supply and 6·3 V for the heater supply. The current taken by the h.t. line is usually fairly small (less than 100 mA) and reasonably constant, and capacitor-input filters are almost always employed. The heater supplies are not rectified. A typical circuit for an a.c. mains power supply unit is shown in Fig. 15.17; a thermionic-valve diode rectifier has been shown to point out that its heater supply is taken from a separate winding on the input transformer. A tapped primary winding is employed so that the unit can be operated from a.c. sources of different voltages. The use of a mains transformer has the advantages of allowing the d.c. output voltage to be greater than the peak value of the supply voltage and of isolating the receiver chassis from the supply, but it has the usual transformer disadvantages of size, weight and cost. To avoid the use of a mains transformer many valve receivers employ a power supply unit that is capable of operating from both a.c. and d.c. mains. Half-wave rectification, followed by a capacitor-input filter is used, see Fig. 15.18. When a d.c. supply is used the unit must be connected with the anode of the diode to the positive side of the supply. The diode then conducts continuously. The heaters of the valves, including the diode rectifier, are connected

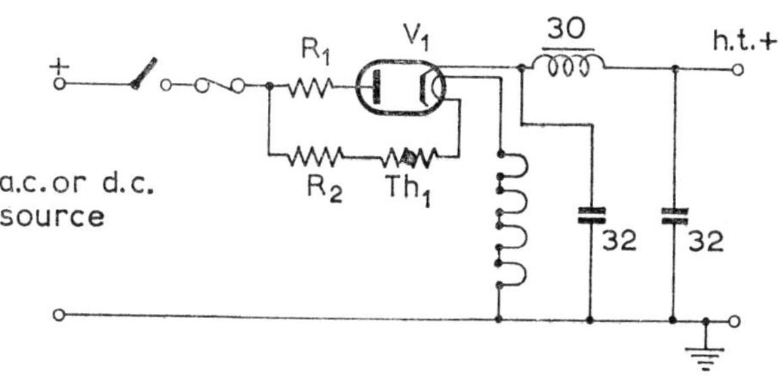

Fig. 15.18

Power supply unit for an a.c./d.c. thermionic-valve radio receiver

in series and the correct heater current is obtained by the correct choice of the value of resistor R_2. Obviously all the valves ought to require the same heater current; if they do not, suitable shunting resistors must be connected across some of the heaters. When the unit is first switched on the heaters are cold and their resistances are low and a large current may flow until the heaters warm up and their resistances increase. To prevent this surge current flowing in the heater chain the thermistor Th_1 is included. (A thermistor is a semiconductor device whose resistance decreases with increase in temperature.) Resistor R_1 limits the peak current passed by the valve. The circuit has the disadvantages of (*a*) no isolation between the mains and the output terminals of the unit and (*b*) the output voltage cannot be greater than the input voltage.

E.H.T. Supplies

Television receivers and cathode-ray oscilloscopes (C.R.O.) employ cathode-ray tubes and these require an extremely high voltage (e.h.t.) of some thousands of volts for their correct operation. Such very high voltages can be obtained in a number of different ways. The e.h.t. voltage can be obtained from the a.c. mains supply by the use of a transformer having a very large turns ratio, or from a mains-fed voltage-multiplying circuit. Both of these methods, however, suffer from the disadvantages of bulky and costly transformers and danger to personnel working on the equipment (because large quantities of electricity are stored in the smoothing capacitor (s)). If a higher-frequency a.c. source were to be employed smoothing would be easier and smaller capacitors could be used; the charge stored in the smoothing capacitor(s) would then be much less.

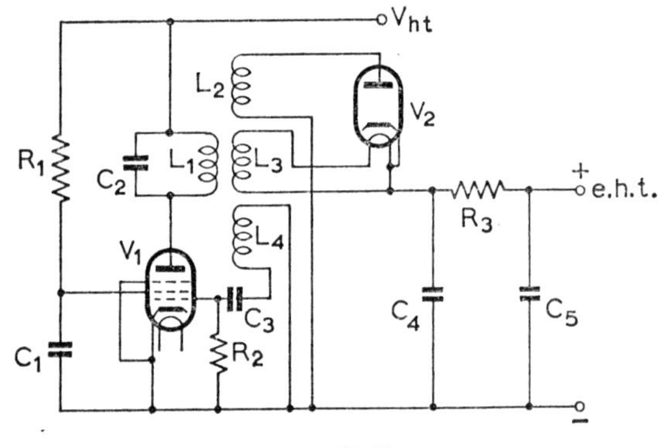

Fig. 15.19

An r.f. oscillator e.h.t. supply unit

Since the current required at e.h.t. voltage by a cathode-ray tube is very small it is possible to employ an r.f. oscillator, frequency some 10–500 kHz, whose output voltage can be stepped up by a transformer, rectified and smoothed, all with physically small components. A possible circuit for a C.R.O. e.h.t. power supply is shown in Fig. 15.19 and employs a tuned-anode r.f. oscillator. The normal anode/grid coupling for a tuned-anode oscillator is via coils L_1 and L_4 and the oscillatory voltage is stepped up by the coupling between coils L_1 and L_2. The stepped-up r.f. voltage is detected by the diode valve V_2, and smoothed by the capacitor-input filter C_4, R_3, C_5.

In a television receiver it is not necessary to provide a special r.f. oscillator because the e.h.t. supply can be derived from the rapid flyback of the line synchronization waveform.* A typical circuit for

---

* Operation of a television receiver requires the application of a sawtooth current waveform shown in Fig. 15.20, to the cathode-ray tube. The waveform is arranged to reduce to zero, or fly back, in a very short time.

a valve television receiver e.h.t. supply is shown in Fig. 15.20. The rapid flyback of the line synchronization current waveform induces a large e.m.f. in winding L_2 ($e = -M di/dt$) and this voltage is rectified by the diode, V_2, and then smoothed in the usual way. The diode heater is fed from a separate winding on the transformer because of the high voltages that might otherwise exist between heater and cathode. A capacitor-input R-C filter has been shown, but often a single capacitance is considered to give adequate smoothing, and even this may not be a physical component but may

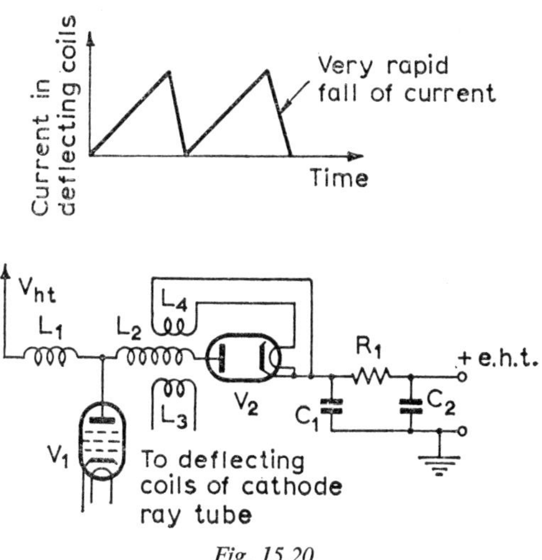

Fig. 15.20

E.H.T. supply for a television receiver

merely be the capacitance of the cathode-ray tube between its inner and outer coatings.

The e.h.t. supply for the cathode-ray tube in transistorized television receivers and C.R.O.s can be obtained either from the transistor versions of Figs. 15.19 and 15.20 or from a transistor d.c. to d.c. convertor.

Conversion of D.C. Supplies

Many equipments, particularly portable ones, are required to operate from a low-voltage d.c. supply and there is a need for d.c./a.c. invertors and d.c./d.c. convertors. A d.c./a.c. invertor is a circuit that converts a d.c. source into a required a.c. supply and a d.c./d.c. convertor is a circuit that converts a d.c. source at one voltage into a d.c. supply at another voltage. A d.c./d.c. convertor consists of a d.c./a.c. invertor followed by a suitable rectifier as shown by the block schematic of Fig. 15.21. Before the advent of transistors recourse

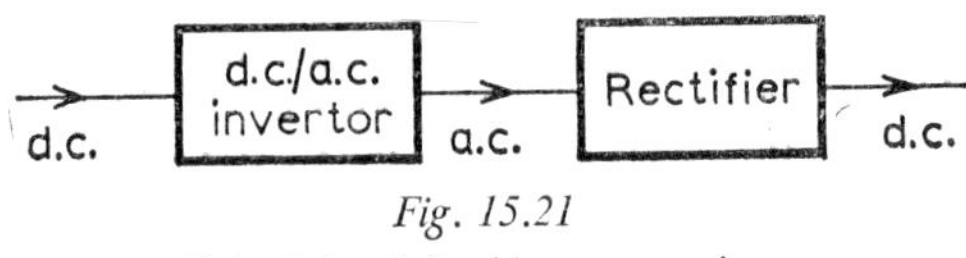

Fig. 15.21

Principle of d.c./d.c. conversion

was had to electro-mechanical devices, such as vibrators, but nowadays transistor circuits are normally employed.

THE VIBRATOR

Fig. 15.22 shows the basic circuit of an electro-mechanical vibrator power unit that may be employed to provide a d.c. h.t. supply from a low voltage battery. When the switch S is closed the coil is energized and the armature is operated. Operation of the armature breaks contact 5 and this disconnects the armature operate circuit;

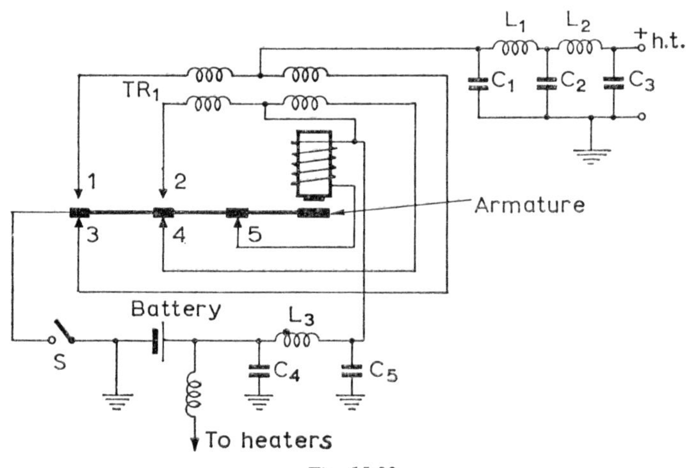

Fig. 15.22

A basic vibrator circuit

the armature releases, contact 5 re-makes and the coil is re-energized. The armature therefore vibrates at a frequency that is adjusted to be in the region of 150–200 Hz. Contacts 2 and 4 making and breaking alternately cause an alternating current at 150–200 Hz to flow in the primary winding of transformer TR_1, and the transformer primary voltage is stepped up to appear across the secondary winding at the required h.t. value. Rectification of the secondary voltage is achieved by the making and breaking of contacts 1 and 3 and the rectified voltage is smoothed by the filter circuit, C_1, L_1, L_2, C_2 and C_3. The heater supply is taken directly from the low-voltage battery, the live line being taken via an r.f. choke to block the passage of r.f. currents.

The vibrator has a number of disadvantages that have led to its obsolescence. These disadvantages are its relatively large size, its high fault liability, its noisy operation, its inefficiency, and its tendency to produce r.f. interference.

TRANSISTOR D.C./A.C. INVERTORS AND D.C./D.C. CONVERTORS

Two types of d.c./a.c. invertor are in common use: firstly, there is the ringing-choke convertor and, secondly, there is the push-pull convertor. The latter type has the advantages of requiring a smaller and cheaper transformer and of giving a higher output power with better voltage regulation, but it has the disadvantages of lower efficiency and of requiring two transistors as opposed to only one.

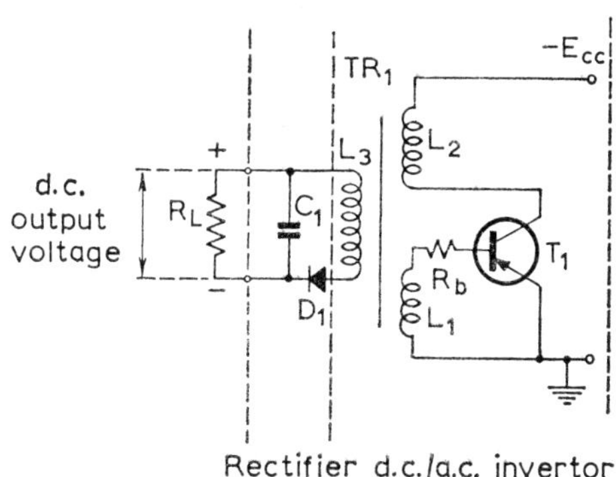

Rectifier d.c./a.c. invertor

Fig. 15.23

The ringing-choke d.c./d.c. convertor

The basic circuit of a ringing-choke d.c./d.c. convertor is shown in Fig. 15.23, and consists of a push-pull d.c./a.c. invertor followed by a half-wave rectifier. When the circuit is first switched on the transistor is biased so that the resistance between its collector and emitter terminals is low. Practically the whole of the supply voltage is then applied across winding L_2 and a current flows in this winding that increases linearly with increase in time. A constant voltage is induced in winding L_1 and so a constant base current is provided for the transistor. The magnitude of this current is determined by the value of the resistor R_b. Diode D_1 is non-conducting during this period. The increase in base current changes the bias (operating) point of the transistor, and immediately this point coincides with the knee of the output characteristic the collector/emitter resistance increases abruptly, and the current flowing in winding L_2 falls. The sudden decrease in the current flowing in L_2 reverses the e.m.f. induced in winding L_1 and this causes the base current to fall rapidly. The transistor then rapidly ceases to conduct, or cuts off. The rapid change in the current flowing in L_2 induces a large voltage in winding L_3 and this voltage rises rapidly, charging capacitor C_1 via diode D_1, until the capacitor voltage has risen to the output voltage. The energy that was stored in the magnetic field has now

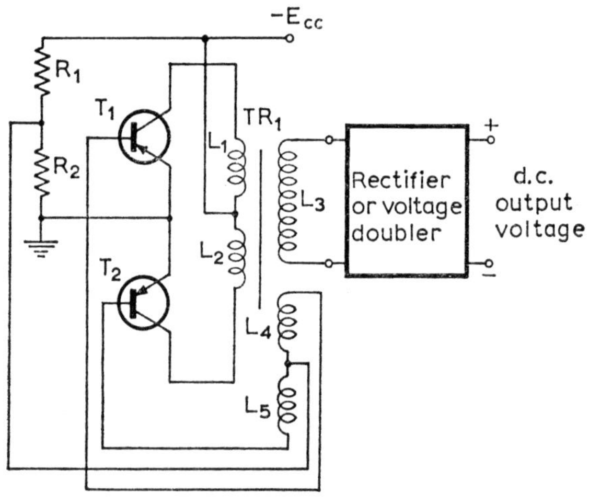

Fig. 15.24

The push-pull d.c./d.c. convertor

been transferred to the dielectric of the capacitor. The voltage across L_2 now collapses (reverses direction) and induces a voltage in winding L_1 that switches T_1 on again and restarts the cycle. The d.c. output voltage is the voltage appearing across capacitor C_1 and it can be smoothed if necessary.

When a high output voltage is required a very high turns ratio is necessary for the transformer and this leads to difficulties in transformer design. The use of a voltage-multiplying circuit is then advantageous.

Ringing-choke convertors are used for low output powers and mainly for d.c. outputs.

Fig. 15.24 shows the basic circuit of a push-pull d.c./d.c. convertor; the block marked "rectifier" may consist of any form of full-wave rectifier or voltage multiplier. When the supply is switched on the potential divider, $R_1 + R_2$, provides a base current for transistors T_1 and T_2 that biases them into a state of conduction. Because of slight unbalances in the circuit, and in the transistors themselves, one transistor conducts more heavily than the other, and this results in the circuit rapidly going into the condition of one transistor fully conducting and the other completely non-conducting. Imagine transistor T_1 to be fully conducting and T_2 to be non-conducting. The collector-to-emitter resistance of T_1 is then small and so practically the whole of the supply voltage appears across winding L_1. This causes the current in L_1 to rise linearly until the point is reached where the transformer core is magnetically saturated. The current then rises suddenly, and rapidly, to try to maintain the same rate of change of flux in the core, and a large voltage is induced in winding L_3. The rise in current is limited by the circuit parameters and when this limit has been reached the rate of change of flux can no longer be maintained and the flux decays to zero. The voltages appearing across the windings collapse and T_1 is switched to the non-conducting state. The current then ceases to flow in winding L_1 and the resultant decrease in core flux induces voltages of opposite polarity in the other windings. Transistor T_2 is then switched to the fully conducting state and the cycle repeats in the opposite direction. The alternating voltage that appears across winding L_3 may be either fed directly to the load or applied to a rectifier as shown. If necessary, voltage multiplying and/or smoothing may be used.

Push-pull convertors are normally operated at a frequency of some 300–2,000 Hz and can produce output voltages of up to several kilovolts from a few volts input.

Exercises

1. Sketch and describe the operation of the following—

   (*a*) power units suitable for the h.t. and heater supplies in an a.c./d.c. broadcast receiver,

   (*b*) a transistorized d.c.-to-d.c. convertor for use with mobile equipment.

   In part (*a*) state suitable component values. (A 1964)

2. Draw circuits suitable for the h.t. power supply of communication-type receivers, using (*a*) half-wave and (*b*) full-wave rectifiers.

   Indicate typical values of the smoothing components.

   If the half-wave rectifier were being used for a television broadcast receiver, what would be the advantage of using a selenium rectifier?

   (A 1963)

3. Draw a circuit diagram of an a.c. mains power supply unit for a broadcast receiver. Give suitable values for the components, and briefly explain its operation. (A 1962)

4. Draw a circuit diagram and explain the operation of an a.c. mains power supply unit, suitable for transmission test equipment such as a signal generator.

What are the advantages of full-wave over half-wave rectification for such applications?   (A 1961)

5. Draw a circuit diagram, and explain the operation, of a power supply unit suitable for the cathode-ray tube in an oscilloscope.

What is the main advantage of an r.f. oscillator e.h.t. generator for such an application?   (A 1960)

6. Give circuit diagrams of h.t. power supplies suitable for the following—

(*a*) a domestic television receiver, using a half-wave selenium rectifier,
(*b*) a communication-type receiver using a hot-cathode rectifier valve.

In both cases show the values of smoothing components used, and account for any differences. What are the advantages of using a selenium rectifier in a television receiver?   (A 1959)

7. What precautions should be taken to avoid objectionable 50 Hz and 100 Hz hum in the output of a mains-operated high-gain audio-frequency amplifier? The amplifier has an input transformer associated with the first stage and a mains transformer is incorporated in the power unit.   (1 1958)

8. Describe the construction of a multiplate metal rectifier suitable for use in the h.t. power supply unit of a radio receiver.   (1 1957)

9. Draw the circuit diagrams of a.c. mains rectifying circuits for supplying h.t. to a receiver, (*a*) for half-wave rectification and (*b*) for full-wave rectification.

Suggest suitable components for the smoothing circuits in the two cases and give reasons for any differences.   (1 1956)

16 *Communication Systems*

Communication systems may be placed into two classes; firstly, point-to-point systems that provide communication, usually two-way, between two points, one of which may be mobile, and secondly, broadcasting systems that provide one-way communication between a central transmitter and a large number of receivers situated within a given geographical area. Point-to-point systems may consist of a single circuit, or of many circuits, and be routed over telephone cable, radio link or a combination of both. Broadcasting systems employ radio propagation for the link between transmitter and receivers, but communication between transmitter and studio, or between different studios, may be either by line or radio link.

The intelligence signals to be transmitted over the various types of communication system fall into one of four classes: (*a*) telephony (speech), (*b*) telegraphy, (*c*) music, and (*d*) television. The first two are employed for the passing of messages and the choice between them is based upon the nature of the message. Telegraphy is generally preferred when there is a large volume of data traffic to be passed between two points and an immediate answer is not required and/or a paper copy of the message(s) is required. Telephony has the advantage of permitting immediate conversation between two persons and is preferable for social purposes and for many business purposes. Music and television signals are mostly employed for entertainment purposes and are well known.

A large number of different types of communication system are in existence and the object of this chapter is to describe briefly, with the aid of block diagrams, the principles of some of the more common.

The G.P.O. Telephone Network

The telephone network of Great Britain is divided into local lines, junctions and trunks. Local lines consist of pairs of wires that connect the individual telephone subscribers to their local telephone exchange, junctions are two-wire circuits that may, or may not, be amplified and connect nearby telephone exchanges together, and trunks are amplified four-wire circuits that connect distant exchanges together.

Long-distance telephone lines are extremely expensive and it is not economically possible to connect every exchange in the network to every other exchange; direct trunks are only provided between two exchanges when justified by the traffic carried. A number of telephone exchanges are therefore grouped together and these exchanges pass all their trunk traffic via a trunk-switching exchange

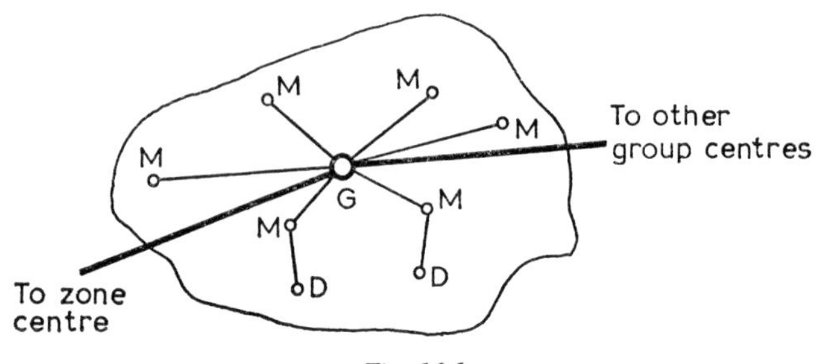

Fig. 16.1

Layout of a group area

G: Group centre
M: Minor exchange having direct access to the group centre
D: Dependent exchange not having direct access to the group centre

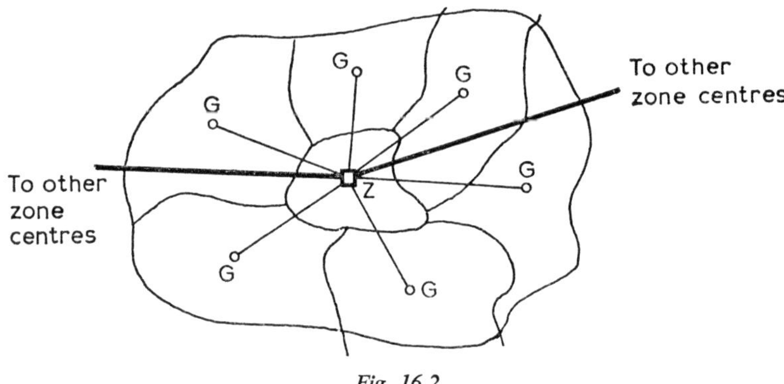

Fig. 16.2

Layout of a zone area

Z: Zone centre
G: Group centre

known as the group centre. Not all the exchanges within a group have direct junctions to the group centre, some exchanges obtaining access to the group centre via another local exchange. A typical group is shown in Fig. 16.1. The group centre is normally located as near the centre of the group area as possible in order to minimize the total length of junction cable required.

It is economically impossible to fully interconnect all the group centres in the country (there are approximately 250 of them) and so some of the group centres are chosen to act as zone centres also. There are 26 zone centres and the function of each of them is to serve as the trunk-switching centre for all the group centres within its zone. A typical zone is shown in Fig. 16.2. The zone centres are almost fully interconnected by low-loss trunk circuits. Whenever justified economically by the telephone traffic, direct routes are provided between two exchanges, for example, between two group centres or between a zone centre and a group centre in another zone.

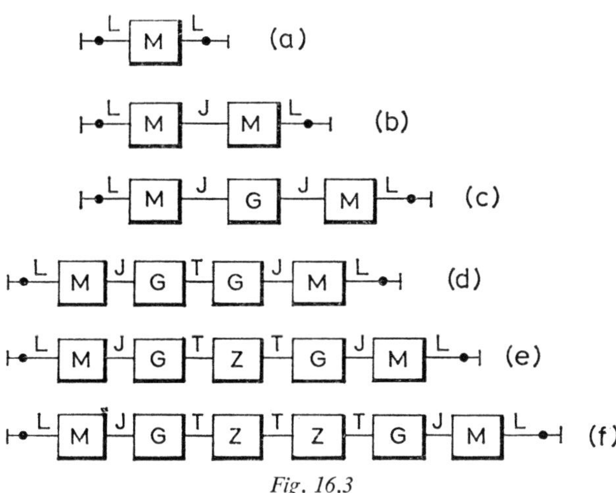

Fig. 16.3

Some possible connections in the telephone network

M: Minor exchange
G: Group centre
Z: Zone centre
L: Local line
J: Junction
T: Trunk

Some of the possible connections that might be set up in the telephone network are shown in Fig. 16.3. The connection between two subscribers to the same exchange is shown in Fig. 16.3*a*, and Figs. 16.3*b* and *c* show two possible connections between subscribers to different exchanges in the same group. Two connections between two subscribers in different groups in the same zone are shown in Figs. 16.3*d* and *e*, and Fig. 16.3*f* shows a connection between subscribers in different zones. Zone-to-zone trunk circuits have a loss of approximately 1·5 dB, group-to-zone trunks have a loss of 3 dB, and group-to-local exchange junctions have a loss somewhat greater than 4·5 dB.

LOCAL LINES

The connection between a subscriber and his local telephone exchange consists of a pair of wires in a telephone cable. Since a large telephone exchange may have up to 10,000 subscribers the local line network can be quite complicated particularly because provision must be made for fluctuating demand. The local line network is provided on the basis of forecasts made of the future demand for telephone service, the object being to provide, on demand, service as economically as possible. Since the demand fluctuates considerably there is

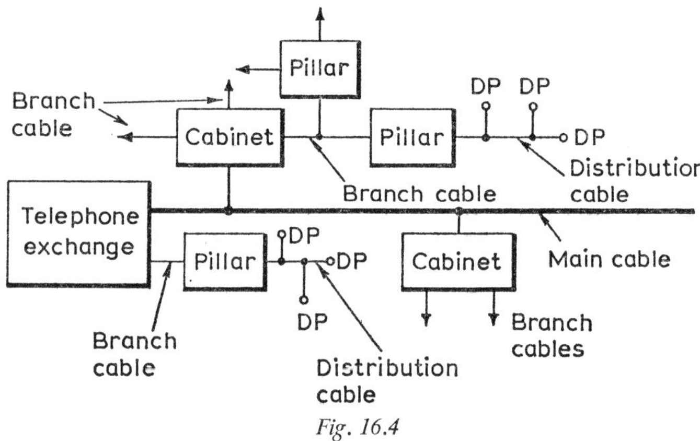

Fig. 16.4

Layout of a telephone exchange area

DP: Distribution point

the problem of forecasting requirements and deciding how much plant should be provided initially and how much at future dates. No matter how carefully the forecasting is carried out some errors always occur and allowance for this must be made in the planning and provision of cable, i.e. the local line network must be flexible. A network must be laid out so that the situation does not arise where potential subscribers cannot be given service in some parts of the exchange area while in other parts spare cable pairs remain. The modern way of laying out a local line network is shown in Fig. 16.4. Each subscriber's telephone is connected to a distribution point, such as a terminal block on a pole or a wall. The distribution points are connected by small distribution cables to pillars, a pillar being a street structure that provides flexibility because it allows any incoming pair to be connected to any outgoing pair. The pillars are connected by larger, branch, cables to cabinets; these have the same function as pillars but are larger. Finally, main cables connect the cabinets to the telephone exchange.

JUNCTION AND TRUNK CIRCUITS

Junctions are employed to interconnect exchanges that are less than about 15 miles apart and trunks are employed to link exchanges even further apart. Junctions and the shorter trunks are generally operated on a two-wire basis and only the shorter junctions do not require amplification. The longer trunks are always worked four-wire.

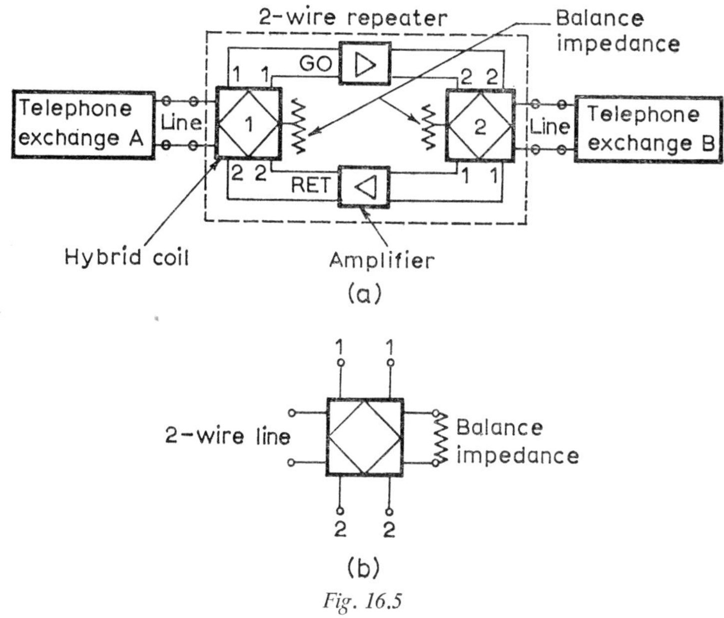

Fig. 16.5

The two-wire repeater

Amplification may be provided in a two-wire circuit, provided the line loss is not greater than 11 dB, by a two-wire repeater inserted as near the middle of the circuit as possible. The repeater may incorporate conventional amplifiers, Fig. 16.5a, or may be a negative-impedance amplifier, Fig. 16.6. The former type consists of two amplifiers—valve or transistor—and two hybrid coils connected as shown. A hybrid coil, Fig. 16.5b, is a device that, when all pairs of terminals are correctly matched, has 4 dB loss between the 2-wire terminals and terminals 1,1; 4 dB loss between the 2-wire terminals and terminals 2,2, and a very high loss between terminals 1,1 and terminals 2,2.

Speech signals reaching the repeater from exchange A divide equally at hybrid coil 1, and the signals arriving at terminals 1,1 are 4 dB below the level at the 2-wire terminals, and are amplified by the GO amplifier before application to terminals 2,2 of hybrid coil 2. The overall gain of the repeater between the 2-wire terminals of the two hybrid coils is equal to the gain of the amplifier minus 8 dB. Typically, the amplifier gain might be 16 dB when an input level of, say, -3 dBm would produce an output level of $+5$ dBm. The signals appearing at terminals 2,2 of hybrid coil 1 are dissipated in the output impedance of the RET amplifier and serve no useful purpose. Since the loss of a hybrid coil between terminals 2,2 and 1,1 is very high, very little energy is fed around the circuit, a necessary condition if the repeater is not to oscillate.

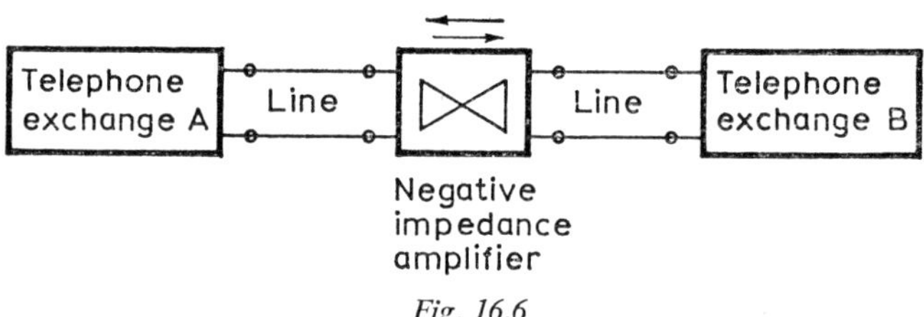

Fig. 16.6

The negative-impedance repeater

A negative-impedance amplifier, Fig. 16.6, is a transistor circuit that provides amplification of signals in both directions by effectively lowering the line resistance. Such an amplifier has a gain variable between 2·5 and 12 dB and is designed for use in conjunction with a particular type of audio telephone cable—naturally the most commonly employed.

The operation of both types of repeater depends upon the extent to which a balance impedance matches the impedance of the two-wire line. For example, if the balance impedances in Fig. 16.5*a* do not accurately match the two-wire line impedance the loss of a hybrid coil between terminals 1,1 and 2,2 is greatly reduced, and some of the energy appearing at the output of one amplifier will be fed into the input of the other. There is then considerable risk of instability and oscillation, known as singing. The impedance balance must be obtained over the band of frequencies to be amplified and it is difficult both to achieve and to maintain. As a result two-wire amplification is not employed nowadays for circuits whose length demands two, or more, points of amplification. Longer lines are always operated on a four-wire basis; however, the obsolete practice of providing long two-wire amplified circuits will be briefly discussed.

Fig. 16.7 shows a two-wire trunk circuit having three two-wire repeaters in tandem. A possible loop path exists in each repeater station and there is an ever present possibility of instability. For the circuit to be stable each of the six two-wire balances must

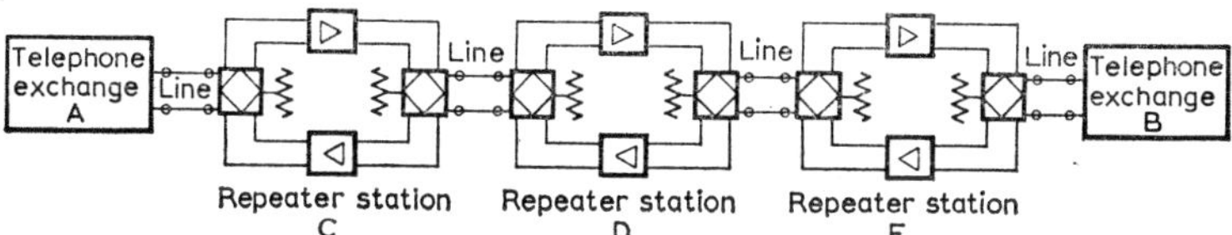

Fig. 16.7

A two-wire amplified circuit

accurately match the associated two-wire line, and this requires complex balancing networks that are difficult to adjust. Because of the instability problem of two-wire circuits their use by the G.P.O. is restricted to short lines of no more than 11 dB loss, for which a single two-wire repeater provides sufficient gain.

All longer trunks are worked four-wire and the basic arrangement of a four-wire circuit is shown in Fig. 16.8. The lines connecting the

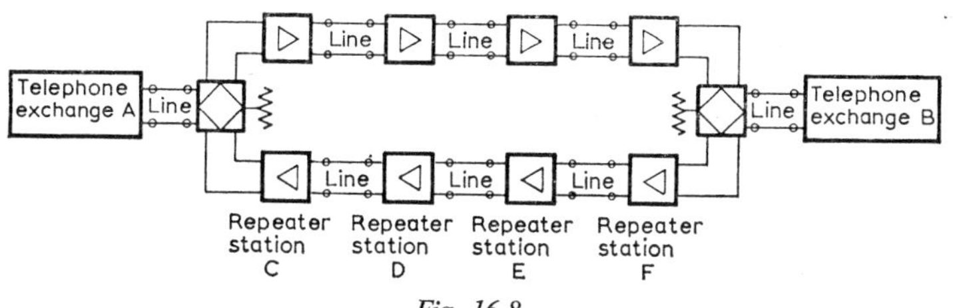

Fig. 16.8

A four-wire amplified circuit

telephone exchanges to the terminal repeater stations are operated two-wire but at these stations the circuit is split into GO and RETURN paths. Only two hybrid coils or terminating sets, as the hybrids used in conjunction with four-wire lines are called, are employed, one at each end of the circuit. There is thus only one possible loop path and the possibility of instability is greatly reduced. Also the risk of instability is not increased by increasing the length of the circuit and/or the number of amplifiers. A four-wire circuit is set up to be stable when the two-wire terminals of the two terminating sets are open-circuited and this practice results in the requirements of the two-wire balance being much less exacting. In the majority of cases a 600 ohm resistor is found to give adequate balance. It is usual to set up a four-wire circuit to have an overall loss of between 1·5 dB and 4·5 dB so that with a 0 dBm test tone (usually 800 Hz) applied to the two-wire terminals of a terminating set, the amplifier output levels, for that direction of transmission, are all $+10$ dBm. The gain of the amplifiers used is nominally 27 dB and so the minimum input level to an amplifier is -17 dBm.

In practice, the level is very often greater than this and the amplifier must then be preceded by an attenuator of suitable loss. Fig. 16.9 shows a four-wire circuit of 3 dB overall loss and gives the levels to be expected at different points in the circuit.

So far the four-wire circuit has been taken as consisting of two pairs in some telephone cables but, in practice, a four-wire circuit

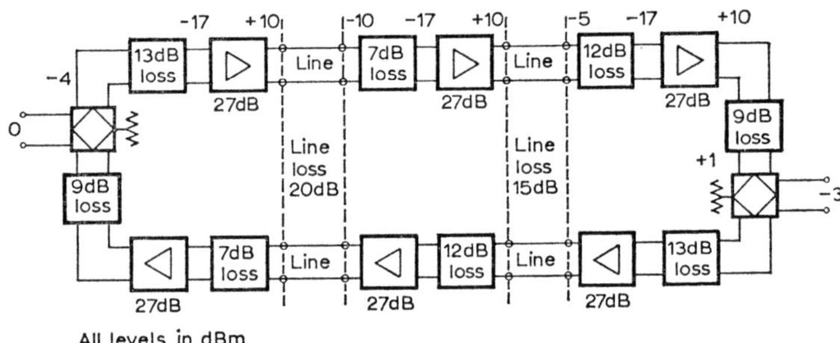

Fig. 16.9

Typical losses and gains for a four-wire circuit

may be routed, wholly or partly, over a single channel in a multi-channel telephony system.

A comparison of the relative merits of two-wire and four-wire circuits shows that while the two-wire circuit is economical in the use of cable pairs—an important consideration—the four-wire circuit requires fewer amplifiers for a given length of line and is *much* easier to set up and maintain.

THE INTERCONNECTION OF JUNCTION AND TRUNK CIRCUITS
The need often arises in a telephone network for trunk and/or junction circuits to be connected in tandem in order to route a call from one exchange to another. Short two-wire circuits are easily switched by manual or automatic means but the problem is somewhat more difficult for four-wire circuits. The automatic equipment required to switch four-wire circuits directly is both complex and costly, and for many years it has been the practice to convert four-wire circuits to two-wire before switching. Fig. 16.10 shows a typical connection that involves two minor exchanges, two group centres and one zone centre. The points marked X denote switching points

in telephone exchanges; such a point may consist of automatic equipment or a telephone switchboard. For s.t.d. calls, of course, all the switching points are automatic. The two-wire switching of four-wire trunk circuits suffers from the disadvantages that (*a*) there is a 4 dB loss through each terminating set and (*b*) the overall loss of each trunk circuit cannot be less than about 1·5 dB because of the instability problem and this means that the switching losses cannot be compensated for. These disadvantages naturally assume greater importance as the number of switching points is increased, and in

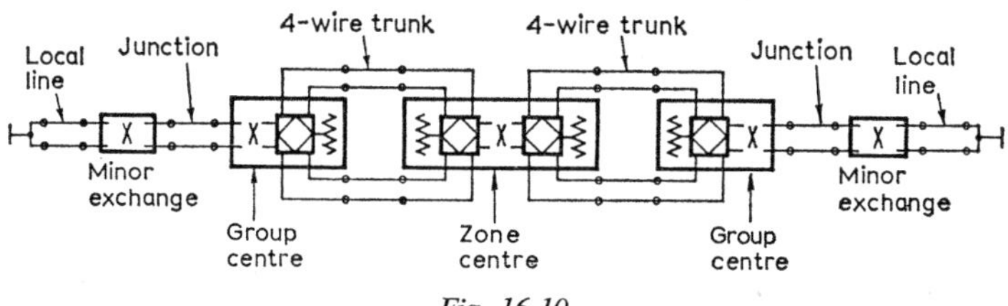

Fig. 16.10

Two-wire switching of trunk circuits

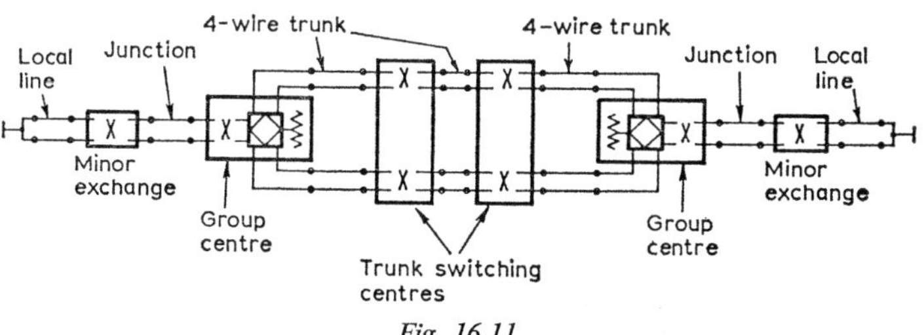

Fig. 16.11

Four-wire switching of trunk circuits

consequence of the introduction of s.t.d. the intention is to switch all connections involving three or more trunk circuits on a four-wire basis. The arrangement of a typical four-wire switched connection is shown in Fig. 16.11 in which switching points are again marked X. It can be seen that the connection is four-wire between the two terminal group centres and that the group centre to local exchange link is worked two-wire. Any losses introduced at the four-wire switching points can be compensated for by increasing the amplifier gains.

INTERNATIONAL CIRCUITS

Many international circuits are routed on multi-channel carrier-telephony circuits passing over submarine cable, radio link, or earth satellite and these will be discussed later. Here the intention is to mention the method employed to interconnect two subscribers in different countries by a four-wire radio link operating in the high-frequency band 3–30 MHz. Fig. 16.12 shows a typical arrangement.

A form of s.s.b. working, known as independent sideband (i.s.b.), is generally employed that enables either two high-quality or four commercial-quality speech circuits to be accommodated in a 12 kHz bandwidth. A subscriber in one country is connected, via the junction and trunk network of that country, to his international telephone exchange. Here an operator takes details of the call and makes contact, via a radio circuit to the required country, with an operator in the distant international exchange. The connection is then completed by the foreign operator. Signals passing between the two subscribers are at audio frequency up to the radio link itself; at the transmitter the signals amplitude modulate a carrier in the h.f. band and the resulting waveform is radiated to the distant receiver. Different frequencies are employed for the two directions of transmission over the radio link to eliminate the possibility of singing around the loop.

Radio Telephony

Radio telephony systems are employed to provide communication between a central, fixed, station and a number of mobile stations, for example, police, taxi-cab, A.A. and ambulance services; and to provide connection to the public telephone system for (*a*) remote locations such as some Scottish islands where only a few telephones are required and the cost of installing a telephone cable would be prohibitive, (*b*) private motor cars, and (*c*) ships at sea. Systems of the first type are classified as private, because they are not given direct access to the public telephone network, and work in the v.h.f. band (71·5–88 MHz and 165–174 MHz) with a bandwidth of 25 kHz.

BASE-MOBILE SYSTEMS

One method of operating a base-mobile radio telephony system is outlined in Fig. 16.13, only one mobile station being shown to simplify the drawing. In their normal position the two receivers are connected to their respective aerials and are ready to receive messages. If either station wishes to transmit, a switch on the transmitter is pressed and the relay S is operated. Contact $S1$ (operated) connects the transmitter to the aerial and transmission may begin. When a speaker has finished he must release his transmitter switch in order to reconnect his receiver to the aerial. The same frequency is used for both transmission and reception since this practice gives the most

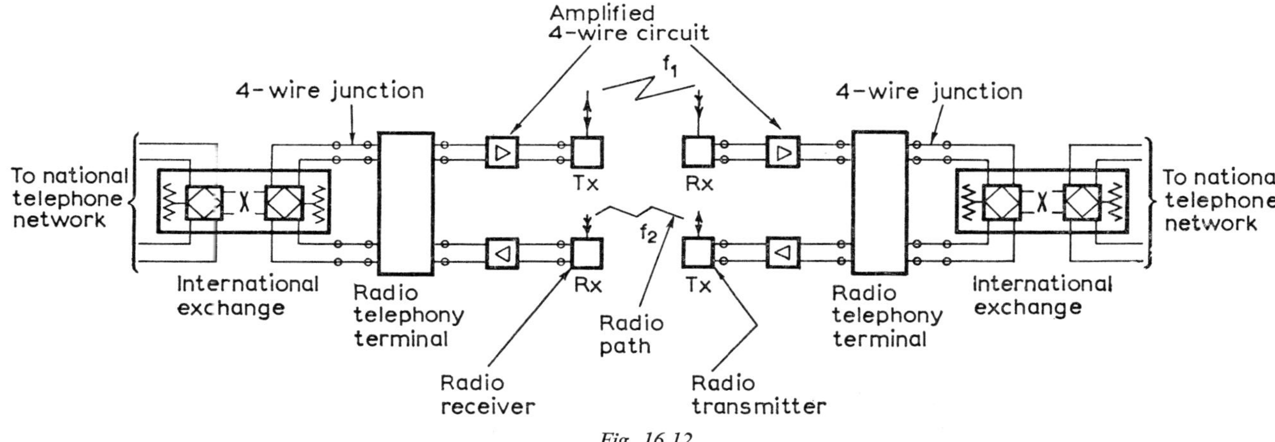

Fig. 16.12 An international telephony circuit

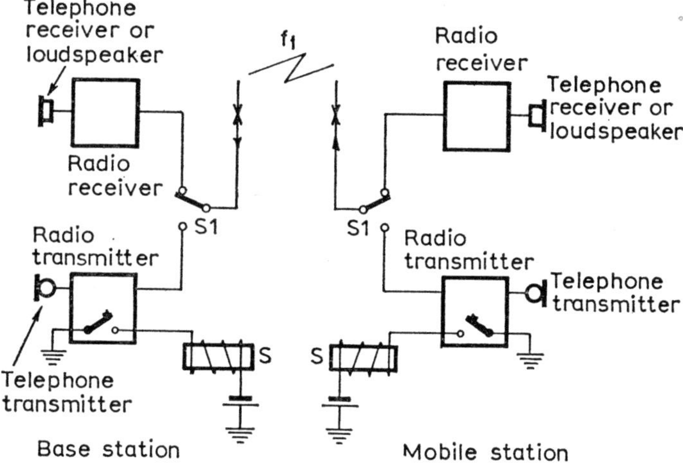

Fig. 16.13

A single-frequency base-mobile radio telephony system

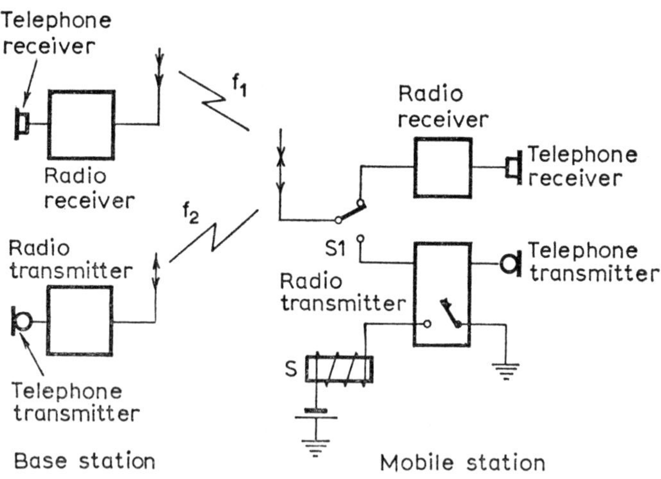

Fig. 16.14

A two-frequency base-mobile radio telephony system

economical use of the frequency spectrum. The use of a single frequency, however, demands wide geographical separation between base stations transmitting on frequencies fairly close to one another.

If the base stations of a number of different services are concentrated within a fairly small area, as is the case in London, for example, it is generally necessary to use a two-frequency method of working. The basic arrangement of a two-frequency radio-telephony system is shown in Fig. 16.14. The base station employs two aerials, one for transmission and one for reception. Base-to-mobile transmissions are on frequency f_2 and the mobile receivers are tuned to this frequency. Transmissions from mobile to base use frequency f_1 and the base receiver is tuned to this frequency. As before, an operator at a mobile station must depress his transmitter switch before speaking.

Both types of system may use amplitude, frequency or phase modulation, but amplitude modulation is the most common.

CAR RADIO-TELEPHONES

Two car radio-telephone services are operated in Great Britain, one covering South Lancashire and the other covering London. The services provide communication between mobile radio stations in motor cars and the public telephone network. The base stations are owned and operated by the G.P.O. but the mobile stations are owned by the private user. Fig. 16.15 shows the block schematic diagram of a connection between an ordinary telephone subscriber and a mobile station in the South Lancashire service.

The ordinary subscriber is connected via the telephone network to a hybrid coil in the Peterloo telephone exchange and the circuit is converted from two-wire to four-wire before extension to Telephone House and the base station. The base station transmits on a frequency f_1 in the band 164·05–164·35 MHz and receives on a frequency f_2 in the band 159·55–159·85 MHz; f_1 and f_2 are 9·5 MHz apart.

For optimum performance the frequency-modulated base transmitter must be fully modulated by low-level speech signals and yet not over-modulated by high-level signals, and so a constant-volume amplifier (c.v.a.) is included. (A c.v.a. is an amplifier that provides a constant-level output for different levels of input signal). The function of the re-radiation suppressor is to prevent received signals being re-radiated because of mismatch between the two-wire line and the hybrid balance impedance. Incoming radio signals operate relay MH and contact $MH1$ completes the receive circuit; if very strong signals are picked up by the auxiliary aerial, relay MM will operate and contact $MM1$ will complete an alternative receive circuit.

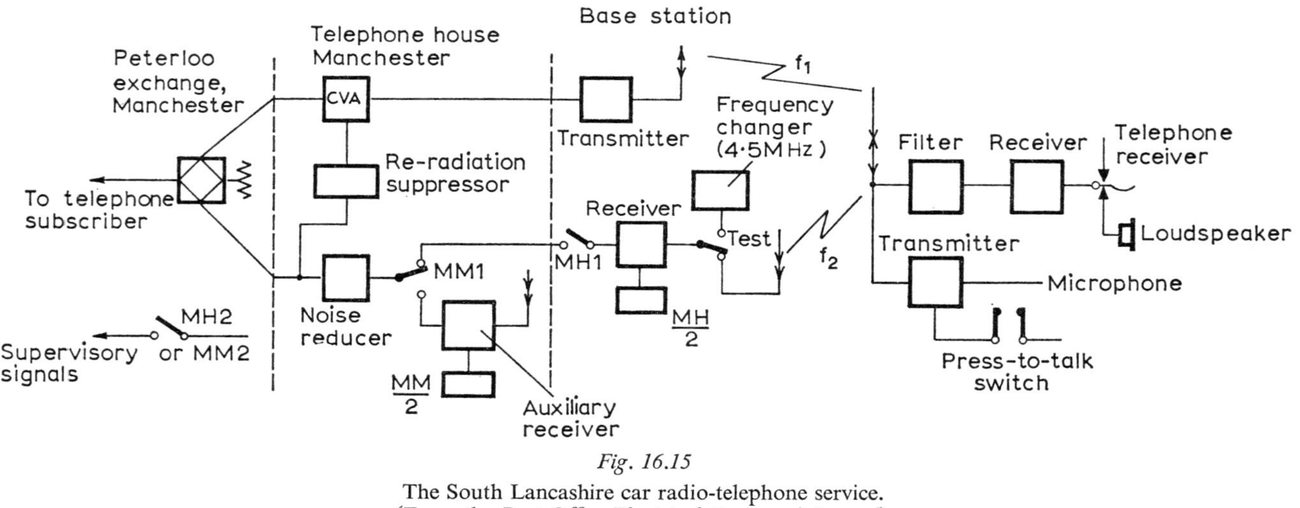

Fig. 16.15

The South Lancashire car radio-telephone service.
(From the *Post Office Electrical Engineers' Journal*)

The noise reducer is a device that reduces the magnitude of low-level noise signals but has little effect on the relatively high-level speech signals. The frequency changer is included merely as a method of testing the base station.

SHIP RADIO-TELEPHONES

Telephonic communication is often required between a telephone subscriber and a ship at sea and the arrangement for setting up such a connection is shown in Fig. 16.16. The telephone subscriber is connected, via the trunk network, with the control centre. The control centre is linked by cable to a number of coastal transmitting and receiving radio stations, each of which transmits or receives to or from a different part of the world. The control centre establishes

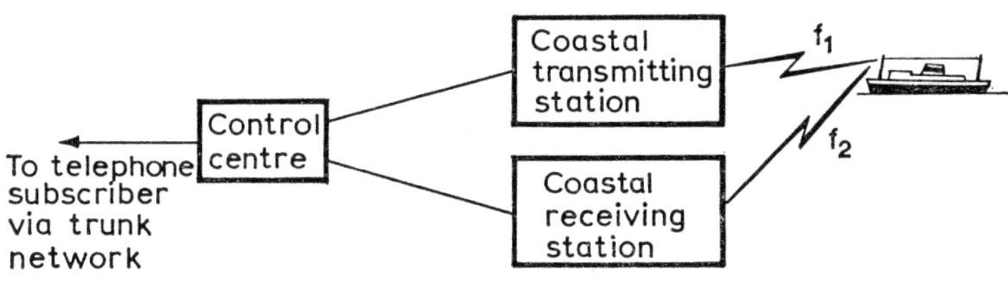

Fig. 16.16

Shore-to-ship telephone connection

the required connection with a distant ship via the appropriate pair of radio stations. To provide good coverage of a particular area of the world each transmitting station transmits on several different frequencies at the same time. The actual frequencies employed depend upon the part of the world concerned but lie in either the h.f. band or the v.h.f. band.

Multi-channel Telephony Systems

The essentials of an f.d.m. multi-channel telephony system have been previously discussed in an earlier chapter. Here the intention is to give an outline description of some of the more common systems employed in Great Britain. Single-sideband amplitude-modulation working is universally used because of the economies in cable frequency spectrum that result and a channel bandwidth of 300–3,400 Hz is usual.

For a number of telephone channels to be simultaneously transmitted along a single pair in a telephone cable it is necessary to employ amplitude modulation to position the various channels in different parts of the frequency spectrum. The number of channels that can be carried by a given cable pair is primarily determined by the loss of the cable and the minimum repeater spacing along the line that can bc economically justified (the greater the line loss the closer line amplifiers must be spaced to prevent the signal level falling below a pre-determined value).

The basic principle of any f.d.m. carrier telephony system is shown by the block diagram of Fig. 16.17; it can be seen that the

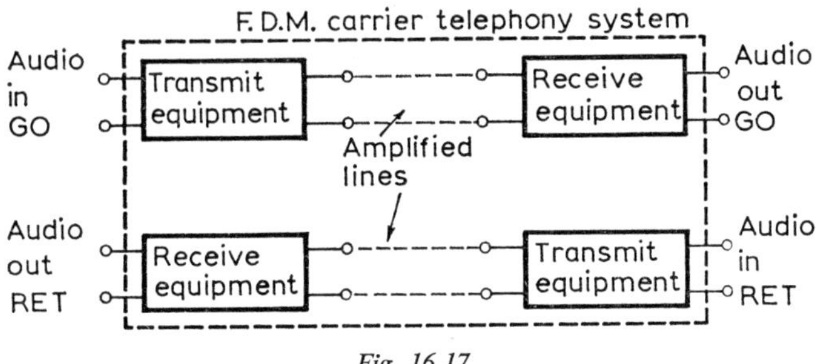

Fig. 16.17

The principle of a multi-channel telephony system

systems are operated on a four-wire basis. The equipment required for the two directions of transmission is the same and so further discussion will refer to one direction only.

THE 12-CHANNEL SYSTEM

The basic building block for most of the multi-channel carrier systems in use in Great Britain is the C.C.I.T.T.* 12-channel group, a block schematic diagram of which is shown in Fig. 16.18. The system has twelve channels that are translated to different positions in the 60–108 kHz frequency band by the channel modulating equipments; the lower sideband is selected in each case. The channel carrier frequencies are spaced at 4 kHz intervals between the highest frequency of 108 kHz and the lowest of 64 kHz. After assembly in the 60–108 kHz band the twelve channels are shifted to the band 12–60 kHz, by modulating a 120 kHz carrier and selecting the lower sideband, before transmission to line. At the receiving end of the system the channels are passed into the group demodulating equipment and restored to the frequency band 60–108 kHz. Filters in the channel demodulating equipments select the frequencies appropriate to their particular channel and the selected signals are demodulated against the appropriate channel carrier frequency to produce audio-frequency signals.

* C.C.I.T.T.: International Telegraph and Telephone Consultative Committee.

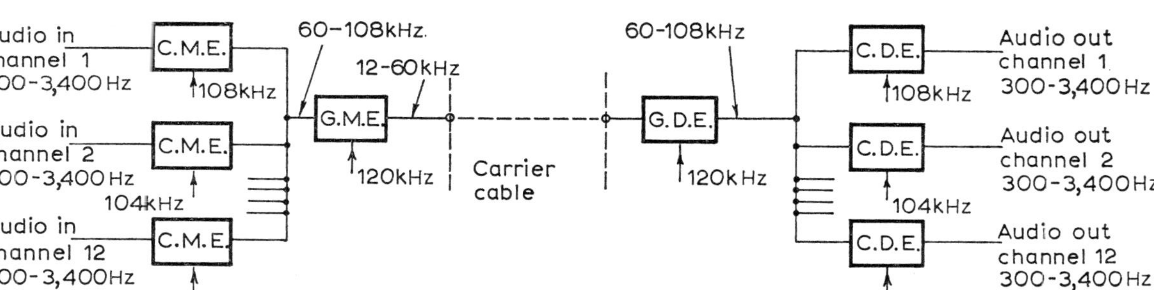

Fig. 16.18

A 12-channel carrier telephony system

C.M.E.: Channel modulating equipment
G.M.E.: Group modulating equipment
G.D.E.: Group demodulating equipment
C.D.E.: Channel demodulating equipment

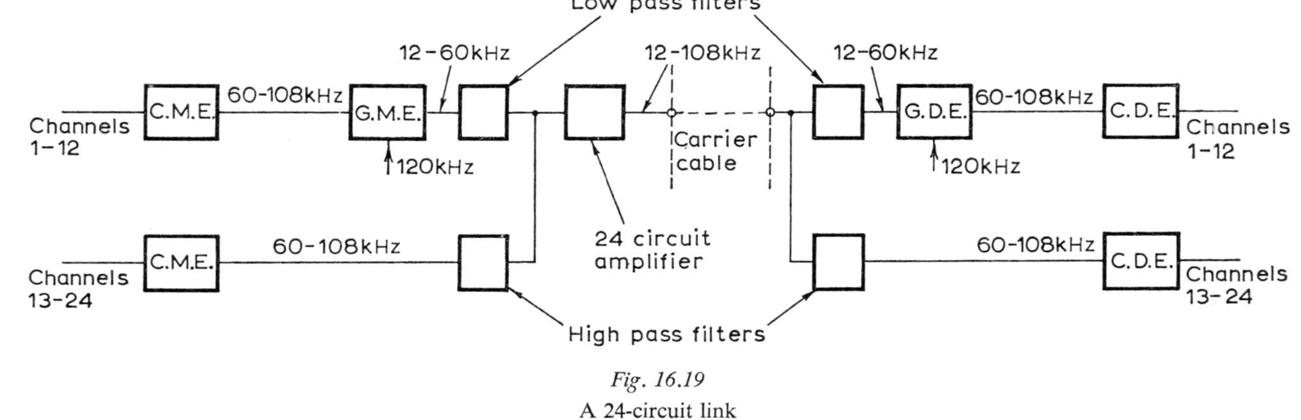

Fig. 16.19
A 24-circuit link

THE 24-CIRCUIT SYSTEM

The 12-channel carrier group is transmitted over pairs in carrier telephone cables that are capable of handling frequencies up to 108 kHz. It is therefore possible to transmit more than twelve channels at once over this type of cable and almost all the 12-channel links in Great Britain have now been converted to 24-circuit working. The method employed to combine two basic 12-channel groups to form a 24-circuit link is shown in Fig. 16.19. The low- and high-pass filters are required at the receiving end to direct incoming signals from the line to the appropriate carrier group, and at the transmitting end to prevent inter-group interference.

OTHER MULTI-CHANNEL SYSTEMS

The principle of the 24-circuit system can be extended to allow the transmission of five 12-channel groups over an improved type of cable. Such 60-circuit systems are known as supergroups and occupy the frequency band 12–252 kHz, but few of them are in use. Providing even more channels than this there are coaxial systems of various kinds, so named because coaxial cable is employed. A coaxial system consists of a number of the basic 12-channel groups combined in an appropriate manner, and a number of different systems are installed in Great Britain. For example, one system provides 600 channels in the frequency band 60–2,540 kHz, and another 2,700 channels in the frequency band 308–12,435 kHz.

EARTH SATELLITES

Long-distance inter-continental circuits routed via high-frequency radio links are not very reliable and do not provide good quality communication. A further disadvantage is the congested state of the h.f. band which limits the number of circuits that can be provided. To overcome these disadvantages a number of multi-channel telephony systems have been provided over submarine cables, the Commonwealth cable for example. Submarine cables are, however, extremely expensive and of limited capacity, and in recent years experiments have been carried out to investigate the possibility of using earth satellites for world-wide telephonic and television communication.

The basic principle of an earth satellite communication system is shown in Fig. 16.20. The ground stations are fully integrated with their national telephone networks and, in addition, the European stations are fully interconnected with each other. Four frequencies are used: the North American station transmits on frequency f_1 and receives frequency f_4, the European stations transmit frequency f_3 and receive frequency f_2. The function of the satellite is to receive the signals transmitted to it, frequency change them (f_1 to f_2 or f_3 to f_4), and then amplify them before transmission to the ground station at the other end of the link.

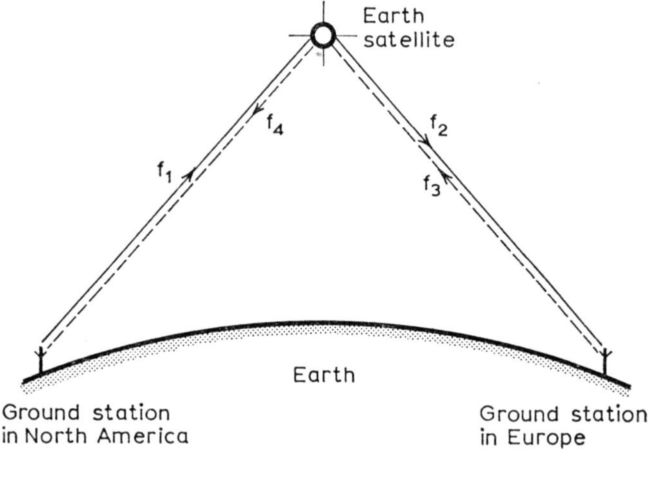

Fig. 16.20

An earth satellite communication system

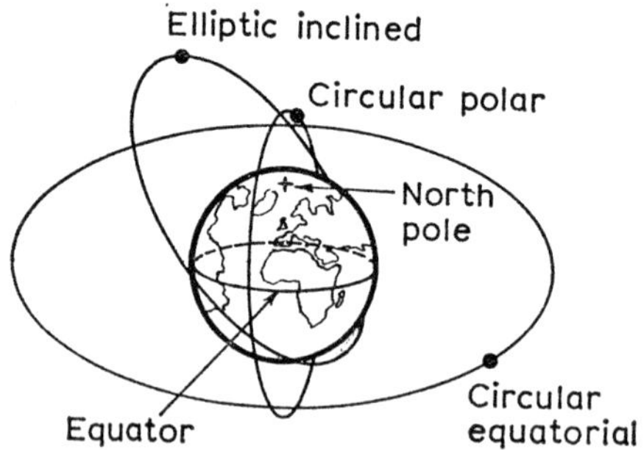

Fig. 16.21

Some possible earth satellite orbits
(From the *Post Office Electrical Engineers' Journal*)

An earth satellite can be placed into a number of different orbits around the Earth; possible orbits include circular and elliptical orbits in the equatorial and polar planes of the Earth, or inclined at some angle to one of these planes. Three possible orbits are shown in Fig. 16.21.

Tests have been carried out with satellites in different orbits at different heights above the Earth, but the latest communication satellite Early Bird I is travelling a circular equatorial orbit at a height of 22,300 miles. This particular orbit is known as the synchronous orbit because a satellite in this orbit travelling in the same direction as the Earth's rotation appears to be stationary above a particular part of the Earth. Early Bird I was the first communication satellite to be employed as an integral part of the transatlantic telephone service, and its performance has been so satisfactory that it has been decided to set up a global satellite communication system using three or four satellites in synchronous orbit.

Telegraphy Systems

Telegraphy is the passing of messages by means of signalling code. Several different codes are in use for different telegraphy systems but the two most common are the Morse code and the Murray code.

In the Morse code characters are represented by a combination of dot signals and dash signals; the difference between a dot and a

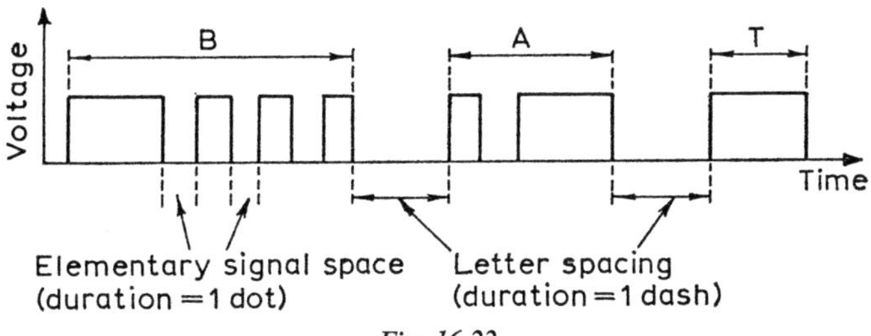

Fig. 16.22

"BAT" in Morse code

dash is one of time duration only, a dash being three times the length of a dot. Spacings between elementary signals, between letters and between words are also distinguished from one another by different time durations. As an example, Fig. 16.22 shows the word "BAT" in Morse code—B is dash and three dots, A is dot, dash, and T is just dash.

The Morse code suffers from the disadvantage that the number of signal elements needed to indicate a character is not the same for all characters and also the signal elements themselves are of different lengths. This makes the design of automatic printing receiving equipment difficult and for teleprinter systems the Murray code is used. In the Murray code all characters have exactly the same number of signal elements and the signal elements are of constant length. Each character is represented by a combination of five signal elements that may be either a mark or a space. In Great Britain a mark is represented by a negative potential or the presence of a tone and a space is represented by a positive potential or the absence of a tone. It is possible to indicate 2^5, or 32, combinations directly by the Murray code but this number is insufficient for general use, because 26 combinations are required for the letters of the alphabet and there are a number of figures and punctuation marks also to be trans-

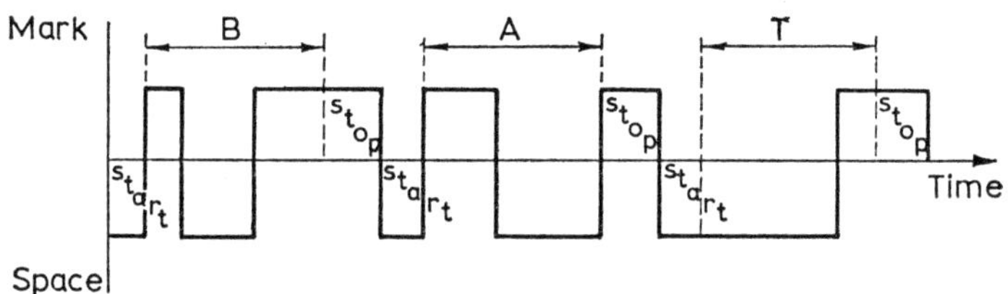

Fig. 16.23

"BAT" in Murray code

mitted. Two combinations, therefore, are used as letter-shift and figure-shift signals and they have the function of setting up the receiving teleprinter to print either figures or letters. This arrangement means that the other combinations can each be used to represent two different characters and so the capacity of the code is greatly increased. Fig. 16.23 shows the word "BAT" in Murray code; to maintain synchronism between the transmitting and receiving mechanisms each character is preceded by a start signal (equal to one space) and followed by a stop signal (equal to one and a half marks).

BANDWIDTH REQUIRED

Telegraph speed is measured in terms of a unit known as the baud. The baud speed of a telegraph signal is the reciprocal of the time duration of the shortest signal element employed. In the Morse code the shortest signal element is the dot; in the Murray code the elements are of the same length. The bandwidth required for the transmission of a signal in Morse code depends upon the number of words transmitted per minute and is not standardized, but generally lies in the range 100–1,000 Hz. Teleprinter transmissions are normally at the rate of 50 bauds and this means that the time duration of a mark, or space, is 1/50 second, or 20 milliseconds. The frequency band required to transmit a teleprinter signal depends upon the characters sent, but the maximum bandwidth is readily determined since it corresponds to the transmission of either R or Y. Both these letters consist of alternate marks and spaces, Fig. 16.24. The time for a cycle of the waveform is 40 milliseconds and so the fundamental frequency of the waveform is 1,000/40 or 25 Hz. A bandwidth of 25 Hz is therefore required to transmit the maximum fundamental component of a teleprinter signal, but it is usual to transmit up to

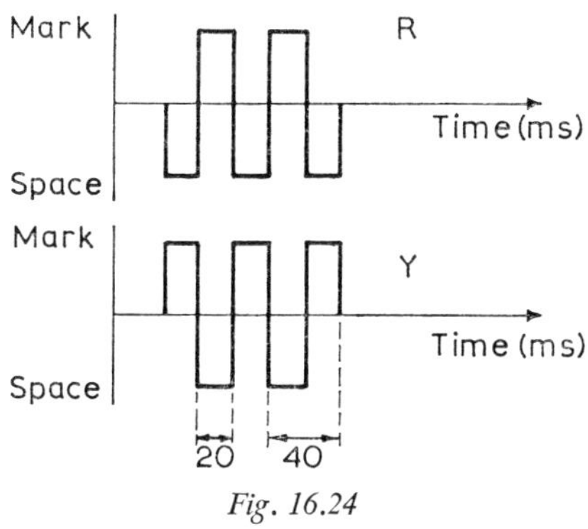

Fig. 16.24

"R" and "Y" in Murray code

about 140 Hz and thereby retain the fifth harmonic. A bandwidth of 0–140 Hz is then required. For any other characters consecutive marks and/or spaces occur in the code and the fundamental frequency is less than 25 Hz.

The use of Morse code telegraphy systems has greatly declined in the last 25 years or so in favour of teleprinter systems using the Murray code. This is mainly because the teleprinter is simple and convenient to use (its operation is very similar to that of a typewriter) and provides a typed copy of the message, at both ends of the circuit if required. Morse systems are still widely employed where the sending or receiving, or both, ends of the circuit are mobile.

SINGLE-CHANNEL TELEGRAPHY

The basic principle of a Morse code telegraph system is shown in Fig. 16.25*a*. When the key is pressed at the sending end of the system a current flows to line and passes through the windings of a telegraph relay at the receiving end. The relay operates and its contact completes a circuit for the buzzer to operate. If the key is operated in accordance with the Morse code a trained operator listening to the buzzer will be able to recognize the message and write it down. Fig. 16.25*b* shows the essentials of a radio telegraphy system. When the sending end key is opened the radio transmitter is switched off and no signal is radiated; when the key is pressed the transmitter radiates a single-frequency tone. If the key is operated in accordance with the Morse code an interrupted continuous wave is radiated and is received by the distant radio receiver. The radio receiver converts the received signal into pulses of direct current that operate the buzzer relay. This method of radio telegraphy has a number of

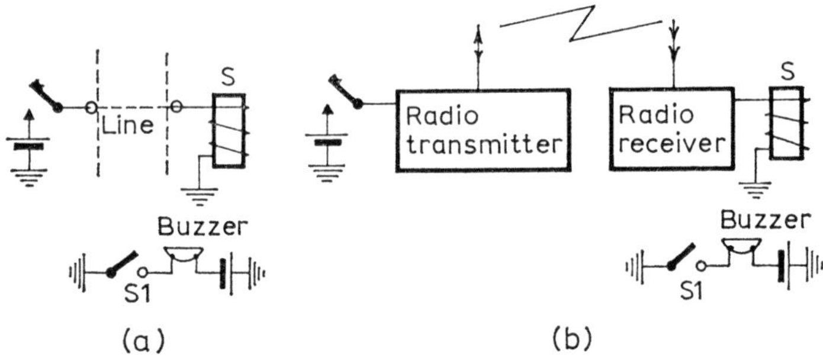

Fig. 16.25

Morse telegraphy systems over (*a*) telephone line and (*b*) radio link

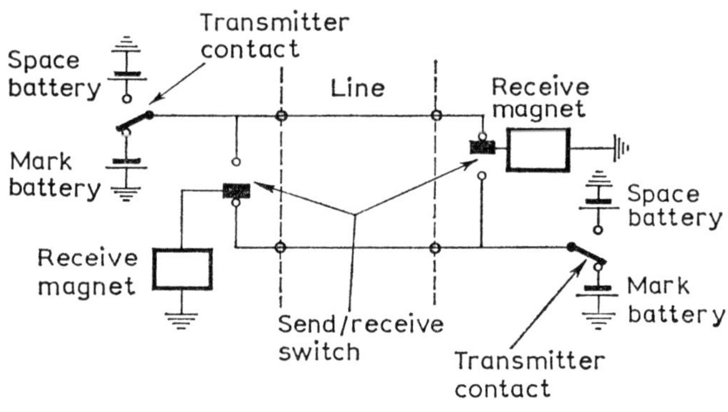

Fig. 16.26

A teleprinter circuit

disadvantages and it is being replaced by a system, known as "frequency shift keying", in which dots and dashes are represented by continuous wave signals of different durations at one frequency, and the spaces between letters, etc. are represented by continuous wave signals of different lengths at another frequency.

Fig. 16.26 shows the arrangement employed for two teleprinters to work to one another. When a teleprinter is sending, its send/receive switch is operated and so its receive magnet is connected to the send line to obtain a local copy of the transmitted message. The transmitter contact normally rests on the mark battery, but operation of the teleprinter keyboard causes the contact to change-over between the mark and space batteries in accordance with the characters to be signalled.

MULTI-CHANNEL TELEGRAPHY

Multi-channel telegraphy systems may be provided using either frequency- or time-division multiplex. The latter was widely used in the early days of telegraphy but nowadays the former is by far

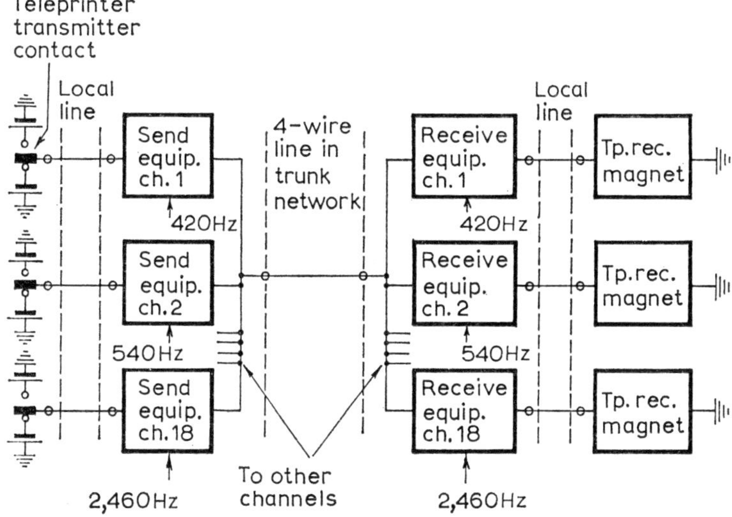

Fig. 16.27

An 18-channel M.C.V.F. telegraphy system

the more common. Consequently, only f.d.m. multi-channel telegraphy will be mentioned here. In principle, a multi-channel telegraphy system is similar to a multi-channel telephony system but the bandwidth of the channels is much less. Most systems are designed to provide 18 or 24 50-baud telegraph channels over an ordinary telephone circuit. Since the bandwidth of these systems lies entirely within the band 300–3,400 Hz they are known as multi-channel voice-frequency (M.C.V.F.) telegraphy systems.

The principle of an amplitude-modulated 18-channel M.C.V.F. telegraphy system is shown in Fig. 16.27. One direction of transmission only is shown, the other direction being identical. The signals transmitted by a teleprinter consist of a series of marks and spaces and these are fed into a channel send equipment; here the marks switch the carrier frequency ON and spaces switch it OFF. The output from a channel send equipment thus consists of a series of pulses of carrier as shown in Fig. 16.28. The outputs of the 18 channels are combined and transmitted to the distant M.C.V.F.

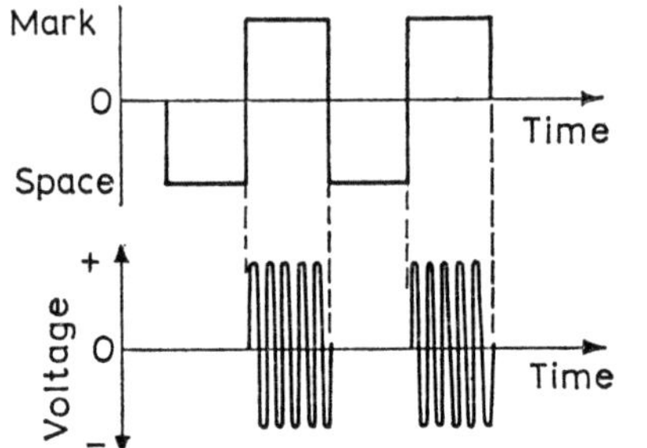

Fig. 16.28

The output signal from a channel send equipment

terminal over a permanent link in the trunk telephone network. This link may be routed, wholly or partly, over one or more multi-channel telephony systems. At the receiving end of the system the various signals are selected by their appropriate channel equipments and demodulated to give a series of marks and spaces again

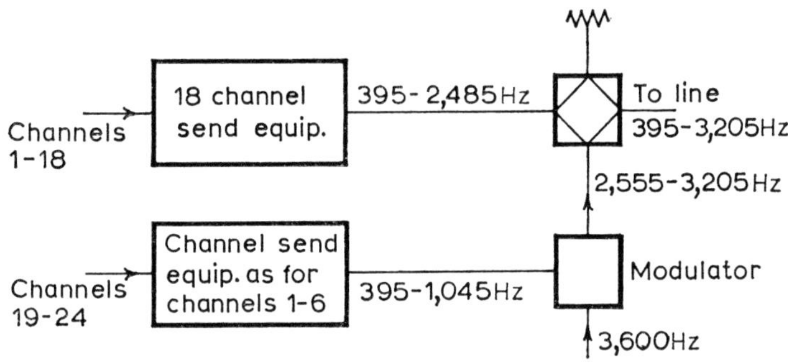

Fig. 16.29

A 24-channel M.C.V.F. telegraphy system

which can be transmitted over a local line to operate the receive magnet of a teleprinter.

Double-sideband amplitude modulation is used, and since the maximum signalling speed is 50 bauds, a bandwidth of 2 × 25 or 50 Hz per channel is required. The channel carrier frequencies are spaced 120 Hz apart starting from 420 Hz and so the bandwidth required for the system is 395–2,485 Hz.

Many telephone lines are capable of transmitting the C.C.I.T.T. recommended bandwidth of 300–3,400 Hz and the use of such a line enables the number of telegraph channels to be extended to 24. The method employed to derive the extra six channels is shown in Fig. 16.29; a similar arrangement is employed at the receiving end of the system. The extra six channels are fed into send equipment identical with that employed for channels 1–6 and the combined output, in the band 395–1,045 Hz, is translated to the band 2,555–3,205 by modulation of a 3,600 Hz carrier.

M.C.V.F. telegraphy channels can be connected in tandem and are used for both point-to-point circuits and as trunks in the national telex service. The telex service is the teleprinter equivalent of the s.t.d. trunk telephone network and it allows subscribers to dial connections with any other subscriber.

PICTURE, OR FACSIMILE, TELEGRAPHY

Facsimile telegraphy is the transmission and reception of still pictures, diagrams, etc. The basic principle of a facsimile telegraphy system is shown in Fig. 16.30.

The picture to be transmitted is mounted on a circular drum that is free to move along a lead-screw as it revolves about it. A small

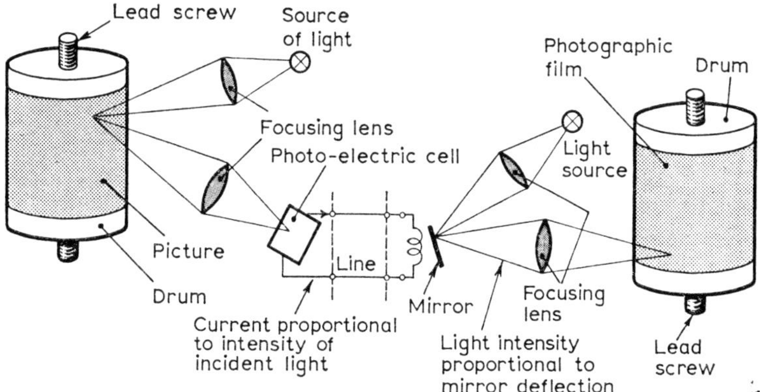

Fig. 16.30

Facsimile telegraphy

spot on the picture is illuminated by the focused light from a light source and as the drum revolves all parts of the pictures are successively scanned by the spot of light. The light reflected from the illuminated spot is focused on to a photo-electric cell that gives a voltage output proportional to the intensity of the light incident upon it. As the picture is scanned the light reflected from the illuminated spot fluctuates according to the nature of the picture and so the current delivered by the cell to the line varies also.

At the receiving end of the system a photographic film is scanned in similar fashion by a light beam whose intensity is proportional to the magnitude of the received current. Light from a light source is focused on the surface of a mirror whose position is a function of the magnitude of the photo-cell current passing through an inductive winding. The intensity of the light reflected from the mirror, and focused on the photographic film, depends upon the angle of the mirror. If the system is correctly adjusted each part of the film is illuminated as it is scanned just sufficiently for the original picture to be reproduced.

The bandwidth required for a facsimile telegraphy system depends upon the speed with which pictures are transmitted and the nature of the pictures, but usually a bandwidth between 0–500 Hz and 0–1,000 Hz is required for direct transmission. Very often the pictures are to be transmitted between points of wide geographical separation and then the signals will require amplification at intervals along the line. It is usual, therefore, for the picture signal to amplitude-modulate a carrier before transmission so that a more suitable part of the frequency spectrum is occupied. The C.C.I.T.T. recommend two carrier frequencies, 1,300 Hz and 1,900 Hz, and assuming these, maximum bandwidths of 300–2,300 Hz and 900–2,900 Hz are required.

Facsimile telegraphy is employed for the transmission of Press photographs but it is not economical, compared with teleprinters, for the transmission of written messages.

TELEGRAPHY SERVICES
Telegrams can be sent by members of the public to any address in the country. A telephone subscriber wishing to send a telegram dials "Telegrams" and is connected to an operator in a local sub-post office or, if nearby, a head post office. The operator writes down details of the message on a telegram form. The telegram is then passed by telephone to a head post office where it is transmitted by tele-printer over the automatic telegram switching network (T.A.S.) to the head post office nearest the destination address. At the receiving head post office the telegram is received on a teleprinter and either given directly to a messenger boy for delivery, or dictated by tele-phone to a sub-post office for delivery from there.

The telex service is a teleprinter-operated service in which sub-scribers are able to dial numbers and obtain direct connection with other subscribers. Automatic switching is used and the telex network, similar in layout to the trunk telephone network, comprises physical local pairs and trunks routed over channels in M.C.V.F. systems. Telex subscribers have automatic access to subscribers in a number of European countries.

Telegrams can also be sent to ships at sea by members of the public and telex subscribers. Fig. 16.31 shows the arrangement employed. The connection between the coastal radio station and

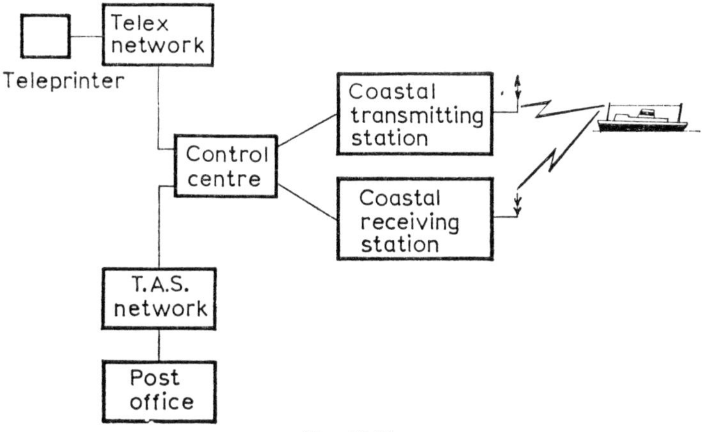

Fig. 16.31

Shore-to-ship telegraphy connection

the control centre is by M.C.V.F. channels, and that between coastal radio station and ship is by radio link. The radio links are operated on six frequencies in the h.f. band, i.e. 4, 6, 8, 13, 17, and 23 MHz and the bandwidth required is between 100 and 1,000 Hz.

Sound Broadcasting

A broadcasting system is one in which programmes of mixed entertainment, news and educational content are made available to a large number of persons operating radio receivers. Since complete coverage of a particular geographical area is required omni-directional transmitting aerials are employed. Fig. 16.32 illustrates the concept of multi-station broadcasting. The programme is radiated by the transmitting aerial as an electromagnetic wave in all directions in the horizontal plane. The wave induces an e.m.f. in every receiving aerial that is cut by its magnetic field. If a receiver is switched on and is tuned to the programme frequency the receiver will amplify and detect the signal and the programme will be received. If a receiver is not switched on, or is switched on but is tuned to another frequency, the programme will not be received.

Programmes radiated at the lower radio frequencies require only a single transmitter to cover the whole country, but at the higher frequencies the range of a particular station is much smaller and it becomes necessary to have a number of transmitting stations situated in different parts of the country. The frequencies used for sound broadcasting have been given in Chapter 2. In the London area, for example, the Light Programme is broadcast on 200 kHz,

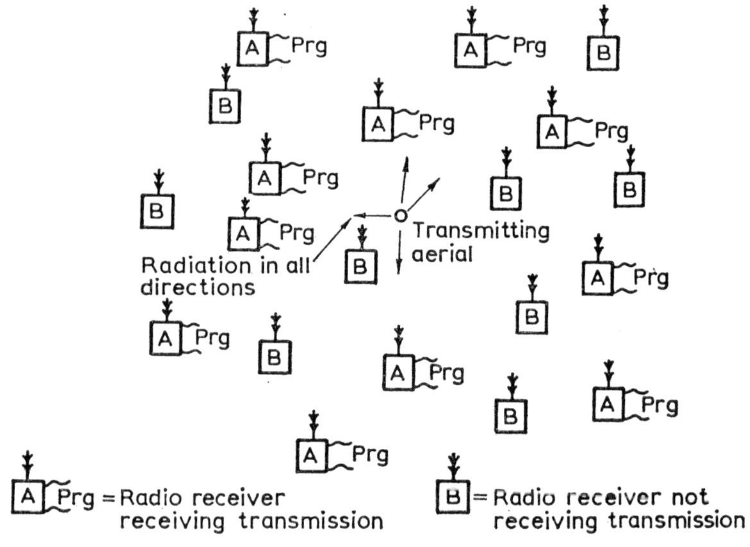

Fig. 16.32

The concept of sound broadcasting

1,214 kHz, 90 MHz and 89·1 MHz, the Home Programme on 908 kHz, 94·4 MHz and 93·5 MHz, and the Third Programme on 647 kHz, 1,546 kHz, 92·4 MHz and 91·3 MHz.

Fig. 16.33 shows how the transmitting stations are connected to the programme control room and thence to the various studios and outside broadcast points.

The programmes are originated from various points and are fed into the programme control room via local tie cables if from an adjacent studio, or via G.P.O. lines if from a distant studio or an outside broadcast. The G.P.O. lines must be capable of carrying, without undue distortion, the frequency band required for music and this means that special lines must be provided by the G.P.O. Programmes are processed as required in the control room and are then fed into the lines, or radio links, linking the control room to the transmitting stations.

Television Broadcasting

In principle, television broadcasting is very similar to sound broadcasting, differences lying in the nature of the signal, and in the transmitting and receiving equipment employed.

A television picture is built up from a number of horizontal lines that are traced out in sequence by a spot travelling on the screen of a

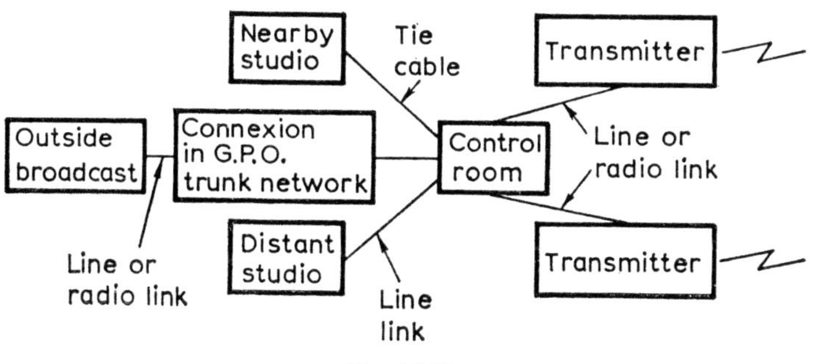

Fig. 16.33

Sound broadcasting arrangements

cathode-ray tube in the television receiver. The various shades of grey, and black and white demanded by a particular picture are created by modulation of the brightness of the spot as it travels. The modulating signal is derived from the television camera that is scanning the original scene in a similar series of lines. A television picture is divided into a large number of small elements of picture, and the television camera transmits an electrical signal at each instant, whose amplitude is proportional to the brightness of the picture element being scanned at that instant. For the picture to be reproduced correctly at the receiver it is essential for the two scans (camera and receiver) to be synchronized. The picture produced by the television camera scanning the picture is therefore accompanied by a number of synchronizing pulses.

The bandwidth required for a television picture signal depends upon a number of factors, such as the number of lines that comprise a picture, the number of pictures transmitted per second, and the durations of the synchronizing pulses. In the British television system either 405 or 625 lines are used and this gives a bandwidth requirement, assuming equal horizontal and vertical resolution, of

3·11 MHz for 405 line and 7·37 MHz for 625 line. For economic reasons these bandwidths are reduced in practice, at the expense of the horizontal resolution, to approximately 3 MHz and 5·5 MHz. The picture signal is used to amplitude-modulate a carrier frequency—in the v.h.f. band for B.B.C.1 and commercial television, in the u.h.f. band for B.B.C.2—and a vestigial-sideband signal is transmitted. A vestigial-sideband signal consists of the carrier component, one sideband (usually the lower) and a small part of the other sideband and economizes in the bandwidth required. For example, a double-sideband 405 line television signal would require a bandwidth of about 6 MHz, but vestigial-sideband working reduces this requirement to less than 5 MHz.

Exercises

1. Draw suitably annotated block schematic diagrams of the following simple line and radio communication systems—

    (*a*) Connections required for an overseas radio-telephone call between two subscribers on different continents.
    (*b*) Connections required for a multi-station sound broadcast from an outside broadcast event.
    (*c*) Connections required for an overseas radio telegraphy call between a shipping company and one of its ships.

    State in each case the approximate carrier frequency and bandwidth appropriate to any radio link involved.　　　　　　　(A 1964)

2. By reference to a block schematic diagram, explain the equipment needed to provide a repeatered audio junction circuit on (*a*) a 2-wire basis, and (*b*) a 4-wire basis. Quote typical losses for each part of the circuit.
    State the relative advantages of 2-wire and 4-wire operation.　　(A 1963)

3. With the aid of a block diagram, explain the connections required for an overseas telephone call between a subscriber on a remote island in one continent and a subscriber in another continent.
    Suggest suitable carrier frequencies which might be used on the national telephone network and the international radio circuit involved.　　(A 1962)

4. State, with explanations, the bandwidths required for each of the following line systems—

    (*a*) for 18 channels of voice-frequency telegraphy, each signalling at 50 reversals per second (bauds),
    (*b*) for 12 telephone channels,
    (*c*) for a high-quality music channel,
    (*d*) for a 405-line television channel.　　　　　　　　　　(A 1961)

5. Show, by means of block schematic diagrams, the equipment needed to provide a repeatered audio junction circuit on (*a*) a 2-wire basis and (*b*) a 4-wire basis.
    Quote typical losses for each part of the circuit, and explain the advantage of 4-wire operation compared with 2-wire operation.　　(A 1961)

Answers to Numerical Problems

2.1. The depth of modulation is $m = V_m/V_c$, where V_m = amplitude of modulating signal and V_c = amplitude of carrier wave.

$$\therefore \qquad V_m = 0.75\, V_c$$

The maximum amplitude of the a.m. wave is $V_c + 0.75\, V_c$ and the minimum value is $V_c - 0.75\, V_c$.

Bandwidth = twice maximum modulating frequency

$$= 2 \times 3{,}400 = 6{,}800 \text{ Hz} \qquad Ans.$$

The lower sideband is from $104{,}000 - 3{,}400$ Hz to $104{,}000 - 300$ Hz, i.e.

$$100{,}600 \text{ Hz to } 103{,}700 \text{ Hz} \qquad Ans.$$

The upper sideband is from $104{,}000 + 300$ Hz to $104{,}000 + 3{,}400$ Hz, i.e.

$$104{,}300 \text{ Hz to } 107{,}400 \text{ Hz} \qquad Ans.$$

2.3(*a*) $m = V_m/V_c = 1.0$ or $V_m = V_c$.

The a.m. wave should have a maximum value of $V_c + V_c$ or $2\,V_c$ and a minimum value of $V_c - V_c$ or 0.

(*b*) $m = V_m/V_c = 0.25$ or $V_m = 0.25\, V_c$.

The a.m. wave should vary between a maximum value of $1.25\, V_c$ and a minimum value of $0.75\, V_c$.

Bandwidth = twice the highest modulating frequency
$$= 2 \times 4{,}500 \text{ Hz} = 9{,}000 \text{ Hz} \qquad Ans.$$

The lower sideband is from $506{,}000 - 4{,}500$ Hz to $506{,}000 - 50$ Hz, i.e.

$$501{,}500 \text{ Hz to } 505{,}950 \text{ Hz} \qquad Ans.$$

The upper sideband is from $506{,}000 + 50$ Hz to $506{,}000 + 4{,}500$ Hz, i.e.

$$506{,}050 \text{ Hz to } 510{,}500 \text{ Hz} \qquad Ans.$$

2.5. The required waveform is shown in Fig. A.1*a*. Note that there are 10 cycles of the carrier wave in a time of 0·2 millisecond compared with 1 cycle of the modulation envelope. If the two signals are merely added to one another the resulting waveform is shown in Fig. A.1*b*.

The frequencies present in the modulated wave are the carrier frequency, 50,000 Hz, the upper side frequency, 55,000 Hz and the

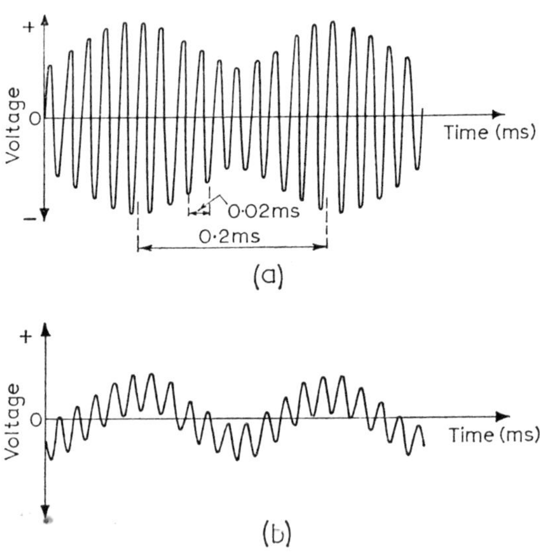

Fig. A.1

lower side frequency, 45,000 Hz. The only frequencies present in the other waveform are the two original frequencies 50,000 Hz and 5,000 Hz.

2.6. The envelopes of the a.m. waves should vary (*a*) between 1·5 V_c and 0·5 V_c and (*b*) 2 V_c and 0.

(*a*) Lower side frequency = 1,000 kHz − 1 kHz = 999 kHz *Ans.*
Upper side frequency = 1,000 kHz + 1 kHz = 1,001 kHz
Ans.

Amplitude of a side frequency = $\frac{1}{2}mV_c$ = $\frac{1}{2}(0·5 \times 1)$ = 0·25 V *Ans.*

(*b*) Lower side frequency = 999 kHz as before   *Ans.*
Upper side frequency = 1,001 kHz as before   *Ans.*

Amplitude of a side frequency = $\frac{1}{2}(1 \times 1)$ = 0·5 V   *Ans.*

2.7(*a*) "High quality" implies v.h.f. frequency-modulated broadcasts in the band 87·5 to 100 MHz.   *Ans.*

(*b*) 12 to 60 kHz.   *Ans.*

(*c*) 1,900 to 2,300 MHz, 3,790 to 4,200 MHz, and 5,925 to 6,425 MHz.   *Ans.*

$$\text{Frequency} = \frac{\text{velocity}}{\text{wavelength}} = \frac{2 \times 10^8}{0\cdot04} = 5 \text{ GHz} \qquad Ans.$$

2.8. $f = v/\lambda$, where f = frequency in hertz, v = velocity in metres/sec, and λ = wavelength in metres.

(*a*) $\lambda = 12\cdot3$ metres,

$$\therefore \qquad f = \frac{3 \times 10^8}{12\cdot3} = 24\cdot4 \text{ MHz}$$

The maximum frequency tolerance in this band is ±30 parts in 10^6.

$$\therefore \quad \text{Frequency tolerance} = \pm \frac{30}{10^6} \times 24\cdot4 \times 10^6 = \pm732 \text{ Hz } Ans.$$

(*b*) $\lambda = 5\cdot5$ metres,

$$\therefore \qquad f = \frac{3 \times 10^8}{5\cdot5} = 54\cdot55 \text{ MHz}$$

The maximum frequency tolerance in this band is ±10 parts in 10^6.

$$\therefore \quad \text{Frequency tolerance} = \pm \frac{10}{10^6} \times 54\cdot55 \times 10^6 = 545\cdot5 \text{ Hz } Ans.$$

Last part:

Frequency variation allowed = frequency $\times$ frequency tolerance

$$\therefore \qquad \pm300 = \text{frequency} \times \pm \frac{30}{10^6}$$

$$\text{or} \qquad \text{Frequency} = \frac{300 \times 10^6}{30} = 10 \text{ MHz}$$

$$\therefore \qquad \lambda = \frac{3 \times 10^8}{10 \times 10^6} = 30 \text{ metres} \qquad Ans.$$

2.9. $\dfrac{\lambda}{2} = \dfrac{3 \times 10^8}{2 \times 45 \times 10^6} = 3\cdot33 \text{ metres} \qquad Ans.$

2.11. Wavelength difference $= \dfrac{3 \times 10^8}{1 \times 10^6} - \dfrac{3 \times 10^8}{1\cdot01 \times 10^6}$

$$= 300 - 297\cdot0297$$
$$\simeq 2\cdot97 \text{ metres} \qquad Ans.$$

2.12(*a*) $\lambda = \dfrac{3 \times 10^8}{600 \times 10^3} = 500$ metres   *Ans.*

  (*b*) $\lambda = \dfrac{3 \times 10^8}{9 \times 10^6} = 33\cdot3$ metres   *Ans.*

  (*c*) $\lambda = \dfrac{3 \times 10^8}{50 \times 10^6} = 6$ metres   *Ans.*

2.13. Maximum value of a.m. wave $= V_c + V_m = 1\cdot5$ V
Minimum value of a.m. wave $= V_c - V_m = 0\cdot9$ V

Adding,   Carrier component $V_c = 2\cdot4/2 = 1\cdot2$ V   *Ans.*
Subtracting,   Amplitude of modulating signal $V_m = 0\cdot6/2 = 0\cdot3$ V

$\therefore$ Depth of modulation $= \dfrac{V_m}{V_c} \times 100\% = \dfrac{30}{1\cdot2}\% = 25\%$ *Ans.*

  Frequency components of the wave are

  (i) Carrier frequency 310 kHz   *Ans.*
  (ii) Upper side frequency 315 kHz   *Ans.*
  (iii) Lower side frequency 305 kHz   *Ans.*

4.2. Turns ratio required $= \sqrt{\left(\dfrac{5{,}000}{3}\right)} = 40\cdot8$ *Ans.*

4.6. Turns ratio required $= \sqrt{\left(\dfrac{4{,}900}{4}\right)} = 34\cdot99$ *Ans.*

5.1. $24 = 20 \log_{10}\left(\dfrac{V_{IN}}{1 \times 10^{-6}}\right)$

$\therefore$ $V_{IN} = 1 \times 10^{-6} \times \text{antilog}_{10}\, 1\cdot2 = 15\cdot85\ \mu\text{V}$ *Ans.*

  Since the input and output impedances are equal, the gain is given by

$$\text{Gain} = 20 \log_{10} \dfrac{V_{OUT}}{V_{IN}}$$

$\therefore$ $\dfrac{V_{OUT}}{V_{IN}} = \text{antilog}_{10}\, 1\cdot5 = 31\cdot62$

and $V_{OUT} = 31\cdot62 \times 15\cdot85 = 501\cdot2\ \mu\text{V}$ *Ans.*

5.2. Total gain $G = G_1 G_2$ and total loss $= 10$ dB.

$\therefore$ Overall gain $G_0 = 20 \log_{10} G_1 G_2 - 10$ dB

Tabulating—

FREQUENCY (kHz)	60	66	72	78	84	90	96	102	108
G	29·8 × 28·2	34·5 × 37·2	38 × 37·6	38·9 × 36·7	38·9 × 36·5	38·5 × 36·7	38 × 38·7	34·7 × 39·1	29 × 29·8
G_0 (dB)	48·52	52·17	53·10	53·09	53·04	53·00	53·35	52·65	48·73
Approx. value	48·5	52·2	53·1	53·1	53·0	53·0	53·4	52·7	48·7

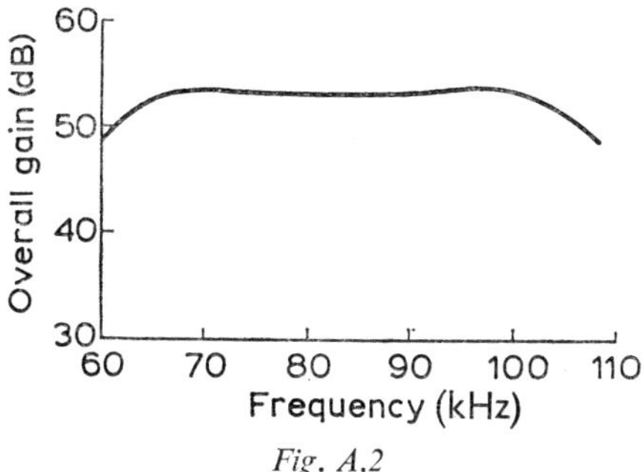

Fig. A.2

The required graph is given in Fig. A.2.

5.3. Total gain $G = G_1 G_2$, and total loss $= 15$ dB.

∴ Overall gain $G_0 = 20 \log_{10} G_1 G_2 - 15$ dB

Tabulating—

FREQUENCY (kHz)	60	76	92	108
G	310 × 330	330 × 345	340 × 380	390 × 325
G_0 (dB)	85·198	86·126	87·226	87·06
Approximate value	85·2	86·1	87·2	87·1

The required graph is given in Fig. A.3.

5.4. $23 \cdot 5$ mW $= 10 \log_{10} \dfrac{23 \cdot 5 \times 10^{-3}}{1 \times 10^{-3}} = +13 \cdot 7$ dBm   *Ans.*

$1 \cdot 25$ W $= 10 \log_{10} \dfrac{1 \cdot 25}{1 \times 10^{-3}} = +31$ dBm   *Ans.*

Fluctuation $= 31 - 13 \cdot 7 = 17 \cdot 3$ dB   *Ans.*

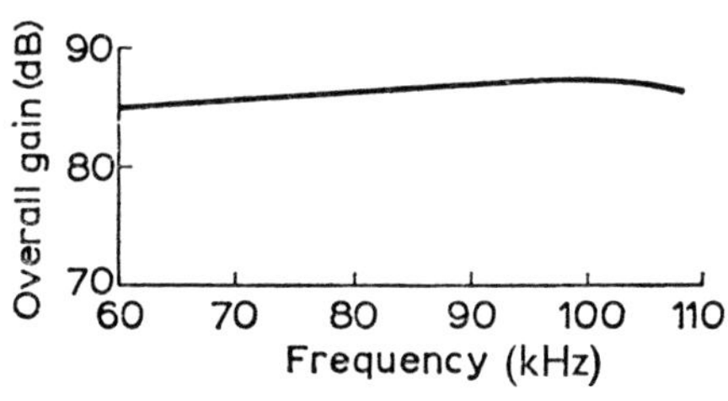

Fig. A.3

7.2. $I_e = I_c + I_b$

where $I_e =$ emitter current, $I_c =$ collector current and $I_b =$ base current.

∴ $I_b = 1 \cdot 0 - 0 \cdot 98 = 0 \cdot 02$ mA   *Ans.*

Current gain $= \dfrac{\delta I_c}{\delta I_b} = \dfrac{0 \cdot 98}{0 \cdot 02} = 49$ *Ans.*

7.6. The collector-voltage/collector-current characteristic is shown in Fig. A.4, and the collector-current/base-current characteristic is shown in Fig. A.5.

The output resistance R_{OUT} of the transistor is given by

$$R_{OUT} = \dfrac{\delta V_{ce}}{\delta I_c}, \ I_b \text{ constant}$$

With $I_b = 80 \ \mu$A and taking equal increments either side of $V_{ce} = -6$ volts (Fig. A.4) (since the I_b curve is linear the size of the increments has no effect on the accuracy), $\delta V_{ce} = 4$ V and $\delta I_c = 0 \cdot 8$ mA.

∴ $R_{OUT} = \dfrac{4}{0 \cdot 8 \times 10^{-3}} = 5{,}000 \ \Omega$ *Ans.*

The current gain is given by $\delta I_c / \delta I_b$ and choosing increments of $5 \ \mu$A either side of $I_b = -10 \ \mu$A (Fig. A.5) gives $\delta I_c = 0 \cdot 65$ mA.

∴   Current gain $= \dfrac{0 \cdot 65 \times 10^{-3}}{10 \times 10^{-6}} = 65$ *Ans.*

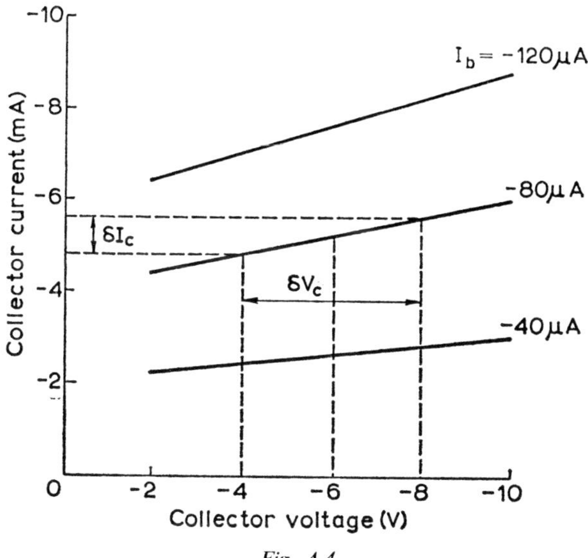

Fig. A.4

$$\delta I_c = 5 \cdot 6 - 4 \cdot 8 = 0 \cdot 8 \text{ mA}$$
$$\delta V_{ce} = 8 - 4 = 4 \text{ V}$$

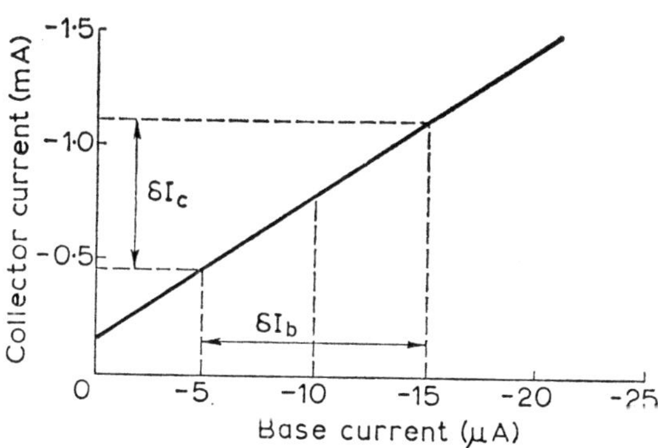

Fig. A.5

$$\delta I_c = 1 \cdot 1 - 0 \cdot 45 = 0 \cdot 65 \text{ mA}$$
$$\delta I_b = 15 - 5 = 10 \, \mu A$$

8.3. The mutual characteristics of the triode valve are shown plotted in Fig. A.6.

(i) With V_a constant at 200 V and taking increments of 0·5 V either side of $V_g = -1$ V gives $\delta V_g = 1$ V and $\delta I_a' = 5$ mA.

$$\therefore \qquad g_m = 5/1 = 5 \text{ mA/V} \qquad Ans.$$

(ii) With V_g constant at -1 V a change of V_a from 200 V to 175 V gives a change, $\delta I_a''$, in anode current of 5 mA.

$$\therefore \qquad r_a = \frac{25}{5 \times 10^{-3}} = 5{,}000 \ \Omega \qquad Ans.$$

$$\mu = r_a\, g_m = 5{,}000 \times 5 \times 10^{-3} = 25 \qquad Ans.$$

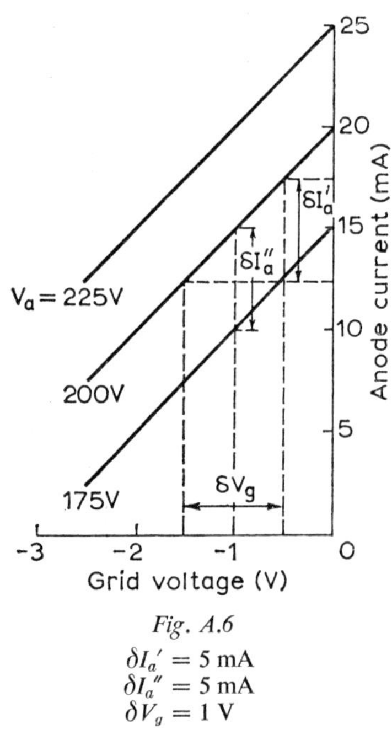

Fig. A.6

$$\delta I_a' = 5 \text{ mA}$$
$$\delta I_a'' = 5 \text{ mA}$$
$$\delta V_g = 1 \text{ V}$$

8.6. The required mutual characteristics are shown in Fig. A.7. From the characteristics,

$$\text{Mutual conductance } g_m = \frac{\delta I_a'}{\delta V_g} \quad (V_a \text{ constant at 250 V})$$

$$= \frac{2 \times 10^{-3}}{0\cdot2} = 10 \text{ mA/V} \qquad Ans.$$

Anode a.c. resistance $r_a = \dfrac{\delta V_a}{\delta I_a}$ (V_g constant at -1 V)

$$= \frac{50}{6\cdot 8 \times 10^{-3}} = 7{,}353\ \Omega \qquad Ans.$$

Amplification factor $\mu = g_m r_a = 73\cdot 53$ $\qquad Ans.$

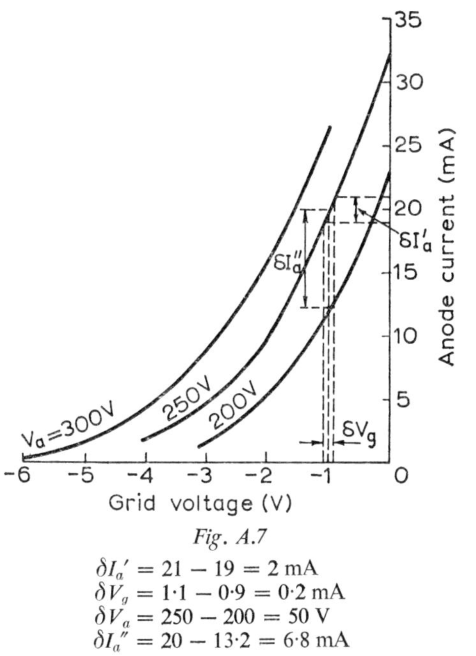

Fig. A.7

$\delta I_a' = 21 - 19 = 2$ mA
$\delta V_g = 1\cdot 1 - 0\cdot 9 = 0\cdot 2$ mA
$\delta V_a = 250 - 200 = 50$ V
$\delta I_a'' = 20 - 13\cdot 2 = 6\cdot 8$ mA

The required anode characteristics are shown in Fig. A.8.

Anode a.c. resistance $r_a = \dfrac{\delta V_a}{\delta I_a'}$ (V_g constant at -1 V)

$$= \frac{10}{1\cdot 3 \times 10^{-3}} = 7\cdot 692\ \text{k}\Omega \qquad Ans.$$

It is clear that the spacing between the curves is not constant and hence the value of g_m will vary considerably with the point of measurement. At $V_a = 200$ V,

$$g_m = \frac{\delta I_a''}{\delta V_g} \ (V_a \text{ constant at } 200 \text{ V})$$

$$= \frac{7\cdot 6 \times 10^{-3}}{1} = 7\cdot 6\ \text{mA/V} \qquad Ans.$$

and at $V_a = 250$ V,

$$g_m = \frac{\delta I_a'''}{\delta V_g} \quad (V_a \text{ constant at 250 V})$$

$$= \frac{10}{1} = 10 \text{ mA/V} \qquad Ans.$$

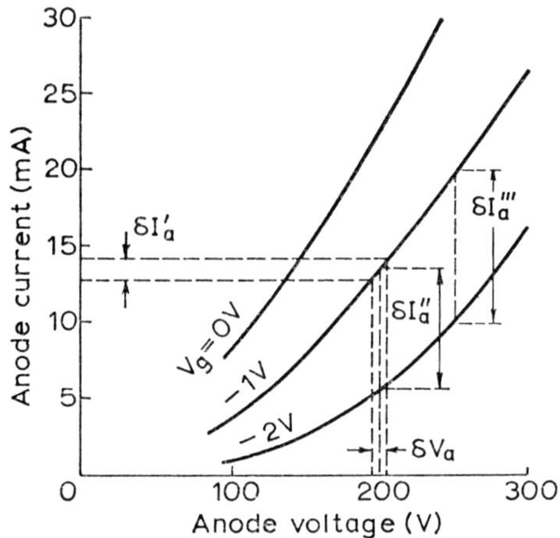

Fig. A.8

$$\delta V_a = 205 - 195 = 10 \text{ V}$$
$$\delta I_a' = 13\cdot8 - 12\cdot5 = 1\cdot3 \text{ mA}$$
$$\delta V_g = 2 - 1 = 1 \text{ V}$$
$$\delta I_a'' = 13\cdot2 - 5\cdot6 = 7\cdot6 \text{ mA}$$
$$\delta I_a''' = 20 - 10 = 10 \text{ mA}$$

Using the first value of g_m, $\mu = 7\cdot692 \times 10^3 \times 7\cdot6 \times 10^{-3} = 58\cdot46$

$$Ans.$$

Using the second value of g_m, $\mu = 7\cdot692 \times 10^3 \times 10 \times 10^{-3}$
$= 76\cdot92$ *Ans.*

8.10. The anode-current/grid-voltage curves are shown plotted in Fig. A.9.

Taking the anode voltage constant at 200 V,

Mutual conductance $g_m = \dfrac{\delta I_a'}{\delta V_g}$ (V_a constant at 200 V)

$$= \frac{8 \times 10^{-3}}{2} = 4 \text{ mA/V} \qquad Ans.$$

Taking the grid voltage constant at -2 V,

Anode a.c. resistance $r_a = \dfrac{\delta V_a}{\delta I_a''}$ (V_g constant at -2 V)

$$= \frac{50}{8 \times 10^{-3}} = 6 \cdot 25 \text{ k}\Omega \qquad Ans.$$

Amplification factor $\mu = r_a g_m$

$$= 4 \times 10^{-3} \times 6 \cdot 25 \times 10^3 = 25 \qquad Ans.$$

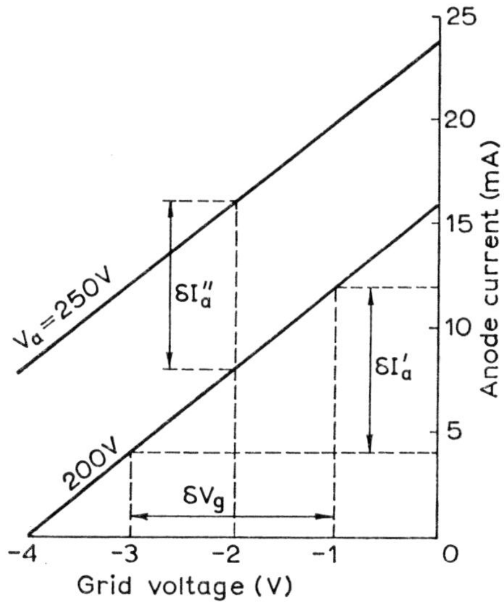

Fig. A.9

$\delta V_a = 250 - 200 = 50$ V
$\delta I_a'' = 16 - 8 = 8$ mA
$\delta V_g = 3 - 1 = 2$ V
$\delta I_a' = 12 - 4 = 8$ mA

9.9. Voltage gain $A_v = \dfrac{\mu R_L}{r_a + R_L}$

$$r_a A_v + R_L A_v = \mu R_L$$
$$R_L(\mu - A_v) = r_a A_v$$

$\therefore$
$$R_L = \frac{r_a A_v}{\mu - A_v}$$

Substituting the given values,

$$R_L = \frac{10 \times 10^3 \times 40}{50 - 40} = 40 \text{ k}\Omega \qquad Ans.$$

The power to be carried by the anode load resistor is the sum of the d.c. and a.c. powers dissipated in it.

$$\text{D.C. power} = (5 \times 10^{-3})^2 \times 40 \times 10^3 = 1 \text{ W}$$

The a.c. power cannot be calculated since the input voltage is not known but the total power will be in excess of 1 W. Hence a power rating of 2 W would provide an adequate safety margin.

9.10. Grid bias voltage $V_g = I_k R_k$ where $I_k = $ cathode current and $R_k = $ cathode resistance.

$$\therefore \qquad R_k = \frac{V_g}{I_k} = \frac{1 \cdot 5}{3 \times 10^{-3}} = 500 \ \Omega \qquad Ans.$$

10.1. $f_0 = 1/2\pi\sqrt{LC}$ where $L = $ inductance, $C = $ capacitance and $f_0 = $ resonant frequency

$$\therefore \quad L = 1/4\pi^2 f_0^2 C = 1/(4\pi^2 \times 1{,}500^2 \times 10^6 \times 1{,}600 \times 10^{-12})$$
$$= 7 \cdot 034 \ \mu\text{H}$$

When C is increased by 4,025 pF to 5,625 pF,

$$f_0 = 1/\ 2\pi\sqrt{(7 \cdot 034 \times 10^{-6} \times 5625 \times 10^{-12})}$$
$$= 800 \times 10^3 \text{ Hz} = 800 \text{ kHz} \qquad Ans.$$

10.2.
$$f_0 = \frac{1}{2\pi\sqrt{(LC)}} = 10^6 \text{ Hz}$$

Let f_0' be the new resonant frequency, then

$$f_0' = 1/2\pi\sqrt{[LC(1 + 21/100)]}$$

$$\therefore \qquad \frac{f_0'}{1 \times 10^6} = \frac{1}{2\pi\sqrt{(L \times 1 \cdot 21C)}} \Big/ \frac{1}{2\pi\sqrt{(LC)}}$$

$$\text{or} \qquad f_0' = \frac{1 \times 10^6}{\sqrt{1 \cdot 21}} = 0 \cdot 909 \times 10^6 \text{ Hz} = 909 \text{ kHz} \qquad Ans.$$

10.3.
$$f_0 = \frac{1}{2\pi\sqrt{(LC)}}$$

and
$$C = \frac{1}{4\pi^2 f_0^2 L}$$

The maximum value of C corresponds to the minimum resonant frequency.

$$\therefore \quad C_{max} = 1/[4\pi^2 \times (500 \times 10^3)^2 \times 150 \times 10^{-6}] = 675{\cdot}3 \text{ pF} \quad \textit{Ans.}$$

The required frequency ratio is $1{,}500/500$ or $3{\cdot}1$ and so the capacitance ratio required to tune to $1{,}500$ kHz is $9{:}1$.

$$\therefore \quad C_{min} = 675{\cdot}3/9 = 75{\cdot}0 \text{ pF} \quad \textit{Ans.}$$

Second section:

$$\text{Capacitance for } 1{,}000 \text{ kHz} = \left(\frac{500}{1{,}000}\right)^2 \times 675{\cdot}3 = 168{\cdot}8 \text{ pF } \textit{Ans.}$$

$$\text{Capacitance for } 2{,}000 \text{ kHz} = \left(\frac{1{,}000}{2{,}000}\right)^2 \times 168{\cdot}8 = 42{\cdot}2 \text{ pF } \textit{Ans.}$$

$$\therefore \quad \text{Required capacitance range} = 42 \text{ to } 169 \text{ pF} \quad \textit{Ans.}$$

10.4. $15 \times 10^5 = \dfrac{1}{2\pi\sqrt{(59 \times 10^{-6}C)}}$

$$\therefore \quad C = \frac{1}{4\pi^2 \times 15^2 \times 10^{10} \times 59 \times 10^{-6}} = 190{\cdot}7 \text{ pF}$$

The new value of capacitance is $290{\cdot}7$ pF and the new resonant frequency f_0 is

$$f_0 = 1/2\pi\sqrt{(59 \times 10^{-6} \times 290{\cdot}7 \times 10^{-12})} = 1{\cdot}215 \text{ MHz} \quad \textit{Ans.}$$

10.5. The capacitance C_s of two capacitors C_1 and C_2, connected in series is given by

$$C_s = \frac{C_1 C_2}{C_1 + C_2}$$

and the capacitance C_p of two capacitors C_3 and C_4, connected in parallel is given by

$$C_p = C_3 + C_4$$

$$\text{Hence} \quad \text{Minimum total capacitance} = 10 + \frac{50 \times 1{,}000}{50 + 1{,}000} = 57{\cdot}62 \text{ pF}$$

$$\text{and} \quad \text{Maximum total capacitance} = 10 + \frac{500 \times 1{,}000}{500 + 1{,}000} = 343{\cdot}33 \text{ pF}$$

$$\begin{aligned}
\therefore \quad &\text{Highest oscillation frequency} \\
&= 1/[2\pi\sqrt{(250 \times 10^{-6} \times 57{\cdot}62 \times 10^{-12})}] \\
&= 1{\cdot}326 \text{ MHz} \quad \textit{Ans.}
\end{aligned}$$

The lowest oscillation frequency
$$= 1/[2\pi\sqrt{(250 \times 10^{-6} \times 343\cdot33 \times 10^{-12})}]$$
$$= 0\cdot5433 \text{ MHz} \qquad Ans.$$

11.1. $\quad f_0 \simeq 1/2\pi\sqrt{LC}$

$$= 1/[2\pi\sqrt{(400 \times 10^{-6} \times 300 \times 10^{-12})}] = \frac{10^7}{2\pi\sqrt{12}}$$

$$= 459\cdot4 \text{ kHz} \qquad Ans.$$

Anode load impedance $R_d = \dfrac{L}{CR}$

$$= \frac{400 \times 10^{-6}}{300 \times 10^{-12} \times 20} = \frac{2}{3} \times 10^5 = 66{,}667 \text{ ohms}$$

Voltage gain $A_v = \dfrac{\mu R_d}{r_d + R_d}$

$$= \frac{10 \times 66\cdot667 \times 10^3}{50 \times 10^3 + 66\cdot67 \times 10^3} = \frac{666\cdot67}{116\cdot667} = 5\cdot715$$

Also, $A_v = \dfrac{V_{OUT}}{V_{IN}}$

Thus, for $V_{OUT} = 1$ V r.m.s.,

$$V_{IN} = \frac{1}{5\cdot715} = 0\cdot175 \text{ V r.m.s.} \qquad Ans.$$

11.2. $C = \dfrac{1}{4\pi^2 f_0^2 L}$

$$= \frac{1}{4\pi^2 \times 1 \times 10^{12} \times 150 \times 10^{-6}} = 168\cdot9 \text{ pF} \qquad Ans.$$

12.1. $m = \dfrac{\text{maximum amplitude} - \text{minimum amplitude}}{\text{maximum amplitude} + \text{minimum amplitude}}$

$$= \frac{1\cdot5 - 0\cdot5}{1\cdot5 + 0\cdot5} = 0\cdot5 \qquad Ans.$$

The frequencies present in the modulated wave are

   (*a*) the carrier frequency 205 kHz   *Ans.*
   (*b*) the lower side frequency 205 − 3 kHz or 202 kHz   *Ans.*
   (*c*) the upper side frequency 205 + 3 kHz or 208 kHz   *Ans.*

12.2. $C = 1/4\pi^2 f_0^2 L$

$\qquad = 1/(4\pi^2 \times 1 \times 10^{12} \times 60 \times 10^{-6}) = 422 \text{ pF} \qquad Ans.$

Now $\quad C \propto 1/f_0^2.$

$\therefore \quad C$ required at 2 MHz is 422/4 or 105·5 pF $\qquad Ans.$
Capacitor range $= 105$ to 422 pF $\qquad Ans.$

12.5. $m = \dfrac{\text{maximum amplitude} - \text{minimum amplitude}}{\text{maximum amplitude} + \text{minimum amplitude}}$

$\qquad = \dfrac{3-1}{3+1} = 0\text{·}5 \qquad Ans.$

12.7. Frequency range to be covered is 10 kHz to 100 kHz and this must be achieved with a variable capacitor of 4:1 range. Since $f \propto 1/\sqrt{C}$ the capacitor gives a 2:1 frequency range.

   Therefore one coil will cover the band 10 kHz to 20 kHz, a second 20 kHz to 40 kHz, a third 40 kHz to 80 kHz, and a fourth 80 kHz to 160 kHz. Therefore four coils are required. $\qquad Ans.$

Index